PLAN-MAKING FO

Plan-making for Sustainability
The New Zealand Experience

NEIL J. ERICKSEN
University of Waikato

PHILIP R. BERKE
University of North Carolina

JANET L. CRAWFORD
Planning Consultants Ltd, Auckland

JENNIFER E. DIXON
University of Auckland

Routledge
Taylor & Francis Group

LONDON AND NEW YORK

First published 2004 by Ashgate Publishing

Published 2017 by Routledge
2 Park Square, Milton Park, Abingdon, Oxfordshire OX14 4RN
711 Third Avenue, New York, NY 10017, USA

First issued in paperback 2017

Routledge is an imprint of the Taylor & Francis Group, an informa business

British Library Cataloguing in Publication Data
Plan-making for sustainability : the New Zealand
    experience. - (Urban planning and environment)
    1.Environmental policy - New Zealand 2.Environmental policy
    - New Zealand - Case studies 3.Sustainable development -
    Government policy - New Zealand 4.Sustainable development -
    Government policy - New Zealand - Case studies 5.City
    planning - Environmental aspects - New Zealand 6.City
    planning - Environmental aspects - New Zealand - Case
    studies
    I.Ericksen, Neil J.
    333.7'0993

Library of Congress Cataloging-in-Publication Data
Plan-making for sustainability : the New Zealand experience / Neil Ericksen ... [et al.].
    p. cm. -- (Urban planning and environment)
    Includes bibliographical references and index.
    ISBN 0-7546-4066-3
    1. Environmental policy--New Zealand. 2. Environmental management--New Zealand. 3.
    Sustainable development--New Zealand. I.Ericksen, Neil J. II. Series.

GE190.N45P56 2004
363.7'00993--dc22
 2003062015

ISBN 13: 978-1-138-25893-8 (pbk)
ISBN 13: 978-0-7546-4066-0 (hbk)

Contents

List of Figures

List of Tables

Notes on Authors

Neil J. Ericksen is Professor and founding Director of the International Global Change Institute at The University of Waikato, Hamilton, New Zealand. His research interests are in evaluating development of national policies for managing resources, hazards, and environment and factors influencing their translation into practice by local government. He is author and co-author of several books and monographs in this general field of planning and governance, and has particular expertise in human response to natural hazards.

Philip R. Berke is a Professor in Land Use and Environmental Planning in the Department of City and Regional Planning at the University of North Carolina, Chapel Hill, USA. He was a Senior Fulbright Research Scholar at the International Global Change Institute, New Zealand in 1993, and is a Research Fellow at the Lincoln Institute of Land Policy, Cambridge, Massachusetts. His recent research focuses on local planning and sustainable development, and the influence of compact urban forms on watersheds. He is co-author of several books and has contributed articles to a variety of environmental policy and planning journals, receiving the Best Article Award in 2000 by the American Planning Association. He is currently preparing the 5th Edition of Urban Land Use Planning (Philip Berke, Edward Kaiser, and David Godschalk, University of Illinois Press).

Janet L. Crawford is a practitioner and Director of Planning Consultants Ltd in Auckland, New Zealand. She has over 25 years of planning experience, first within local government and then as a consultant to both the public and private sectors. Her main interests are environmental conflict resolution, plan writing and plan implementation and she is an established teacher and trainer in these areas.

Jennifer E. Dixon is Professor and Head of Department of Planning, University of Auckland, New Zealand. Before taking up an academic position, she worked for several years as a practitioner. Her research interests are planning practice, environmental assessment, and urban intensification. She has contributed a number of articles and book chapters in these fields. She was formerly President of the New Zealand Planning Institute (1996-98).

Foreword

Since the enactment of the Resource Management Act in 1991 the Parliamentary Commissioner for the Environment (PCE) has taken an active interest in its implementation. Through the mid 1990s a number of investigations focused on consultation processes, compliance with consents, the effects of 'side agreements' (consent purchasing) and the management of such matters as amenity values and urban vegetation. More recently, the PCE has focused on whether or not the Act is enabling communities to achieve the desired environmental outcomes. All too frequently it is not, and the reasons often lie in the quality of the plans and the breadth and depth of understanding of what is really needed to improve environmental sustainability.

The RMA, as has been said many times, is an innovative piece of environmental legislation. However, it was introduced at a time when sustainability concepts were only just beginning, in the immediate post Bruntland phase, to enter the wider public policy discourse of the western world. This was a difficult context for a major paradigm shift, a new way of thinking about how we live within environmental limits (i.e. sustainably). Given the lack of understanding, and conflicting understandings, as to just what sustainability really means in the RMA context it is not surprising that communities the length and breadth of New Zealand are still struggling to develop and implement effective plans.

Plan-making for Sustainability: The New Zealand Experience by Neil Ericksen and his team makes a substantive contribution to our understanding of the quality of RMA plans. The research literature on plan making, and implementation, for sustainability is limited not just in New Zealand but also internationally. This is surprising given the magnitude of what is trying to be achieved through environmental plans and sustainability policy in general in many countries. *Plan-making for Sustainability: The New Zealand Experience* provides robust insights into why RMA plans have not lived up to expectations. The study puts a spotlight on the magnitude of the shift from an activities based planning system to an effects based one against a backdrop of market liberalization and restructuring of local government. In organizational terms, making such a large transition in planning approach, on a canvas of significant governance restructuring and societal changes, necessitates robust legislation, central government capacity building and investment in local capacity to create good plans.

Neil Ericksen and his team have provided evidence that these conditions were not met and provide suggestions for remedying them. In doing so, they are making an important contribution to global understanding of the effectiveness of different approaches to environmental policy making. The work joins that of recent seminal studies such as *Green Society: The Paradigm Shift in Dutch Environmental Politics (2002)* in which Peter Driessen and others have proposed that advancing environmental sustainability necessitates employing sociology, policy and political sciences to assist the design of effective policies. They have also highlighted that overt faith in managing environment effects through ever more sophisticated technical solutions has limitations.   Increasingly it is evident that advancing environmental sustainability requires countries and communities to define what they are trying to sustain.   Then, via new relationships between government, the market, and civil society, they need to develop mechanisms that will lead to major redesign of production systems and changes in our consumption patterns. In short, environmental sustainability will only be advanced from deep within the values, beliefs, and behaviour systems of our societies. Hopefully the next generation of RMA plans will be built on such understandings.

Dr J. Morgan Williams
Parliamentary Commissioner for the Environment, New Zealand
October 2003

Preface

This book is about planning for environmental sustainability with particular reference to the innovative approach taken by New Zealand in the late 1980s, including its *Resource Management Act* (RMA) 1991. The RMA transformed the traditional land use planning system into an environmental management regime designed for the purpose of 'promoting the sustainable management of natural and physical resources'. The focus of the book is on the quality of plans produced under this new planning regime and factors of governance influencing that quality.

Several fields of scholarship are drawn upon, including land use planning, environmental science, political science and systems thinking. Consequently, many books and articles have informed the research for this book. Principal among these is *Environmental Management and Governance* (May, et al., 1996), which not only provided theoretical underpinnings for our evaluation of the effectiveness of the RMA as a devolved and co-operative mandate, but also prompted our focus on its effectiveness as evidenced by the state of practice. The need for this focus was reinforced because, since Pressman and Wildavsky's seminal work, *Implementation* (1973), scholars have struggled to develop methodologies for evaluating plans and the effectiveness of their implementation. Faludi, Healy, Kaiser and Godschalk, among others, have written on this topic. The literature on plan quality and implementation is, however, limited and little deals with plan-making for sustainability. Further, our book joins the ranks of those that chronicle the failures and successes of the 'New Zealand experiment'. One of the more influential has been Jane Kelsey's *The New Zealand Experiment: A World Model for Structural Adjustment* (1995), which was critical of the market-led reforms initiated in 1984 by the left-wing government of the day. Boston et al., in *Public Management: The New Zealand Model* (1996) describe and critique main features of the public-sector reforms. Harris and Twiname in *First Knights: An Investigation of the New Zealand Business Roundtable* (1998), critically examine the influential role of New Right businessmen in shaping the reforms.

These works dealt briefly with environmental management as affected by the reforms, if at all, leaving in-depth studies for others. Memon's *Keeping New Zealand Green: Recent Environmental Reforms* (1993) provided a useful description of the changes in the planning mandate and their underlying philosophy. Years later, he helped examine its impact in Memon and Perkins (eds) *Environmental Planning and Management in New Zealand* (2000). An early review was provided by Bührs and Bartlett in *Environmental Policy in New Zealand* (1993) and soon after the first empirical study of the RMA as a devolved and co-operative mandate was carried out by May, et al. (1996).

While our book owes much to what has gone before, there are many ways in which it extends scholarship and therefore makes a unique contribution to the literature on both plan-making for sustainability internationally and the New Zealand reform movement in particular. Much of this contribution is due to the comprehensive and integrated nature of the research. The fact that New Zealand is a small country means that it is possible to see the workings of the environmental planning system as a whole. Thus, in addition to developing a new method for assessing variation in plan quality that is suitable for New Zealand and capable of modification for use in other countries, this book delves deeply into the ways in which governance and capacity influence the whole planning system, including plan quality. It also scrutinises 'sustainability' as defined in the mandate, identifying shortcomings and their implications for effective resource management. The breadth and depth of analysis provides a compelling explanation of the gap between the early expectations for, and the reality of, plan-making in New Zealand under the RMA, and raises deep-seated concerns that poor governance will badly compromise its intent — a quality environment. The New Zealand experience should therefore provide useful lessons for other countries contemplating effects-based environmental planning within the context of a devolved and co-operative system of governance. (See for example Sumits and Morrison, 2001.)

This book covers the period 1991-2000, the first decade under the RMA. The methods on which the research was based were critiqued by professionals from private and public enterprise at Peer Review Group workshops in Auckland, Wellington and Christchurch, as were the findings subsequently. Recommendations from the research were summarized in a Report to Government in early 2001 (Ericksen, Crawford, Berke and Dixon, 2001), and distributed to key ministers and their departmental heads, regional and district councils, and various end-user groups with a view to influencing public policy and work programmes.

Thus, since finishing the research on which this book is based, many of the issues raised for, and recommendations made to, Government for improving the environmental planning and governance system have been considered by relevant agencies. The second generation of plans ought to be much improved as a result.

This book, like the research itself, was a collaborative effort. Neil Ericksen had overall responsibility for the structure and style of writing for the book, but more specifically for: writing the Introduction, chapters on governance (1, 3 and 4), and two cases (9 and 12). He also contributed to Chapter 2, the one on Māori interests (5), and the Conclusion. Philip Berke, principal theoretician and methodologist, wrote Chapters 6, 7 and 8 (plan quality in regional and district councils), and contributed much to the writing of Chapters 1, 2 and 5 (the theory and practice of mandates and plan-making, and Māori interests). His international perspective was valuable in placing New Zealand's experience into the world context. Using her expertise as a planning practitioner in New Zealand, Janet Crawford had an ongoing role of critically reviewing what others produced, while also writing chapters on Making Plans (2) and the Tauranga case (11). She also worked with Neil Ericksen on the Far North case study (9) and the Introduction and Conclusion. Jennifer Dixon, with input from Neil Ericksen, wrote Chapter 10, a case study on Queenstown Lakes District, contributed to Chapter 5 (Māori interests) and reviewed Chapters 6 and 7 in light of field interviews.

Acknowledgements

Research for this book could not have been done without the support of a large number of individuals and organizations. Of critical importance were staff and politicians in regional and district councils who graciously filled in questionnaires and participated in lengthy interviews. Together with representatives from many stakeholder groups in our four case study councils, staff and councillors willingly shared time, insights, and resources during our extended stays in the field. Likewise, staff in the Ministry for the Environment and Department of Conservation and other central government agencies provided information through documents and interviews. Over 70 staff from private and public agencies gave feedback in peer review workshops on research design, methods, and outcomes.

Important to the research effort were our research assistants, Sherlie Gaynor and Audrey Aird, who undertook content analysis of the policy statements and plans for the plan quality evaluations in Chapters 6-8, and written material for Chapters 9-12. Adhir Kackar, a post-graduate student, Department of City and Regional Planning, University of North Carolina at Chapel Hill, assembled the plan quality databases and helped with analyses. Claire Gibson of the International Global Change Institute (IGCI), University of Waikato, assembled secondary databases, organised field research, evaluated plans and provided valuable editorial assistance. Max Oulton, Senior Technician in the Department of Geography, The University of Waikato, and Wanda Ieremia-Allan, IGCI Information Manager, produced illustrations for the book.

The book gained a great deal by being reviewed in part, or as a whole, by several scholars and practitioners. For this we sincerely thank: Dennis Bush-King, Sarah Dawson, Lindsay Gow, Tom Fookes, Doris Johnston, Rene Kampman, Steve Markham, John Martin, Peter May, John Mitchell, Lynley Newport, Dennis Nugent, Andy Ralph, Carl Reller, Ray Salter, and Sue Veart. Sarah Michaels kindly allowed us to draw on her working paper on regions prepared for an earlier project.

In 2000, we invited Andrew Mason to be our editor and the book is all the better for his help. His involvement led to reorganisation and re-writes of the book and helped to shape the findings into a more interesting story.

In addition to assistance from these specific individuals, we benefited from comments by colleagues and practitioners on papers presented at various conferences in New Zealand and overseas, and in reviews of journal articles. A substantial portion of Chapter 8 and a figure and table in Chapters 1 and 2, respectively, appear in an article by Philip Berke, Janet Crawford, Jennifer Dixon, and Neil Ericksen, 'Do cooperative environmental planning mandates produce good plans? Empirical results from the New Zealand experience', *Environment and Planning B: Planning and Design*, 26(1999): 643-664 under copyright by Pion. Three illustrations and related text in Chapters 5 and 7 appear in an article by Philip Berke, Neil Ericksen, Janet Crawford, and Jennifer Dixon, 'Planning and Indigenous People: Human Rights and

Environmental Protection in New Zealand', *Journal of Planning Education and Research*, 22, no. 2 (Winter 2002) copyright by Sage.

Financial support for the research was from the Public Good Science Fund of the Foundation of Research Science and Technology, New Zealand, under contract to Massey University (MAU504; MAU604) and the University of Waikato (UOW504; UOW606), each with sub-contracts to the University of North Carolina at Chapel Hill, USA, and Planning Consultants Ltd, Auckland. This research built upon comparative research (United States, Australia and New Zealand, 1992-95) led by Professor Peter May, University of Washington, and funded by the National Science Foundation (USA). The contents of this book are not necessarily endorsed by any of these institutions.

Glossary of Māori Terms

haka	war dance or challenge
hapū	family or district groups or communities
hui	meetings
iwi	tribal groups
iwi authority	a term describing all bodies which represent iwi
kaitiakitanga	ethic of guardianship
kaumatua	male elder
kawanatanga	right of governance
kuia	female elder
marae	open space in front of a meeting house
pākehā	person of European descent
papakainga	general term for Māori housing on Māori land and marae
rangatiratanga	full chieftainship, the authority of the hapū or iwi to make decisions and control resources
raupo	wetlands
rohe	geographical territory of an iwi or hapū
rūnanga, rūnaka	a council of senior decision-makers, covering one or more iwi
tangata whenua, takata whenua	people of the land, Māori people
taonga	valued resources, assets, both material and non-material
urupā	burial place
wāhi taonga, wāhi taoka	sites of special significance
wāhi tapu	special and sacred places
wai	water
whānau	extended family

Source: Drawn from Boast (1989), King (1988), and Parliamentary Commissioner for the Environment (1998)

Abbreviations

CA	Court of Appeal
CEO	Chief Executive Officer of government agency
CHEPA	Coastal Hazard Erosion Policy Areas
DP	District Plan
DoC	Department of Conservation
EBOP	Environment Bay of Plenty (regional council)
FNDC	Far North District Council
FTE	Full-time equivalent staff
GIS	Geographic Information System
KRA	Key Result Areas
LGA	Local Government Act
MAF	Ministry of Agriculture and Fisheries
MfE	Ministry for the Environment
MoC	Ministry of Commerce
NWASCA	National Water and Soil Conservation Authority
NZCPS	New Zealand Coastal Policy Statement
PCE	Parliamentary Commissioner for the Environment
QLDC	Queenstown Lakes District Council
RCEP	Regional Coastal Environmental Plan
RCP	Regional Coastal Plan
RMA	Resource Management Act
RMD	Resource Management Directorate (in MfE)
RPS	Regional Policy Statement
SD	Standard deviation
SMF	Sustainable Management Fund
SOE	State of the environment
SRA	Strategic Results Areas
TA	Territorial Authority (district council)
TasDC	Tasman District Council
TDC	Tauranga District Council
TDP	Tauranga District Plan
TDS	Technical Services Division
TRMP	Tasman Resource Management Plan
WBOPDC	Western Bay of Plenty District Council
WBOPDP	Western Bay of Plenty Proposed District Plan

Introduction

From Rio to RMA: Great Expectations

Since the Stockholm Conference on the Human Environment in 1972, a score of major United Nations conventions and commissions have been established. These are like signposts on the international community's journey in search of solutions for protecting the Earth's threatened environments in the face of ever-increasing consumption and pollution of its natural resources. The journey is through both time and space and deals not only with intergenerational issues, but also with problems of scale from global to local, including the globalization of trade and its social, economic and environmental impacts.

Today, concerns about global environmental sustainability are even more acute than they were 30 years ago. A key challenge over the intervening decades has been for countries to develop policies that prompt local communities to plan for sustainable development in ways that result in quality environments. In the 1980s, as environmental issues were popularized through slogans like 'Think global, act local' some countries developed mandates (i.e. statutory goals) for dealing more effectively with the environmental harms caused by resource development in their communities by adopting a more integrated approach to sustainability. These efforts were enhanced by the Environment and Development Conference in Rio de Janeiro (RIO) in 1992, which adopted not only conventions for dealing globally with the adverse impacts of climate change and loss of biodiversity, but also a charter for local action through Agenda 21.

Around the time of RIO, some countries, like the Netherlands (primarily urban and industrial) and New Zealand (primarily rural and natural resources), were already leading the way with quite innovative approaches to environmental planning. These sparked considerable international interest. In this book, we focus on New Zealand's approach to sustainability and environmental planning.

The New Zealand Government adopted its new environmental planning mandate in 1991 after three years of development and extensive public consultation and discussion. It was called the *Resource Management Act* (RMA). The RMA replaced not only a comprehensive planning statute, but also a range of topic-oriented environmental statutes that are typical of the approach used in many other countries.

The purpose of the RMA was to promote the sustainable management of natural and physical resources through a devolved and co-operative system of governance. At its heart the RMA required individuals or groups to internalize (i.e. absorb) the environmental costs of their use, development and protection of natural and physical resources. Linked to reforms of the economy and central and local government, the RMA provided for a hierarchy of effects-based environmental policies and plans, strengthened local government structures, and gave central government a capacity-building role. Nationally and internationally, there were great expectations for this new environmental planning model. After ten years on the road, have these been fulfilled?

That is the question behind this book. We wanted to know whether this ambitious and sophisticated model of environmental planning was being effectively implemented.

There was some cause for doubt. By the mid-1990s, as the new local policies and plans became publicly notified, there was a considerable public backlash against the RMA, in contrast to the optimism that accompanied its introduction. Media headlines of the day included: 'System overload — industry staggers under the pressure'; 'Unrealistic standards demanded by councils'; 'Some council plans lack "certainty and clarity"'; 'Clean, green and expensive — the high cost of NZ's dollar-driven resource law'.

The pervasive nature of discontent was encapsulated some years later by Kim Hill, a popular national radio interviewer, who observed that the RMA was 'sitting like a blob in the middle of the New Zealand psyche' (Hill, 2002). It seemed as though many of the less than four million New Zealanders living in their South-West Pacific country (see Figure I.1), similar in size to the United Kingdom or the State of Colorado, valued high quality environments, but railed against processes and costs for achieving them.

We confined our study of planning under the RMA to the 1990s. This enabled us to trace the influence of the 'New Right' era of the Labour and National Governments on its implementation. This influence ended as the left-leaning Labour-Alliance Government took the country into the new millennium.

A Need for Change?

Fifty years ago, New Zealand had evolved into a parliamentary democracy with a centrally directed mixed economy. The state owned much of the natural resources, such as land and forests, while others, like rivers, lakes and the ocean, were in the public domain. Coming out of depression and war (1930s and 1940s), there was a great need for growth and development. The Government led the opening of new frontiers and strongly influenced development in order to promote its social and economic objectives. However, the rapid loss of natural heritage and dramatic transformation of the landscape strengthened environmental consciousness (McCaskill, 1973; Powell, 1978; Searle, 1975; Wilson, 1982). Legislation was enacted to control development and activities (through statutory town and country planning) on the one hand, and environmental impacts (through statutes aimed at specific environments) on the other.

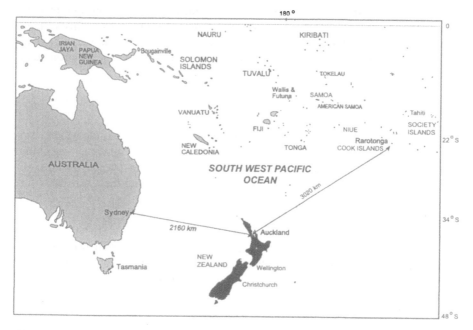

Figure I.1   Location of New Zealand in the South-West Pacific Ocean

Top-down and Piecemeal

By the 1980s, incremental changes to institutional arrangements for dealing with growth management and environmental impacts had resulted in a plethora of central resource management agencies. They tended to operate in a top-down fashion and each had multiple and overlapping functions (e.g. policy and operations and/or development and conservation) underpinned by a multiplicity of statutes. The changes also resulted in a complex and often contradictory system of publicly funded grants and subsidies that confounded sound economic and environmental outcomes. Sandrey and Reynolds (1990) provide useful examples from the agricultural industry.

Too Much Red Tape

Another set of concerns had also emerged by this time. Resource developers, including the Government, complained about the length of time for gaining permission to use and develop resources. These 'red-tape' impediments to development applied as much to large-scale projects as to small-scale local activities. Simplified and faster processes were wanted. There were also loud calls for central government to decentralize decision-making to local levels where the effects of activities occurred and innovative solutions to problems might emerge. While land-use planning and water and soil conservation already provided some measure of decentralized decision-making, central control remained strong. There was dissatisfaction with the environmental outcomes of this system, because developmental goals in central government agencies took precedence over conservation and protection goals. (See for example, Ministry for the Environment, 1988; 1990; and Wilson, 1982.)

Māori Renaissance

Finally, by the late 1980s, the indigenous people of New Zealand, the Māori, had had their interests in resource development and environmental planning heightened. Recent legislative changes, such as in the *State Owned Enterprises Act* (1986) and *Treaty of Waitangi Act* (1975), had improved their prospects for both resolving historical grievances against the Crown[1] for wrongful land acquisitions, and ensuring that their values and interests were acknowledged in contemporary resource development and conservation activities.

In summary, when resource management law reform got under way in 1988, there was broad agreement that planning was handicapped by a system of fragmented, unco-ordinated, and overlapping statutes that was costly to administer, and that the plethora of existing resource law should be integrated into a single statute and serviced by fewer, more efficient, agencies.

Conflicting Philosophies for Change

When the reform movement accelerated in the mid-1980s, an unholy alliance formed between environmental and business interests calling for radical change. Both wanted decentralization of, and accountability in, government and more rational use of grants distributed by central government for developing or conserving resources. While they shared a view of the need to reform not only the system of governance, but also the environmental and economic systems, the philosophies underpinning their respective positions were markedly different (Bührs and Bartlett, 1993).

The 'New Right' Agenda

Business wanted government to be downsized and less regulatory, in order to facilitate resource development through an unfettered market. Among the advocates of economic (and administrative) reforms were leaders of some major

---

[1] 'Crown' refers to the system of government established in New Zealand as a constitutional monarchy. The head of state is the Queen (Queen Elizabeth II of England), represented in New Zealand by the Governor-General. As head of state, the sovereign is the nominal source of law. In practical terms, however, the power of the monarch has largely been superseded by Parliament (McDowell and Webb, 1998: 110).

government agencies and private corporations. For them, adverse environmental effects caused by market failure in the use and development of resources would be best solved not through government regulation, but by creating a set of property rights that would bring the externalities (i.e. the environmental costs of resource use) into the market system. For the public sector, they argued that agencies administering public-owned resources should be corporatized, before being sold. Where this led to private monopolies, legislative reforms would encourage future competition. What is more, they wanted public agencies to be run by a series of contractual arrangements that made their operations more transparent, efficient and effective in achieving specified goals. This perspective is encapsulated in Harris and Twiname (1998).

The 'Greenies' Agenda

Environmentalists wanted controls on development to conserve natural resources and protect the environment, and sought to separate environmental management from developmental responsibilities within government agencies. Advocates for environmental (and administrative) reform were by and large adherents of a devolved participatory democracy. Most environmentalists, and supporters of centre left ideology, saw a need for a combination of measures to bring about environmentally sensitive, resource-sustaining developments, including government regulation, public information campaigns, environmental education, financial incentives and so on. They opposed initiatives that both privatized public assets and diminished the bureaucracy in a manner that threatened democracy, such as through rigorous application of public choice theory. One prominent environmental group (the Maruia Society) did, however, come to embrace creating a set of property rights that would bring the externalities caused by resource use and development into the market system.

Reforming the Regime

The principal reforms in response to the issues being debated by these various pressure groups occurred under what was assumed to be a centre-left Labour Government (1984-90). The process showed the schizoid character of this Government (a centre-left political party implementing 'New Right' policies), a condition that soon tore it apart. While economic and administrative reforms set out on the 'New Right' path, environmental and local government reforms followed a social democratic path. Future tensions seemed inevitable. As the blueprint for change emerged more starkly in 1988, a revolt within the ruling party caused the pace of change to slow (see for example, Boston et al., 1991). However, the shift to the 'Right' intensified once again after the conservative National Party came into power in late 1990, while the local government and environmental reforms continued much as planned earlier. While all parties wanted a greater role for sub-national government in decision-making relevant to national mandates, political parties argued over which model to follow — decentralization (where control was retained at the centre) or true devolution (where power was in local government).

New Environmental Ethos

The reform of environmental administration occurred within wider economic, public sector and administrative reforms at central, regional and local levels. All were driven by the New Right ideology of monetarism and market liberalism, and included consideration of economic instruments in policy implementation, commercialization of the public sector (i.e. profits ahead of public-good objectives), corporatization of government agencies (i.e. public agencies trading as businesses) and user-pays strategies for government goods and services.

The overall aim was to get government out of business activities and business practices into government activities (Bührs and Bartlett, 1993; Martin and Harper, 1988). To a large extent, this meant leaving resource development to the private sector and having sub-national government control the environmental effects of the use, development, and protection of not only land, but also water, air, and coastal resources, through environmental planning and management. The intention was to further devolve decision-making from central government to regional and local government, where problems occur. Greater efficiency and democracy would result, it was thought, because local decision-makers would have better information about local problems and solutions and could better ascertain citizens' preferences than decision-makers in central government (Bassett, 1987; Elwood, 1988; McKinlay, 1990; Martin, 1991; Martin and Harper, 1988; Palmer, 1988).

Central government agencies   By 1988, major government departments and ministries had been restructured into functionally based agencies. Having both commercial and environmental functions within one agency had resulted in the former dominating the latter. These functions were therefore separated and each placed into primarily single-function agencies (Ericksen, 1990). The elements of conservation and environmental protection and enhancement within the former multi-function resource agencies (e.g. the New Zealand Forest Service, Department of Lands and Survey, and Ministry of Works and Development) went into three new environmental agencies: the Department of Conservation, Ministry for the Environment and Parliamentary Commissioner for the Environment.

The Department of Conservation was charged with administration of public lands held for conservation, advocacy and operations; the Ministry for the Environment was to focus on environmental policy advice and implementation; and the Parliamentary Commissioner for the Environment was to act as Parliament's environmental watchdog (*Environment Act*, 1986). The Ministry for the Environment was to be the lead agency for implementing legislation aimed at environmental planning locally, although the Department of Conservation was given the lead role for implementing environmental policies affecting the coastal marine area (Ministry for the Environment, 1987).

Local government agencies The 1989 reform of local government resulted in 255 regional and local entities (plus 430 special-purpose ad-hoc authorities) being reduced to 93 new units, consisting mostly of 13 regional councils and 74 local councils, including one unitary authority (Gisborne). In 1993, one region was converted to three dual function unitary authorities (Nelson, Marlborough and Tasman), leaving 12 regional councils and four unitary authorities (Figure I.2).

Changes to the *Local Government Act* (1974) required managerial principles of transparency and accountability to be applied to these councils, including the separation of policy-making from service delivery and/or regulatory roles.

The rationale for having new regional councils was to create an organizational framework to carry out integrated resource management within a coherent region, the functions and duties being made clearer once the *Resource Management Bill* (1989) became law in 1991 (Ministry for the Environment, 1991). A key feature of the new regional councils was that their political boundaries were drawn to coincide with the watershed boundaries of the previous catchment boards and regional water boards. The aim was to enhance the prospects for effective management of renewable resources that require regional co-ordination.

The 74 new local councils included 60 district councils (mostly rural or suburban), 13 city councils and one county council (Chatham Islands), each of which would be required to produce new district plans when the *Resource Management Bill* (1989) became law. These local councils consolidated the many separate and independent functions of previous local authorities.

Three-tiered Structure

By 1990, the Government had created a three-tiered structure for administering environmental planning, involving central, regional and local government. On the one hand, the new administrative system was underpinned by notions of devolution, flexibility, co-operation, partnership, participation and capability-building. On the other hand, it was to operate on the competitive principles of free-market liberalism. Through this contradiction, the new agencies would be required to implement a new planning regime under the RMA. This would include the development of new plans by the new councils aimed at achieving sustainable management and dealing with a wider range of resources and environmental concerns than plans produced under the previous planning regime.

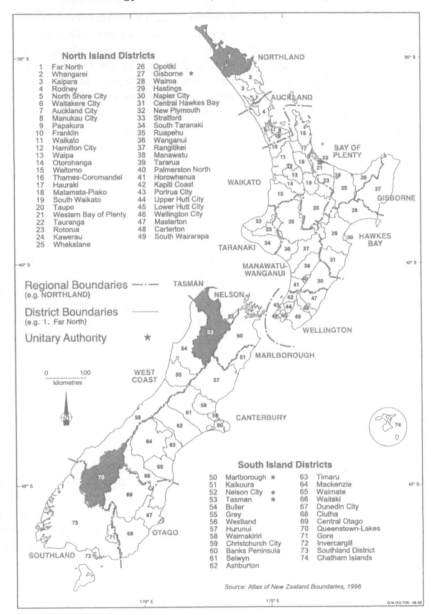

Figure I.2 The location of regional and district councils in New Zealand
(including the four dual function unitary authorities, which are denoted
by an asterisk in the district council list. Darker shading indicates the
four case study districts detailed in Part 4 of the book)

Source: Atlas of New Zealand Boundaries, 1996

The RMA Planning Approach

Restructuring of the resource development and environmental agencies of central government took place between 1985 and 1987 and the reform of local government between 1988 and 1989, but the latter occurred in tandem with the resource management law reforms. The two reforms were under the control of the Minister for the Environment (Geoffrey Palmer), also Deputy Prime Minister. From 1988-90, over 60 statutes and regulations were combined into one statute: the RMA.

Key elements of environmental planning that were drawn into the RMA were the *Town and Country Planning Act* (1977) — concerned mainly with land-use planning — and the *Water and Soil Conservation Act* (1967) — concerned with controlling water quality, flooding and drainage, and soil erosion. Also drawn into the RMA were the project-oriented *Environmental Protection and Enhancement Procedures* developed by the Commission for the Environment in 1973, which had gained statutory recognition in the *National Development Act* (1979), legislation aimed at fast-tracking government projects stalled by multiple statutes and environmental protestors.

Sustainable Management and Development

By 1991, the Government had an overall strategy for sustainable development of natural resources, in which the RMA was a key element. The Government's three-tiered strategy is shown in Figure I.3. There, the main provisions of the RMA are listed down the left side of the diagram as they relate to national, regional and local areas of governance, including the preparation of policy statements and plans. Down the right side of the diagram are various policy instruments ranging from legislation for resources not covered by the RMA, to economic instruments, guidelines, codes of practice, education and research (New Zealand Government, 1991; Ministry for the Environment, 1992).

Two key points emerged from the Government's 1991 sustainable development strategy. First, national and sub-national policies and plans were not the only means by which the Government's environmental goals would be reached. Second, while the single purpose of the RMA was the sustainable management of natural and physical resources, political considerations narrowed its focus to renewable resources. But even there, fisheries, and energy were omitted, along with all non-renewable resources, such as minerals. These were to be dealt with in separate legislation, as indicated by the right side of Figure I.3.

IMPLEMENTING SUSTAINABLE DEVELOPMENT
The Policy Instruments Available

THE NATIONAL RESPONSE

NATIONAL AGENDA FOR THE IMPLEMENTATION OF SUSTAINABLE DEVELOPMENT

RESOURCE MANAGEMENT ACT

- National Policy Statements (only coastal plan is mandatory)
- National Environmental Standards
- Call-in of Projects of National Significance
- Water Conservation Orders
- Regulations (other than National Environmental Standards)
- Amendments to the RMA

OTHER INITIATIVES

- International initiatives and involvement
- Crown-Owned Minerals Act
- Financial policies (taxation, grants, duties, etc)
- Economic instruments (such as transferable quotas)
- Property rights
- Other legislative powers
- Education legislative powers
- Education, information guidelines, codes of practice and research
- Departmental policies

THE REGIONAL RESPONSE

LOCAL GOVERNMENT ACT

CORPORATE PLANNING • MISSION STATEMENT AND CORPORATE GOALS

RESOURCE MANAGEMENT ACT

- Regional Policy Statements (mandatory)
- Regional Plans (including rules)(only coastal plan is mandatory)
- Resource consents
- Enforcement powers

OTHER INITIATIVES

- Other legislative powers
- Education, research guidelines
- Codes of practice
- Other regional functions

THE DISTRICT RESPONSE

LOCAL GOVERNMENT ACT

CORPORATE PLANNING • MISSION STATEMENT AND CORPORATE GOALS

RESOURCE MANAGEMENT ACT

- District Plans (including rules)(mandatory)
- Resource consents
- Enforcement powers

OTHER INITIATIVES

- Other legislative powers
- Education, research guidelines
- Codes of practice
- Other regional functions

Figure I.3   RMA and other policy instruments in the New Zealand Government's agenda for sustainable development of natural and physical resources

Source: adapted from MfE, Information Sheet No. 1, June 1992

Policy Intent of RMA

A mid-1990s book on *Environmental Management and Governance*, the research for which was led by political scientist Peter May, compared New Zealand's approach to environmental management with that of New South Wales (Australia) and of Florida (United States of America). The main aim was to characterize the mandate of each country (e.g. centralized or devolved) and then to examine its implementation. Of interest was the effectiveness of intergovernmental processes within each country aimed at facilitating implementation through the hierarchy of governments (e.g. central, regional, and local).

Analysis of the RMA by May et al. (1996) revealed six unique characteristics that distinguished New Zealand's approach to environmental planning. These are:

1. the RMA requires regional and local councils to manage the use, development and protection of natural and physical resources in ways that enable people and communities to provide for their social, economic and cultural well-being and their health and safety, while at the same time achieving three environmental objectives. These are:
    *   to ensure resources meet the needs of future generations;
    *   to safeguard the life-supporting capacity of air, water, soil and ecosystems; and
    *   to avoid, remedy or mitigate the adverse environmental effects of activities on the environment.
2. to achieve these goals councils have to prepare plans for assessing the environmental effects of the use and development of resources in their areas.
3. regional councils have the role of integrated natural resource management to counter the tendency of district councils to promote parochial interests.
4. strong emphasis is placed on public participation in plan-making, which includes provisions for conflict resolution through a public hearing process.
5. special attention is placed on participation of New Zealand's indigenous people, the Māori, to uphold their rights in the guardianship and stewardship of the land and resources. The legislation specifically requires that local government uphold the Treaty of Waitangi, which guarantees Māori rights.
6. emphasis is placed on a co-operative approach by central agencies to achieve local government compliance with goals of the RMA.

In essence, the policy intent of the RMA is to provide an overarching framework for integrated and sustainable management of renewable resources. It presents a rational, comprehensive and systematic approach to decision-making. The four main components of the system are:

1. policy analysis and plan preparation (by central, regional and local government);
2. assessment of environmental effects (at policy and operational levels);

3.   decision-making on proposed developments (by local government commissioners and, if need be, the Environment Court); and
4.   monitoring of policy implementation and consents (by regional and local government), which leads to ongoing review of and amendments to plans.

Key provisions of the RMA important to our study are in Appendix 1 to the book.

Effects-based Planning

In developing the RMA, the policy-makers considered that complex environmental problems are interconnected and require comprehensive and integrated solutions that can best be derived through local planning.   Their version of planning, however, emphasized controlling adverse environmental effects from the use and development of renewable resources, rather than, as hitherto, controlling the resource use activities themselves.   Thus, the intent of the RMA was to create council plans that provide a flexible resource management framework.   The limits of the environmental effects of activities would be defined, but the activities themselves would be unspecified, giving opportunity for market innovation and technological change to define them.   This would gain maximum environmental benefit with minimum environmental regulation of the market.

Evaluating Planning Under the RMA

In this book we examine the effectiveness of New Zealand's new environmental planning regime to the year 2000 by evaluating the quality of policies and plans prepared by councils under the RMA, and assessing the intergovernmental processes and influencing factors that conditioned their development.   Questions guiding our research included: do local plans produced within a devolved, co-operative system of governance achieve national goals while at the same time provide policy solutions that meet local aspirations; do local councils have the capability to make high-quality plans that meet the environmental goals of central government; does central government have the commitment to effectively build the capability of local councils to make high-quality plans; and what impact has Government reluctance to clarify its Treaty partnership had on the responsiveness of councils to Māori interests when preparing policies and plans?

Mandates and Planning Approaches

In Part I on approaches to planning and governance, we characterize the RMA in greater detail.   It enables us to introduce key terms and concepts and to summarize a little of the planning theory behind the RMA.   Some readers may wish to skip over this material and go directly to Part II.   They should, however, note that Chapters 1 and 2 provide the conceptual basis for Parts II and III of the book, respectively.   In Chapter 1, we outline the theoretical range of mandates and review some of them through international examples. This allows us to place the RMA mandate in relation to others. We then offer a theoretical framework for assessing variation in plan quality between the entities of local government, with particular reference to New Zealand. In Chapter 2, we briefly review the range of approaches for making plans, describe the rational-adaptive approach to plan-making adopted by the RMA, and then, drawing on international research on planning, develop criteria for evaluating the quality of regional policy statements and district plans in New Zealand. The various methods used to carry out the investigations are summarized in Appendix 2, while the plan coding protocol for evaluating plans in a consistent and reliable fashion is included as Appendix 3.

Devolution and Co-operation

In Part II we develop the principles of mandate design explained in Part I by investigating New Zealand's version of intergovernmental planning (central, regional, and local) within a 'co-operative' framework. Such a mandate includes regulatory goals (rather than the prescribed contents typical of command mandates) and relies heavily on the organizational capability of local government to do the mandated planning. Like many other countries, New Zealand has wide variation in local capability across rural and urban councils. Where local capability is lacking, the challenge for central government is to provide leadership in building it up in a facilitative and non-interventionist manner. If central government fails in this task, local governments may view the national mandate with apathy or worse, as abusive, due to central government neglect.

     We examine in turn the extent to which: central government has carried out its role of capability builder for sub-national government under the RMA (Chapter 3); regional councils have been successful partners to local councils in advancing environmental planning (Chapter 4); and Māori interests under the RMA have been considered by central government as Treaty partner, and implications for successful sub-national planning (Chapter 5).

Plan Quality and Influencing Factors

In Part III we give special attention to how well plans integrate the effects-based approach to environmental management required by the RMA, rather than, as previously, simply regulating land use activities. Testing the quality of effects-based plans proved a major challenge. The planning literature is surprisingly narrow when it comes to defining plan quality, and non-existent for effects-based plans. Formulating suitable criteria for evaluating plan quality under the RMA will obviously help in understanding plan-making and making plans better. The framework developed in Chapters 1 and 2 was applied nationwide and the results for regional policy statements and district plans are discussed in Chapters 6 and 7 respectively. Some key issues identified in Part II are also considered in these chapters, including the responsiveness of councils to Māori interests.

We also examine the organizational capability of local government for plan-making, which is essential to carrying out the technically and politically complex tasks required for effects-based plan preparation. We focus on how local government decision-makers can enhance local capability to prepare high-quality plans. Indeed, national goals, even when mandated by central government legislation, can be successfully implemented only when there is local capability to effect change through planning. The links between local and regional plans and factors influencing their quality, such as political commitment and council capability, are tested in Chapter 8.

District Council Case Studies

In Part IV we focus on four local cases, chosen for the relatively good quality of their district plans. They are the Far North District (Chapter 9); Queenstown Lakes District (Chapter 10); Tauranga District (Chapter 11); and Tasman District, which is a unitary authority with both regional and district functions (Chapter 12). The quality of the plans in these four districts, as evaluated against the eight plan-quality criteria (Chapters 2 and 7), depends a great deal on the interplay of various factors within the plan-making steps, as well as the implementation strategies of national and regional government. In each case, crucial to the plan quality outcome is the effectiveness of 'research' and 'consultation' in achieving environmental and community 'fit'. Each case exemplifies a major theme dealt with earlier in the book (e.g. devolution, managerialism, integration, partnership) and one matter of national importance identified in Part II of the RMA (e.g. significant natural areas, outstanding landscape values, coastal margins, and iwi interests).

Main Findings and Lessons Learnt

The Conclusion pulls together the overall findings of our research by structuring them according to four themes: promoting environmental sustainability in New Zealand; why resulting plans are so weak; shortcomings in governance; and the high price paid for risky reforms. From this review stemmed 20 recommendations in five inter-related sets specific to New Zealand (summarized in Annex C.1, Conclusion).

PART 1
APPROACHES TO
PLANNING AND GOVERNANCE

Chapter 1

Planning Mandates:
From Theory to Practice

When national governments promulgate planning mandates, assumptions are made about the capabilities (commitment and capacity) of agencies in the intergovernmental hierarchy to implement them. 'Commitment' is the willingness of key people in national, regional and local agencies to promote goals in the mandate, while 'capacity' is the resources (like funds and expertise) available to agencies for achieving the goals. The levels of commitment and capacity will influence the implementation effort and hence the degree of compliance with the mandated goals. Thus, if the national agencies responsible for environmental planning are properly resourced and committed, the sub-national agencies will have their capabilities enhanced. Overall, this ought to result in good processes for preparing and implementing quality plans. There are, however, different types of planning mandate with distinctly different assumptions about capability and the processes and provisions for achieving the best results. These differences affect the plan-making process and hence the quality of resulting plans.

In this chapter we outline the theoretical range of mandates, from coercive to co-operative, for achieving capability, and illustrate them through some international examples. We then place New Zealand's RMA in relation to the theory and practice, and offer a theoretical framework for assessing variation in plan quality.

Range of Mandates

A mandate is a policy instrument that governments use to help implement their broad policy goals. In a parliamentary system, as in New Zealand or the United Kingdom, the mandated goals of central government influence the operations of sub-national government, like regional and/or local councils. In a federal system, like the USA and Australia, the mandated goals of federal government influence the states, and those of the states influence local councils. In this discussion, we use national and sub-national governments when referring to the differing levels, except for New Zealand, which has central, regional and local governments, the latter two typically called territorial authorities (TAs).

In developing a national (or state) mandate, assumptions are made about the behaviour of people and agencies, and their readiness to sacrifice self-interest for the common good. A pessimistic view of behaviour leads to mandates that are coercive and command-driven from the centre, whereas an optimistic view of behaviour leads to mandates that are devolved from the centre and carried out in co-operation with local governments. Useful explanations of coercive and co-operative mandates in environmental and land-use planning are provided by May and Handmer, 1992; Burby and May et al. (1997) and May et al. (1996).

Coercive Mandates

A coercive mandate assumes that sub-national governments are not committed to the goals of the national government (even though they may well have the capability to comply) and are therefore unlikely to co-operate in achieving them. The primary role of the national government is inducing sub-national governments to adhere to its regulatory standards. The national government does this by, for example, monitoring local plans and regulations for their consistency with national objectives and enforcing their provisions through the use of penalties or sanctions when sub-national governments fail to comply.

The centrally driven and coercive mandate is therefore interventionist, and capability-building is of only secondary concern. Since sub-national governments are assumed to be not committed to national goals, they cannot be trusted voluntarily to develop innovative plans and measures for managing growth and environmental harms. Instead, the national government prescribes not only the plan development process, but also the content for local plans. This is done to ensure that sub-national government will meet the nation's regulatory standards. It is expected that implementation of this mandate may be compromised by political and economic considerations that weaken the monitoring and enforcement effort of the national government.

Co-operative Mandates

A co-operative mandate assumes that sub-national governments are committed to the regulatory goals of the national government and willing to co-operate in achieving them, but do not necessarily have the capability to comply fully. The primary role of national government is to enhance not only intergovernmental co-operation, but also the capability of sub-national governments so that they are better able to comply. National government does this by, for example, technical and financial assistance, education programmes, guidelines and consensus-building forums. This in turn reinforces commitment to the nation's regulatory goals.

The devolved and co-operative mandate is therefore facilitative rather than interventionist. While the national government may prescribe the planning processes sub-national governments must follow in developing plans, it refrains from prescribing the means by which they are to achieve the desired goals and regulatory outcomes. That is, instead of the national government prescribing the content for local plans, it is assumed that sub-national government can devise the best means for reaching the mandated national goals. This concession to sub-national government aims to foster flexibility in dealing with local problems in the hope that it will stimulate innovative solutions. Sub-national governments that successfully meet this challenge are rewarded by gaining good planning outcomes in terms of growth management and environmental threats, whereas the laggards and recalcitrants will in time learn by their mistakes and through further capability-building by national government. Thus, a relatively slow and uneven uptake of the national mandate by sub-national government is to be expected.

Mandate Form and Control

A finer distinction between mandated policies can be made by focusing on form and control, each of which can be characterized by a continuum (May, 1993; Ericksen, 1994). Mandate form has 'programmatic policies' emphasising planning processes and standards at one end of its continuum, and 'prescriptive policies' prescribing or proscribing specific activities at the other. Mandate controls for influencing local government action have co-operative provisions that facilitate action at one end of its continuum and coercive provisions that punish inaction at the other (see Figure 1.1).

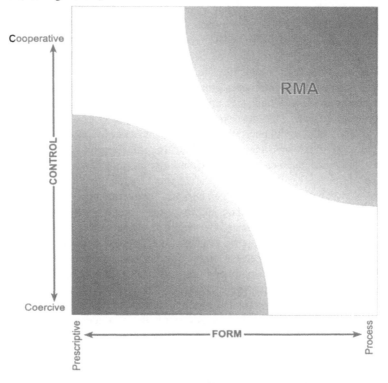

Figure 1.1 A schema illustrating the range of mandate designs based on form (process to prescription) and control (co-operative to coercive) (The location of New Zealand's RMA indicates a primarily co-operative-process mandate)

In theory, on the basis of form and control, mandates or policies could be mixed in various combinations, such as process/co-operative, process/coercive, prescriptive/co-operative, and prescriptive/coercive (see Figure 1.1). Selecting the type of mandate for implementing policies, environmental or otherwise, reflects the political ideology and preferred governance style of the ruling party, modified by electoral considerations.

Implementing Policy Goals

Governments can use various policy instruments or mechanisms to help implement their policy goals. McDonnell and Elmore (1987) provide a useful four-fold typology of mechanisms for translating national or state policy goals into local action.

The first mechanism is 'system-changing', which involves the transfer of authority among individuals and agencies for the purpose of altering the delivery system of public goods and services, such as changing from a command to a devolved regime in order to improve planning outcomes. The reforms in New Zealand (as summarized in the Introduction) exemplify a major system change from central to devolved control. The second mechanism is the 'mandate', like the RMA, which provides the rules aimed at producing compliance by governing the actions of individuals and agencies, such as through use of various coercive/ prescriptive or co-operative/process measures. The third is 'inducements', which transfer resources to individuals or agencies for particular actions, with the aim of fostering commitment. The fourth mechanism is 'capacity building', which is the transfer of resources for the purpose of investing in material or human capital.

Because of their importance in facilitating plan preparation in sub-national government, we define these last two mechanisms in more detail below.

Inducements   Two assumptions lie behind inducements. Without additional funding, certain valued things, such as the development of environmental plans, will not be produced with the desired frequency or consistency. Transferring funds is a main means for eliciting performance. One issue is how much variation in production of the thing of value is tolerable and how narrowly to prescribe the way the funds are to be used and for what — thus, for plan-making, how much variation in quality across jurisdictions is acceptable. The effectiveness of inducements increases when the capacity exists to produce things, like plans, that policy-makers value and when preferences and priorities support the production of those things.

Capacity-building   Like all investment in material or human capital, capacity-building carries the expectation of future returns, which may or may not be certain, tangible, measurable and immediate. In general, specific individuals and institutions gain in the short-term, while society at large gains in the long-term. Capacity-building may be instrumental to both mandates and inducements.

Two assumptions underlie capacity-building. One is that, without immediate investment, society will not realize the material, intellectual or human benefit, such as from the development of environmental plans. The other is that longer-term benefits are desirable in their own right or are instrumental for other important purposes. Usually, capacity-building responses deal with fundamental failures of performance, but they are also important when introducing a new mandate, such as for environmental planning. A major problem for policy-makers is that the results of capacity-building are intangible and uncertain. Success is challenged because there are tensions between mobilising resources for the future and the immediate production of value (McDonnell and Elmore, 1987).

How do Planning Mandates Influence Plan Development?

Political science and public administration have contributed greatly to understanding how top-down mandates are implemented, especially since Pressman and Wildavsky's (1973) ground-breaking book, *Implementation*. Most studies in this genre focus on the political behaviour of those involved in implementing social and economic policy. The objective is to explain why the process may or may not have gone awry (Googin et al., 1990; Younis, 1990). While planning can profit by insights from this research, there has been little parallel inquiry into the implementation of planning mandates aimed at guiding land-use and development patterns in communities in general, and quality environmental planning in particular. Talen observes that 'Planning needs to develop its own brand of [implementation] research that is sensitive to the physical, spatially referenced side of planning — specifically, the making of plans that seek to guide the future development of cities' (1996: 248).

A few studies have begun to address the implementation of planning mandates, but until recent publication of works by Dalton and Burby (1994), May et al. (1996) and Burby and May (1998), how a 'devolved and co-operative' planning mandate influences local plan quality was largely unknown. Earlier research almost always examined the influence of mandates for regulating sub-national land-use patterns and the planning process, but not on the plan itself (DeGrove, 1992; Healy and Rosenberg, 1979; Innes, 1992).

European Community

Some work suggests that specific characteristics of planning mandates do affect plan quality in different ways. Faludi (1987) found that planning in the Netherlands was successful in achieving local plan compliance with national goals because of the co-operative emphasis of the national mandate. A key pre-condition for success was that local governments were willing and capable of achieving national goals.

But he found national planning in Britain to be more coercive and thus not as successful in achieving local compliance. The mandate stipulated detailed standards for achieving national policy goals that were backed up by sanctions if local governments did not comply. As a result, the lack of local discretion stifled innovation and experimentation, and caused a weak sense of ownership of plans by local governments and a low willingness to carry out the mandate.

Cullingworth (1994) reached similar conclusions in his comparative evaluation of Dutch and British planning regimes. But, like Faludi, Cullingworth based his exploratory case studies on limited empirical observation of the links between national planning mandates and local responses through plans. They must, therefore, be viewed as speculative about how different planning mandates influence local plans.

United States

A study by planning scholar Ray Burby and political scientist Peter May (1997) drew attention to the impact that the various characteristics of planning mandates have on the quality of local plans. These included the clarity of the goals in the mandate, incentive funds for local planning, the authority to coerce and assistance for building local capacity and commitment (or willingness) to plan. They were used to classify the planning mandates of various states in the US along a continuum from coercive to co-operative. A series of articles from this study examined the influence of these characteristics on the quality of local plans and on the stringency of regulations for managing development (Berke and French, 1994; Berke, Roegnik, Kaiser and Burby, 1996; Burby and Dalton, 1994; Dalton and Burby, 1994). A major finding was that state mandates had a significant impact on plan quality, and that the stronger mandates had a greater effect. Strong mandates contained clear goals and a combination of coercive characteristics (stringent and frequent use of sanctions) and co-operative characteristics (strong commitment, capacity-building and adequate funds).

Deyle and Smith (1998) extended the characterization of mandates provided by Burby and May in a study of plans produced under the coercive planning mandate of the State of Florida. They found that plan compliance was influenced by the manner in which the state's administrative agency pursued its responsibilities, including how well it monitored and enforced local compliance with mandated goals. Another major finding was that local councils selectively implemented the Florida planning mandate. Only certain highly salient goals were pursued, while less salient ones were ignored. This was not based on objective policy criteria, but depended on political feasibility.

New Zealand

As previously noted, the RMA was characterized as a co-operative mandate in an early study by May et al. (1996). They studied the potential intergovernmental influences on planning and the style of implementation policies of the RMA, because with few plans available for detailed study they could not focus on plan quality. Later, Berke, Dixon and Ericksen (1997) applied Burby and May's conceptualization of the characteristics of the planning mandate in a comparative assessment of seven regional plans in Florida and eight in New Zealand. Given the small sample size, this study did not, however, examine the influence of each mandate characteristic on regional plans. Nor did their study examine the influence of the mandate on district plan quality. But now that there are enough publicly notified plans, we can address this issue. More recently, Kerr, Claridge and Milicich (1998) looked at devolution both from a general theoretical standpoint and the perspective of the RMA with particular regard to residential land use and kiwi habitats (Claridge and Kerr, 1998), while Murphy and McKinley (1998) focussed on the strength of partnerships expected under devolution and beyond the RMA.

Characterizing New Zealand's RMA

The substantial discretion given to regional and local councils (i.e. sub-national government) in developing environmental plans, and the facilitative role of central (i.e. national) government, characterizes the RMA as a devolved co-operative mandate. Below, we assess four main features of the RMA: regulatory goal, devolution, co-operation and capacity-building, and process and prescription. We then consider RMA policies and plans in relation to those required of local councils by other statutes.

Regulatory Goal

Promoting the sustainable management of natural and physical resources under the RMA means that, when people or groups develop and use resources, local government is to apply planning provisions to help safeguard the quality of air, water, soil and ecosystems, and future resource needs. Part of the goal is therefore to ensure that users internalize the environmental costs of development.

Devolution

A partly devolved mandate, the RMA does specify powers reserved for the national government. For example, the Minister for the Environment has overall responsibility for implementing the RMA, although the Minister of Conservation has responsibility for the coastal area. At the Government's discretion, the Ministry for the Environment may draft national policy statements and regulations reflecting the matters of national importance specified in Part II of the Act (see Appendix 1). In contrast, the Department of Conservation is mandated by the RMA to prepare a national policy statement for the coast.

Regional and district councils are given authority to adopt plans with regulatory rules. Each regional council is mandated to prepare a regional policy statement to help guide integrated development and use of resources across the districts within its region, and to prepare a regional coastal plan. Both had to be done within two years of the RMA becoming law. Developing regional plans for other resource topics is optional. The district councils were required to develop district plans for their areas within five years of the due date for review of previous district scheme plans. Both regional and district councils must review their operative policies and plans at least once in ten years.

The RMA requires that plans lower in the hierarchy not be inconsistent with those at higher levels (or adjacent jurisdictions). This is a much milder dictate than plans having to be consistent, and therefore provides considerable discretion for lower-level innovation.

Co-operation and Capability-Building

While the RMA clearly provides a hierarchy of mandated and discretionary policies and plans, there is no concomitant hierarchy of governance. Rather, as reinforced by a decision of New Zealand's highest court, regional and district councils are to work in a co-operative partnership in pursuing the goals of the RMA (Court of Appeal 99/95, 4 July 1995).

While the national government can become involved in local affairs if activities badly compromise the national interest, such intervention is expected to be rare. Rather, the RMA enables the national government to carry out several capability-building activities to help councils achieve the mandated goals. These activities include grants or loans to councils, providing information about plan development, and technical assistance, through reviewing sub-national plans, explaining why revisions are needed, negotiating differences and generally encouraging councils to comply with the RMA.

Process and Prescription

While the RMA specifies a step-by-step process by which the policy statements and plans must be developed by councils (see Figure 2.2), it does not prescribe their content. Instead, councils are free to use as much or little regulation as the local circumstances allow. The policy guidelines provided by the Ministry for the Environment state that councils are 'free to develop their own approaches as long as they achieve the outcomes specified in the Act and follow the specified process'. This can mean using measures other than rules in the plan, such as education and financial incentives.

There are, nevertheless, some coercive elements in the RMA, apart from mandated policies and plans, and meeting deadlines when notifying and reviewing them. Councils must also gather information, conduct research and hold public hearings into policies, plans and applications made under the RMA. There are sanctions, but no incentives promoting compliance and adherence to the mandated goals, but their application by councils is discretionary. Several other prescriptive tools are also discretionary. Clearly, while the RMA can be characterized as a devolved and co-operative mandate, it does have some prescriptive and coercive features, as earlier illustrated by its location among the theoretical range of mandate types in Figure 1.1.

Non-RMA Policies and Plans

In addition to the environmental planning required under the RMA, councils were mandated under other statutes to prepare other types of plans. These aimed to achieve the managerial principles underpinning the local government reforms. Legislation also enabled councils to prepare other sorts of plans to help achieve desired social, economic and environmental goals.

For example, the *Local Government Amendment Act No. 2* (1989) required councils to undertake a publicly notified annual planning and reporting cycle, through the preparation of an annual plan and annual report on performance. The annual plan had to include the significant objectives, policies and activities of the council, forecast annual expenditure on the proposed activities and indicate funding sources, including policy on property taxation.

Later, the *Local Government Amendment Act No. 3* (1996) required councils to produce every three years a long-term financial asset management plan looking ten years ahead, in order to show how funding in relation to activities, such as roading, may be achieved in the long term. In the meantime, many councils continued to prepare non-mandatory strategic plans to assist the overall co-ordination of their policies.

Thus, an issue for councils was how to integrate these various planning instruments, and where the regional policy statements, regional plans or district plans required under the RMA fit in relation to other documents. Figure 1.2 depicts the relationship between the various planning documents developed by councils, and places sustainable resource management in local government under the RMA within the wider context of sustainable development. The diagram therefore links the functions required of councils under the two legislative reforms, the RMA and the *Local Government Act* (LGA). It also identifies resources as comprising not only natural and physical resources, but also people, money, property and goods.

Sustainable development has three elements: social development (of people); economic development (of money, property and goods); and resource development (of natural and physical resources). Social development and economic development are anthropocentric (i.e. focused on human needs and wants), whereas resource development is ecocentric (i.e. focused on ecosystems or the environment).

Policies and plans comprise: community planning and management through processes provided in the LGA such as annual plans; economic planning and management through the LGA processes via annual plans (which also allow for funding of RMA-related actions) and strategic financial plans; and natural and physical resources planning through the processes of the RMA.

The arrows in the diagram suggest integration between the three main components of the system. Central to all of the processes is public participation in decision-making and monitoring of policies, plans and outcomes. Local government may also use non-statutory tools to achieve social, economic and environmental goals, such as strategic planning and economic instruments.

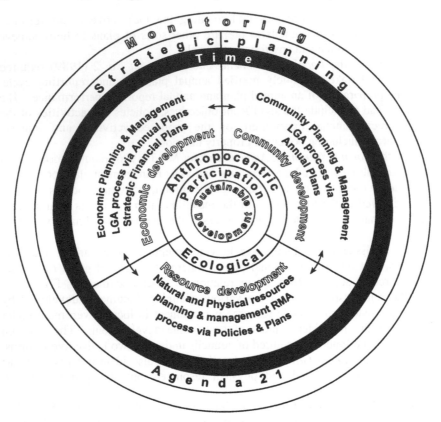

Figure 1.2   Location of policies and plans in sustainable community development

Source: Adapted from Ernest New and Associates, 1993: 11

A Theoretical Framework for Assessing Variation in Plan Quality Between Councils

We present below a theoretical framework for explaining variation in the quality of plans. It draws particularly on the initial evidence provided by Dalton and Burby (1994) for a theory highlighting the important role of planning mandates and plan quality in an intergovernmental planning context. Our adaptation of the Dalton and Burby theory is presented in Figure 1.3 by relating it to planning under a co-operative mandate in New Zealand. The diagram highlights the crucial role of national, regional and local governments by showing that the national planning mandate must be filtered through these partners when preparing plans. In addition, we have posited that the capability of regional and local councils to plan, and hence the quality of their plans, is affected by the availability of resources. Thus, the theory shows how the national mandate, and government agencies that help implement it, influences the capability of regional and district councils to plan, and how regional and districts councils react to their own contextual conditions when producing plans.

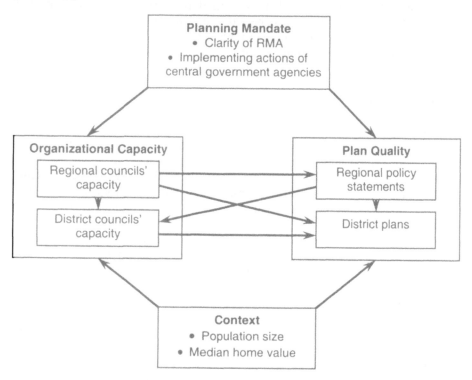

Figure 1.3   Theoretical framework linking mandate, organizational capacity, plans and context

Source: Berke, et al., 1999: 647

The Influence of Intergovernmental Actions

We examined two variables seen as useful by staff who prepare regional and local plans: clarity of provisions in the national mandate, and usefulness of actions by national government agencies in helping them to implement the mandate (see Planning Mandate panel in Figure 1.3). We expected that, if the goals and provisions in the national mandate are clearly understood, district and regional councillors are likely not only to support their intentions, but also be more willing to expand their staff capacity to prepare plans. Furthermore, council planners are then better able to infuse their intent into the regional policy statements and district plans they must prepare (May et al., 1996).

National government agencies are responsible for helping regional and local councils to implement the national mandate in the preparation and review of their plans. We expected that, if the actions of these agencies are perceived as useful by staff and politicians in regional and local councils, councils are likely to be more certain about what they must plan for, and thus more likely to commit staff resources to the plan. We also expected that, when these agencies give clear signals about what makes a high-quality plan, good plans will be achieved, and hence greater compliance with the national mandate (Burby and May et al., 1997; Deyle and Smith, 1998).

The Influence of Local Actions

We expected that staff capacity, as measured by the number of planners assigned, would have a potentially important effect on regional and local plan quality (Berke et al., 1996; Dalton and Burby, 1994). When more staff are available, more attention is given to the actions needed to create high-quality plans, including clarifying issues, developing a strong fact-base, eliciting goals from citizen groups, selecting policies that best fit local conditions and co-ordination with other organizations (see Figure 1.3).

We also expected that activities in regional councils would influence the capability of local councils to prepare district plans through policy direction and technical support. As the capability of regional councils improves, they have more opportunity to facilitate local planning by providing technical assistance, transferring information and supporting consensus-building efforts (DeGrove, 1994). As well, strong regional policy statements can communicate, educate and provide policy guidance to ensure that district plans are not inconsistent with regional documents.

Finally, we posited that any central government initiatives to advance planning in regional and local councils have to recognize the role of key local factors in influencing the ability of councils to produce quality plans. Previous studies indicate that the availability of financial resources to support planning is important (Berke, 1995; Burby and May et al., 1997; Fishman, 1978). We used median house value and population size as an indication of the wealth of councils (see context panel in Figure 1.3).

Chapter 2

Making Plans: From Theory to Practice

This book assumes that plans make a difference and thus warrant serious investigation. Generally, the plan guides decision-makers in managing community growth, land use change, resource development and/or environmental quality. It helps in reaching a consensus about land use and environmental policy among diverse interest groups (Kaiser and Godschalk, 1995). The plan typically promotes co-ordination among the multiple agencies in government. Policy-makers at higher levels of government view local plans as a way of addressing national and/or regional matters that are often beyond the purview of individual local governments. Decision-makers in local governments view the plan as a tool to build local knowledge about current and anticipated problems and trends. Plans should therefore lead to policies that are sensitive to local conditions, desires and needs.

Despite these widely recognized benefits, plans and planning have been subject to both political denigration as impediments to growth and development, and powerful academic critiques, as simply the production of a paper document to meet the requirements of a mandate, to avoid sanctions and/or to qualify for funds (Baer, 1997; Wenger, James and Faupel, 1980). Thus, plans become nothing more than a formalized series of words on paper with local governments having little commitment to their implementation. Issues are also raised about the best approach to use when making plans, ranging from an emphasis on top-down rationality to bottom-up participation.

In this chapter we briefly review the range of approaches for making plans, describe the approach adopted by the RMA, and then, drawing on international research on planning, develop criteria for evaluating the quality of regional policy statements and district plans in New Zealand. Finally, we consider various factors that influence the plan-making effort in local councils, including the organizational capability to plan and intergovernmental activities introduced in the Introduction.

Range of Plan-Making Approaches

Approaches to plan development can be seen as falling along a continuum. At one end are plans that are produced primarily by applying 'rationality', in which plans reflect scientific thought and methods. At the other end are plans that are produced primarily by a consultative 'participatory' process in which plans represent the art of the politically possible. The approach adopted in a particular country is, of course, influenced by the planning mandate — itself a reflection of prevailing political ideology. It is therefore highly likely that some sort of hybrid of the two extreme approaches, rational and participatory, will be used.

Rational Approach

Plans that result from applying scientific methods are prepared by following a rational sequence of steps:

1. generate factual information on broad conditions and trends;
2. develop a long-range vision and set goals;
3. evaluate costs and benefits of all possible policy alternatives;
4. adopt preferred alternatives that provide optimal achievement of goals;
5. implement the plan; and
6. monitor outcomes and evaluate the plan to assess revisions needed.

In this model, experts play a leading role in writing a 'blueprint' for the future of a particular community, resource or environment. The effectiveness of plans can be measured and policies adjusted by a process of continuing improvement through feedback. Thus, rational planning is 'both a normative theory in that it advocates a particular format for making planning decisions and a descriptive theory in that it describes the steps that most planning processes attempt to follow' (Kaiser, Godschalk and Chapin, 1995: 37).

This rational approach is assumed to rely on a scientific process in which facts and logic hold sway. Critics note that it has exorbitant demands for information, artificially separates stages of decision-making and has unreasonable expectations for considering possible alternatives (Kaiser, Godschalk and Chapin, 1995). While facts and logic are very important in the planning process, in reality other factors normally apply. For instance, in developing a vision, setting goals or choosing methods, community values ought to be influential and can be obtained by various participatory methods. Thus, realistically, little will be achieved from a plan that lacks community endorsement, and hence political circumstances must be acknowledged.

Participatory Approach

At the other end of the continuum is a model that emphasizes public participation through consultation. This approach recognizes that power, political will and processes are important. Decisions are made through incremental adjustments to existing policy, and the process is not comprehensive, linear and orderly, but best characterized as 'muddling through' (Braybrooke and Lindbloom, 1963). This is not to say that incrementalists are necessarily proponents of this approach, simply that it has elements of incrementalism. In this model, a good plan is one that is politically possible because it is acceptable to the politicians and their constituents. Advocacy is a vital skill for planners, who must be political to be effective. The participatory approach may also encompass 'critical theory which insists on processes for open communication, including critiques of plans, among all affected interests'. In this, information-sharing techniques are used, including dispute resolution, in order to achieve consensus about plan-making (Kaiser, Godschalk and Chapin, 1995: 39).

Nevertheless, despite reliance on political processes, planning ought to be informed by research and analysis. As a consequence, the community's preferred policy options and desired outcomes will be tempered by the influence of factual information and technical feasibility, and because, realistically, little will be achieved by a plan that lacks rational thought and facts.

Rational-Adaptive Approach

Where the planning approach involves an iterative relationship between research and analysis on the one hand, and public consultation and participation on the other hand, the model is characterized as 'rational-adaptive' (Figure 2.1). 'In this model, plan-making is primarily a rational analysis- and design-based activity, whereas plan implementation is primarily an incremental administrative- and political-based activity' (Kaiser, Godschalk and Chapin, 1995: 38-39). Some rational and adaptive plan development techniques are listed in Table 2.1.

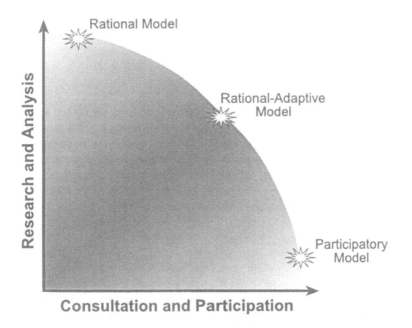

Figure 2.1   **Range of planning approaches** (Where increasing public participation (x axis) and increasing research (y axis) are maximized, a rational-adaptive model is denoted, exemplified by the RMA. At the extremes of each axis lie the rational and participatory models)

Table 2.1   Rational and adaptive plan development techniques

Rational Techniques	Adaptive Techniques
Data analysis	Public participation and discourse
Trend projection	Consensus-building and conflict resolution
Supply and demand derivation	Monitoring and problem solving
System modelling	Impact analysis and mitigation
Goal and objective statements	Capital budgeting and project review
Plan design	Plan evaluation and amendment

Source: Kaiser, Godschalk and Chapin, 1995: 39

Developing a plan using a rational-adaptive approach involves a series of steps, the outcomes of which are, at key stages, tested by consultation with the stakeholders in the affected community. Thus, for plans dealing with environmental management, scientific methods provide data that informs policy development while public participation shapes the policy and builds commitment to its implementation. Each aspect (rationality and participation) moderates the selection of policy options and desired outcomes. Our analysis of the RMA shows that it exemplifies this synthesis.

Plan-making Under New Zealand's RMA

In this section, the main activities and steps in plan development under the RMA are outlined in some detail, since they are fundamental to making good rational-adaptive plans. The focus is primarily on the plan-making process in regional and local councils. The process applies equally well to regional policy statements, regional plans, and district plans, but for brevity we use 'plans' throughout this section. First, though, because of its importance, we return briefly to the hierarchy of policies and plans and review what they are intended to achieve.

Hierarchy of Policies and Plans

Although only one national policy statement (the coastal plan) was mandated in the RMA, the reason for having these in the hierarchy of policies and plans was to help guide regional and local councils on issues of national importance, such as the protection of significant natural areas and highly valued landscapes, and the promotion of Māori interests under the Treaty of Waitangi.

As already noted, regional councils were assigned a critical role, reflected in their foremost responsibilities:

1.   the identification of significant regional resource issues that transcend district council boundaries;

2.   the development of regional policies and plans that promote integrated
     management across environmental media (air, water and land).

Because they are closest to the day-to-day land use decision-making, local
councils were given the critical role of:

1.   developing plans for guiding land use and subdivision;
2.   controlling the impacts of land use and resource development activities for
     the purposes of avoiding, remedying or mitigating adverse environmental
     effects (see Appendix 1).

Instead of prescribing specific uses, the district plans should enable a wide range
of land uses provided the effects can be managed sustainably.
     While regional policy statements may include issues, objectives, policies and
methods, they do not contain regulatory rules simply because they are intended as
guides to the rule-based regional and district plans. Regional plans help to control
the adverse environmental effects of various resource uses, such as taking or
diverting of water from a stream or earthworks on a piece of land subject to
erosion. It is expected, however, that the plans will also emphasize non-regulatory
means for managing adverse environmental effects.

Key Presumptions

The RMA presumes that land use activities are allowed unless otherwise specified
in district plans. In other words, resource developers can do anything on land
unless there is a rule in the district plan that says they cannot. The reverse applies
to water, air, and coastal marine resources because, unlike land, ownership is
vested in the Crown. In other words, resource developers can carry out activities
related to water, air and the coast only if they are expressly permitted by, or do not
contravene rules in, a regional plan. This difference reflects the principles of
individual property rights for land resources and common property rights for
water, air and coastal marine resources.
     Some functions are discrete to regional and district councils while others are
overlapping, such as natural hazards and hazardous substances, one of several
aspects an amendment to the RMA sought to clarify (see *Resource Management
Amendment Act* [1993]). In addition, the distinction between 'controlling the use
of land' for various purposes by regional councils and 'controlling the effects of
land use' by local councils was subtle and has caused confusion, if not friction,
between some district and regional councils (see Chapter 4).

Plan Preparation Process

The RMA identifies matters that need to be considered in the preparation of plans.
It mandates a rational-adaptive model because 'research and analysis' and
'consultation and participation' are carried out iteratively through a series of
normative 'steps in plan development'.   In this model, there are many

opportunities for feedback between the two main activities and between steps. Figure 2.2 outlines the model schematically. It follows that there are salient relationships between these plan-making activities and steps and the plan quality criteria on which the systematic analysis of plans is based (see Chapters 6-8). Our four case studies identify and explain the importance of these relationships and examine how well 'environmental fit' and 'community fit' are pursued by councils in their plan-making (see Chapters 9-12).

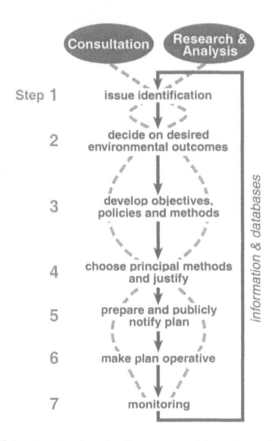

Figure 2.2   Main steps in plan development and how they relate to the activities of research and analysis and to consultation

Research and analysis   In preparing effects-based plans, key issues and shortcomings in knowledge need to be identified and desired environmental outcomes for the area determined. Early research and analysis are essential if the development of objectives, policies and methods for avoiding, remedying or mitigating adverse environmental effects is to be properly informed. Data on which to base subsequent monitoring of the effectiveness of the plan is also necessary. Thus, sound research and analysis are required to ensure that the plan has good 'environmental fit'.

Consultation and participation The second activity running through the plan-making process is that of consultation. The RMA requires councils to consult relevant government agencies, iwi (tribal), and other stakeholders during plan preparation. Consulting stakeholders early to identify issues is an important way of ensuring that the broad objectives of the plan are acceptable to the community at large.

Consultation continues throughout the plan preparation process, and includes not only the objectives and policies, but also the methods and/or rules to be adopted in the plan. Consultation thus ensures that the resulting plan has good 'community fit'. This requirement builds a political constituency in support of the plan's interpretation of what sustainable management of natural and physical resources means for the area.

Main steps The steps in plan preparation shown in Figure 2.2 are based on the elements of policy statements and plans and the process for preparing and changing plans specified in the RMA (Parts 4 and 5 and First Schedule). In carrying out these steps, the paramount consideration is defining sustainable management of natural and physical resources in terms of the circumstances of the region or district. In Step 1, the 'issues' in resource use and development are identified and, desirably, prioritized. Well-defined issues underpin Step 2, which is to think through the 'desired environmental outcomes'. Step 3 is for plan-makers to develop the 'objectives, policies, and methods' for meeting the desired outcomes. Step 4 is to 'select the main methods' and 'justify' them, including any regulatory rules. When writing the plan, the main reasons for adopting the objectives, policies and methods have to be given, along with the 'anticipated environmental results' of their implementation. It is inferred from the RMA that, to be effective, this cascade of elements in the plan must be internally consistent, as illustrated in Figure 2.3.

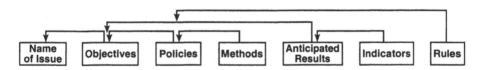

Figure 2.3 Cascade of elements, including feedback, required in plan-making under the RMA

The RMA also requires councils to establish a 'monitoring regime' (section 35). Environmental outcomes, and the means for monitoring them, should be specified in the plan so that the effectiveness of its policies and methods can later be evaluated (sections 62, 67 and 75). More broadly, this enables the overall effectiveness of the plan to be reviewed in the long term and, if necessary, the plan to be amended so that the desirable environmental outcomes are achieved (Step 7, Figure 2.2). This is shown by the feedback loop in Figure 2.3 which links indicators to anticipated results and thence to the objectives for a particular issue.

Additional matters In keeping with a devolved co-operative mandate, the RMA requires plan-makers to ensure there is adequate 'horizontal' and 'vertical integration' across neighbouring councils and between levels of government (sections 62, 67 and 75). It does this by requiring councils to consider other management plans and strategies prepared under other Acts, both internal (e.g. annual and strategic plans) and external (e.g. conservation, heritage, pest management, and transport strategies). It also requires plans to include procedures for considering environmental issues that cross geographical and jurisdictional boundaries. At a minimum, councils must ensure that their plans are not 'inconsistent' with policies and plans higher in the hierarchy.

Given the emphasis on participation in the RMA, the plan's readability is important, because its 'accessibility' to diverse stakeholders can significantly influence community acceptance. The RMA does not, however, specify how plans should be organized and presented. The RMA is sophisticated, and the resulting plans will be complex. This presents a considerable challenge to those responsible for preparing the plan. Consequently, an important measure of plan quality in our study is user-friendliness.

Making Plans Operational

Once the plan has been written, it is required to be 'publicly notified' and therefore open for submissions from the public and statutory authorities (Step 5 in Figure 2.2). Submissions are summarized by the council, and further submissions invited supporting or opposing the matters raised in the primary submissions.

The council then embarks on hearing submissions, and makes decisions. Following notification of these decisions, submitters can then lodge references (i.e. appeals) to the Environment Court, whose decision is final (except on points of law, which can be appealed to the High Court).

Once all references are dispensed with, the council publicly notifies that the plan is 'operative' (Step 6 in Figure 2.2). This public process can take many months, even years. Until that happens, transitional arrangements apply. (Nearly all plans considered in the research for our book were not yet formally operative.)

Plan Implementation

Once operative, the plan is implemented primarily through the resource consents (permitting) process, which is regulatory. At the same time, the council is able to implement complementary strategies, such as facilitating the development of alternative methods (i.e. education programmes, economic instruments), as well as co-ordinating its implementation with other instruments (fiscal strategic plan, annual plan and so on). The process of choosing and justifying the principal methods for achieving the desired environmental outcomes (Step 4 in Figure 2.2) expressly encourages the use of an optimal mix of policies in preference to reliance on regulation.

Assessing Plan Quality

Given that the rational-adaptive model mandated by the RMA combines the best of both rational and participatory approaches to plan-making, it could be argued that it ought to produce high-quality environmental plans and, ultimately, high-quality environmental outcomes. We set out to test the first part of this proposition in a systematic and rigorous fashion, taking into account not only the content of the plans, but also the context within which they were developed. An important part of the research was developing a methodology for testing the quality of plans. We identified eight criteria derived from researching the international literature, peer review by New Zealand practitioners and professional experience (Dixon et al., 1997). This next section reviews the literature and its influence on the conceptualization of the methodology, before describing the eight criteria themselves.

Conceptual Definitions

A difficulty in evaluating the quality of plans is to know how to represent them for testing. There have been several attempts to define the characteristics of plan quality. One line of work has attempted to define a set of normative criteria that must be followed to ensure that plans and local government decisions reflect community-wide goals (Healy, 1993; Kent, 1990). Kent (1990) argues that the key characteristics of good plans include clear policies, maps that display the spatial intent of the policies, and land use design. These characteristics will ensure that plans communicate, effectuate and advise. This view tends to emphasize traditional physical techniques for plan-making.

Kaiser, Godschalk and Chapin (1995) defined plan quality criteria that lend themselves to empirical analysis. The criteria included:

1. a thorough fact-base that documents local conditions and issues, and guides selection of alternative solutions;
2. a comprehensive set of goals that represent the shared local vision of a liveable community; and
3. policies as a general guide for action to achieve goals.

In explaining the response of 180 local governments to reducing the long-range threat from natural hazards, Raymond Burby and his colleagues provided empirical support for this conceptualization of plan quality (see Berke and French, 1994; Berke et al., 1996; Dalton and Burby, 1994; Burby and May et al., 1997). Among various findings, they determined that better-quality plans promote more extensive use of land use controls in hazardous areas. Berke (1994) extended this conceptualization by assessing how well regional governments respond to natural hazards in New Zealand, and then worked with Burby and May, et al. (1997) in examining plans under state mandates.

Building on prior work, Baer (1997) reviewed different ways in which plan quality is defined, including citizen critiques, best alternative plan policy selection, evaluation of the influence of plan policies on environmental outcomes, and research and professional assessment. Drawing on these variants of plan evaluation, Baer extended the Kaiser, Godschalk and Chapin (1995) definition of plan quality by outlining the most comprehensive set of evaluation criteria then derived. The criteria included, for example, the clarity of the plan's purpose, the relevance of identified issues, internal consistency, explicitness of procedural actions specified in plans to achieve desired objectives (e.g. how actions will be initiated, administered and enforced, who is responsible for implementation, how the resources required for implementation will be provided) and assignment of responsibility for implementation.

Quality Criteria for RMA Plans

We further extended Baer's conceptualization in our examination of the quality of regional policy statements and district plans in New Zealand. The criteria and corollary criteria we used are given in Table 2.2.   The eight criteria are:

1.   interpretation of the national mandate;
2.   clarity of purpose;
3.   identification of issues;
4.   the quality of the facts base;
5.   internal consistency;
6.   integration with other plans and policy instruments;
7.   provisions for monitoring and responsibilities; and
8.   organization and presentation.

Table 2.2   Criteria for evaluating plan quality

---

1.   *Interpretation of the Mandate:* Articulation of how a legislative enabling provision is interpreted in the context of local (or regional) circumstances.
    1.1   Is there a clear explanation of how the plan implements key provisions involving matters of national importance, Treaty of Waitangi, duties to assess costs and benefits, and duties to gather information and monitor?
    1.2   Is there a clear explanation of the functions of a district plan, as required by key legislative provisions?
2.   *Clarity of Purpose:* Articulation of a comprehensive overview, preferably early on, of the outcomes the plan attempts to achieve.
    2.1   Does the overview consist of a coherent explanation of environmental outcomes?
    2.2   Does the overview contain a discussion of social, cultural and economic matters affecting those environmental outcomes?

Table 2.2   Continued

3. *Identification of Issues:* Explanation of issue in terms of the management of effects.
   3.1   Are issues clearly identified in terms of an effects-based orientation?
4. *Quality of Facts-Base:* Incorporation and explanation of the use of factual data in issue identification and the development of objectives and policies.
   4.1   Are maps/diagrams included? Do the maps display information that is relevant and comprehensible?
   4.2   Are facts presented in relevant and meaningful formats?
   4.3   Are methods used for deriving facts cited?
   4.4   Are issues prioritized based on explicit methods?
   4.5   Is cost/benefit analysis performed for main alternatives?
   4.6   Is background information/data sourced/referenced?
5. *Internal Consistency (of Plans):* Issues, objectives, policies, and so on are consistent and mutually reinforcing.
   5.1   Are objectives clearly linked to issues?
   5.2   Are policies clearly linked to certain objectives?
   5.3   Are methods linked to policies?
   5.4   Are anticipated results linked to objectives?
   5.5   Are indicators of outcomes linked to anticipated results?
6. *Integration with Other Plans and Policy Instruments:* Plans should integrate key actions of other plans and policy instruments that are produced within the agencies or by other agencies.
   6.1   How clear is the explanation of the relationship of each mentioned policy/policy instrument of the plan under study?
   6.2   How clearly are cross-boundary issues explained?
7. *Monitoring:*   Plans should include provisions for monitoring and identify organizational responsibility.
   7.1   Are provisions for monitoring the performance of objectives and policies included in the plan?
   7.2   Are the specific indicators to be monitored identified?
   7.3   Are the organizations responsible for monitoring and providing data for indicators identified?
8. *Organization and Presentation:* Plans should be readable, comprehensible and easy to use for both lay and professional people.
   8.1   Is a table of contents included (not just a list of chapters)?
   8.2   Is a detailed index included?
   8.3   Is there a user's guide that explains how the plan should be interpreted?
   8.4   Is a glossary of terms and definitions included?
   8.5   Is there an executive summary?
   8.6   Is there cross-referencing of issues, goals, objectives and policies?
   8.7   Are clear illustrations used (e.g. diagrams, pictures)?
   8.8   Is spatial information clearly illustrated on maps?
   8.9   Are individual properties clearly delineated on maps?

Source: Berke et al., 1999: 646

Many of the criteria also constitute steps in the plan-making process (e.g. criteria 3, 4, 5 and 7). The specific language of the corollary criteria was adapted to reflect the rational-adaptive approach to plan-making mandated by the RMA.

We examined how well policy statements and plans in New Zealand use these criteria, and then determined how plans can be improved. We also filled a gap in the planning literature by applying these criteria in a more comprehensive and systematic evaluation than prior efforts.

Operational Milieu of Plan-Making

The making of plans cannot be examined in isolation from their operating milieu. Figure 2.4 puts the plan development processes explained above into a wider organizational and operational context. This includes: intra-organizational factors within the plan-making agency, such as institutional structures and procedures; inter-organizational factors, such as the flows of resources from national to local agencies; and the broader economic and public sector policy settings that influence plan-making agencies. All these factors influence the plan-making process and hence the quality of plans. They also influence the plan implementation and hence the quality of environmental outcomes.

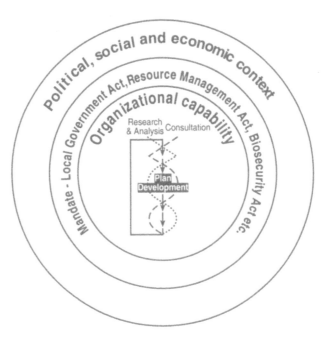

Figure 2.4   Main factors in the operating milieu affecting plan development

Organizational Capability for Plan-making

Critical to the success of the RMA planning regime is having the 'capability' to implement its policy intentions. This includes plan-making not only in regional and local councils, but also in agencies of national government. The RMA was, after all, introducing a national mandate and the Government was responsible for ensuring that its agencies had the requisite capabilities for satisfactorily implementing it.

Building from Above

In moving more forthrightly away from centrally directed planning, intergovernmental co-operation and partnership implied that the Government would provide the necessary resources to its agents (especially the Ministry for the Environment) for building capability in regional and local councils. This was despite new economic policy signalling that financial grants and subsidies would be phased out. Providing financial assistance was all the more important because the RMA's new effects-based planning approach required new skills and resources. At a minimum, this government assistance would include timely production of national policy statements and standards and guidelines that explained to councils the philosophy and practice of the new national mandate for sustainable management.

Ominously, the passage of the RMA through Parliament was accompanied by not only assurances that resources for its implementation would be adequate, but also that it would not be costly. Further, the Government's economic and administrative policies clearly signalled that its agencies would have to do more with less funds. In this light, government resources for building capability at any level looked to be under threat from the outset. How important would this become as the Ministry for the Environment and Department of Conservation set about implementing the RMA, and councils embarked on preparing new policies and plans? Would it affect the ability of the central agencies to offer timely guidance and support to the councils? Would it affect the preparation of plans and their quality? These key questions are explored in later chapters.

Building from Below

Regardless of the Government's commitment to building sub-national capability, councils had a responsibility to ensure that they were adequately resourced for developing plans. This included building appropriate staff levels and skills and providing sufficient time and money for competently making plans. For the newly created regional councils as partner to the local councils within their areas, this was an especially difficult challenge, given the very tight two-year statutory deadlines for producing regional policy statements and coastal plans, and limited resources.

For both regional and local councils, providing the resources for preparing new plans would be affected also by the Government's requirements for organizational and managerial change. The separation of policy-making (e.g. strategic planning, district plan preparation and asset planning in the districts; and policy and corporate services in the regions) from regulatory services (e.g. district plan administration, building consents, and pest control in the districts, resource use and biosecurity in the regions), and service delivery meant that plans would be prepared in the policy-making sections of councils independently of those who administer the plans. This separation could adversely affect plan development unless the planning team expressly devised ways to co-ordinate inputs from staff in other relevant sections.

The complexities of the new mandate for plan-making presented a particular set of challenges to councillors and staff. First, the relationships between council politicians and staff were important in guiding the development of the plan, which is an expression of council policy. Thus, the committee structure was relevant for building council 'ownership' of the plan. The organizational possibilities ranged from a planning committee matched to the functional sections of a council through to the whole council acting as the 'planning committee' and working with planning staff. More important was the willingness of elected members to demonstrate commitment to and ownership of the plan.

Second, the demands of writing new effects-based plans and carrying out the necessary consultation required significant additional funding. The provision of new resources would depend on councillors' understanding of the demands of the new mandate.

Finally, the process of plan-making was not new to local government. However, the particular combination of facts-based research and public consultation demanded by the new mandate suggested that councils needed both rigour and a clear sense of strategy on the one hand, while on the other maintaining flexibility to meet democratic expectations as they occur over time. In these circumstances, two questions take on considerable importance for our study, and are addressed most directly in Part IV. Did council leadership allocate adequate resources for plan preparation, and how influential were organizational arrangements and decisions on making good plans? These questions are also explored in later chapters.

PART 2
INTERGOVERNMENTAL
PLANNING IN NEW ZEALAND

Chapter 3

Central Government: Walking the Talk

Did the Government provide clear policy direction for the RMA, and how well did it do? Did it adequately fund its environmental agencies to implement the Act, and how well were they able to build capability in regional and local councils? These questions are explored in this chapter. We examine how central government went about implementing its 'policy goal' of internalizing the environmental costs of the use, development, and protection of natural and physical resources, and the main 'policy instruments' for doing so: system change, mandate, inducements and capacity-building.

Changing Policy Intentions

Development and implementation of the RMA were marked by uncertainty about policy direction, especially on governance. This is explored here through three interrelated issues:

1. the locus of control in the intergovernmental relationship (devolution or decentralization);
2. tensions and contradictions between policies and practices; and
3. the influence of managerialism and cost-cutting on inducements and capacity-building, and hence implementation of the new planning regime.

Locus of Control

By the 1980s, most political parties believed that central government was too dominant and interventionist in local affairs, and that increased power of decision-making should be accorded to local government. How this might be achieved was highly debatable. The possibilities ranged from decentralization to devolution. 'Decentralization' involves the 'delegation' of power and authority from national to sub-national levels of government, but ultimate responsibility remains at national level. 'Devolution' involves the 'transfer' of power and responsibility from national to sub-national levels. The key tests are that sub-national organizations derive authority from statute, are run by locally elected members and empowered through 'powers of general competence' to raise revenue and make laws without reference to central government (Boston et al., 1996: 163-164; Bush, 1995: 301).

Devolution allows for intergovernmental co-operation and implies a trustee relationship or partnership between national and sub-national agencies in achieving given ends, whereas decentralization involves a delegate or principal-agent relationship with more power and responsibility retained centrally. Both, however, require clear legislative signals from central government about national goals and objectives. As well, where 'system change' is considerable, both decentralization and devolution require central government to provide 'resources' for 'building capability' in sub-national government to ensure that there is commitment to and capacity for the national mandate. To achieve this, central government must ensure that its own agencies have the necessary resources and capability.

Towards devolution under Labour The Fourth Labour Government came to power in 1984 with a policy goal of smaller central government and stronger local democracies. The means by which this could be achieved were publicly discussed, including changing the locus of control for decisions that affect local communities away from central government (Martin and Harper, 1988; The Officials Co-ordinating Committee on Local Government, 1988a, 1988b, 1988c; State Services Commission, 1988).

When the Labour Government retained power in the 1987 elections, some policy coherency was evolving. The blueprint for reform of *Government Management* (December 1987) integrated reforms for restructuring, management, democracy and finance. Among social democrats (centre-left 'wets') in the Government was an emerging view that it should develop, through devolution, a true partnership with local agencies, including councils, in delivering services and funding in, for example, social welfare, health, education, Māori development and resource management (Boston et al., 1996; Martin and Harper, 1988).

On paper, one of the best examples of devolution to emerge from the 1980s reforms was in resource management and environmental planning. (The other was for Māori development — see Chapter 5.) As we have already seen, this was achieved by coupling reforms of local government and resource management law to create a new environmental administration for implementing a new RMA. By the mid-1990s, the reforms had resulted in different sectors of government being at different points on the spectrum, such as social welfare (decentralization) and resource management (devolution) towards each end, and education and Māori development more in the middle (Boston et al., 1996: 164, 180).

For resource management, Government was able to build on a reasonable measure of devolution and co-operation under the previous planning regime. The Minister of Local Government (Michael Bassett) saw the comprehensive changes to the structure and management of local government as making it 'a true partner in the governance of the country', and the transfer of resource management functions from central government a case of 'true devolution' (Bassett, 1989; Boston et al., 1996: 183; Martin, 1991: 268). A key feature would be the creation of an intermediate tier of regional government (see Chapter 4).

While generally supporting devolution, the political arm of local government wanted a blueprint for the sweeping reforms to see how they would work in practice — particularly in power sharing and revenue raising (State Services

Commission, 1988). Central government had produced discussion papers on these and related topics, to which local government and other interested parties, including the public at large, responded but the work remained incomplete (e.g. The Officials Co-ordinating Committee on Local Government, 1988a, 1988b, 1988c, 1989). The Government considered it more important to complete restructuring of local government before the local body elections in late 1989, and ahead of the *Resource Management Bill* (introduced in August 1989) becoming law (Ministry for the Environment, 1991a). The latter was especially important because it provided for the 'transfer of functions, power, and responsibility from central to local government for resource management purposes'. By election time, however, only the structural changes were complete, together with some legislative provisions through amendments to the *Local Government Act* (1974), and some principles of taxation and possibilities for finance. The Government believed there was enough in current law to allow the system to operate while more work was done on other aspects of accountability and funding (The Officials Co-ordinating Committee on Local Government, 1988b).

The local government fraternity was enticed into accepting rapid implementation of the reforms through two prospects: amalgamation of existing councils into fewer, larger units would greatly improve their ability to meet both old and new functions; and increased powers of general competence — the ability to make laws and raise taxes — the hallmark of a devolved system. Under the existing regime, local government could carry out only those functions for which it had explicit statutory authority, although few of the functions were mandatory. While local government was a creature of Parliament, central government had rarely intervened in its affairs, leaving councils a wide range of choice in supporting various services for their citizens (Bush, 1995; Boston et al., 1996). However, the sustainable management of renewable resources through the RMA was one of the few mandatory functions of local government, along with civil defence and public health (Boston et al., 1996; Kerr, Claridge and Milicich, 1998).

Towards decentralization under National The devolution policies of the Labour Government had already met resistance from the increasingly powerful liberal democrats (centre-right 'dries') within its ranks when the National Party won power in the elections of late 1990. Mainly market liberals in economics, they favoured decentralization because it more readily enabled central government to retain control over achieving national goals, especially economic ones (Bushnell and Scott, 1988: 32). Even more preferable from this perspective would be the return to the market of publicly provided activities in which the Government did not have to be involved. Inside the Government, staff in the Treasury dominated this view, while outside the Government it was channelled through members of the newly created Business Round Table, the powerful lobby group drawn from the chief executives of major corporations (Harris and Twiname, 1998).

Thus, an important policy shift occurred when the National Party assumed power. Being composed almost exclusively of liberal democrats, the new Government halted the devolution process, but intensified the market liberalism and managerialism begun by liberal democrats within the previous Labour

Government. Fiscal prudence and discipline were the catchwords of the day. As we shall see, this had important consequences for the resource management mandate, especially in central government's commitment to building and funding capability among its partners in local government and iwi.

Although Labour's plan for devolution to iwi was quickly revoked by the new National Government (see Chapter 5), its *Resource Management Bill* was passed in late 1991, but with some major changes, such as the omission of non-renewable resources, like minerals, and hazardous substances. A definition of the 'meaning of effects' was added (section 3).[1] As noted in the Introduction, some resource management functions previously carried out by central agencies (e.g. for water, soil and town planning) were transferred to local government. Matters of national importance that had been identified earlier in the *Town and Country Planning Act* (1977) were to be considered when making local plans.

Nevertheless, central government retained important powers and responsibilities. First, it supervised the RMA's implementation through the Ministry for the Environment (generally) and Department of Conservation (coastal marine), and could produce statutory national policy statements and monitoring standards and non-statutory guidelines for regional and local councils to follow (sections 24 and 28). Second, central government retained direct responsibility for allocation of energy, minerals, fisheries and coastal marine resources, although not for the local environmental effects of their development and use. Third, it could intervene in local government decisions through call-in procedures for matters of national importance (sections 24(c) and 140; and 141-150).

The power to intervene in local government affairs was increased in 1992 when the new Minister of Local Government (Warren Cooper) further amended the *Local Government Act* (1974). This enabled a Minister of Local Government to initiate a review of a council if its statutory obligations (resource management or otherwise) were not being met and significant mismanagement or deficiency in management and decision-making was evident. By then discussions on powers of general competence for local government had long ceased.

Tensions in Policy and Practice

The shift in policy emphasis away from devolution and towards decentralization had important consequences for implementing the new planning regime, because the system of governance was neither truly devolved nor truly decentralized. Rather, it emerged as a hybrid, which generated many tensions, even contradictions, over policy and practice under the RMA.

---

[1] When introducing the Bill to Parliament, the new Minister for the Environment (Simon Upton), emphasized environmental effects-based planning and the need for alternatives to regulatory instruments, including plans, for achieving environmental bottom lines. He reiterated this theme time and again as the purpose and principles of the RMA got publicly debated in the years that followed.

For example, while the intergovernmental structure for implementing provisions of the Act made it a devolved co-operative mandate, ministerial powers and the absence of a free hand for councils in raising finances to carry out their functions meant it was more like 'an exercise in decentralization' (Boston et al., 1996: 170; Reid, 1999: 166). That is, government governs, sets the regimes and takes the responsibility, including getting good local government, by prescription or other means, including intervention.

Further, although central government made it clear that local government was to engage in integrated resource management through effects-based planning that focused on environmental bottom lines, it did not alter the purpose and principles in Part II of the RMA to become consistent with that view, especially sections 5 and 6 (see Appendix 1). Thus, many in local government, and beyond, saw scope in Part II for continuing with land-use activities-based planning that the integration of environmental, social and economic issues enabled under the previous planning regime (*Town and Country Planning Act* [1977]).

Moreover, while the framework for resource management in the RMA assumed intergovernmental co-operation, its implementation (to judge by the behaviour of central government) was neglectful. On paper, there were to be national policy statements and standards to guide local government policies and plans on matters of national significance. In practice (except for the compulsory national coastal policy statement) the National Government resiled from national policy statements, not by eliminating relevant provisions in the RMA, but by failing to promote and resource their preparation. There was a view that local councils could work it out variously for themselves. Central government would not, therefore, be seen as interventionist, and it would also avoid the considerable cost of policy development. Consistent with market-liberal philosophy, it was acknowledged that some councils would fail, an acceptable outcome akin to business failure, regardless of the environmental costs.

Finally, a devolved system and co-operative mandate assumes that central government will ensure that sub-national government has the capability for implementing the national mandate by providing adequate inducements and resources. For that to happen, it must first resource its own implementing agencies. Instead, central government embarked on savage cost-cutting throughout the system, accompanied by the entrenchment of 'managerialism'. That is, it applied to public agencies the same set of principles that commonly apply to business, including management skills, to gain results, contestability and contracts. This had important implications for implementation of the RMA — a co-operative mandate — as discussed below.

Managerialism and Cost-Cutting

More than anything else, governance in the 1990s was driven by the efficiency principles of managerialism and principal-agent contractual arrangements, accompanied by cost-cutting (Boston et al., 1996; Halligan, 1997). This had important implications for all government sectors, including the new environmental planning regime. In central government, operations would be made more efficient through implementing a range of managerial principles recognized in the *State Sector Act* (1988) and *Public Finance Act* (1989). Ministries and departments were to operate as in private business. They were required to describe objectives through the specification of outputs and outcomes, to which were later added Strategic Result Areas (SRAs) through Key Result Areas (KRAs), and to set overarching goals and strategic priorities. Each year, their activities had to be specified and purchased in advance by the respective ministers, and later reported on annually to Parliament in terms of outputs achieved. All this change required agency staff to come to grips with radically new ways of thinking and doing business. And, it took for granted that the RMA would be effectively implemented through the intergovernmental system of planning.

Cut the fat   From the outset, the new National Government warned, 'fat would be cut' from the public service, in order to make it more efficient. This quest was epitomized in mid-1991 by the self-proclaimed 'Mother of All Budgets' of the new Minister of Finance (Ruth Richardson) — evoking the 'Mother of All Wars' proclaimed earlier in the year by Iraq's President Saddam Hussein during the Gulf War. It was as if the severe budget cuts would help facilitate the adoption of managerialism in Government agencies, including regional and local councils (see Chapter 4). It was in this context that the RMA was to be implemented. Seemingly, the emphasis lay on cost-cutting, not capability-building.

Operational and cost efficiencies in government agencies have accrued from managerialism. In his review of public sector restructuring globally, Bangura (2000) says that the decentralized management reforms in New Zealand are generally regarded as the most revolutionary and successful. However, many local policy analysts point to disadvantages. For example, splitting the organization along functional lines may well increase transparency and accountability, but in the absence of strong counter-mechanisms it jeopardizes much-needed cross-sector interaction — a very important factor in developing integrated resource management policies and plans (see Chapters 9-12). As another example, 'the industry sector approach to the management of services in councils where resources like water and roading are being split off to the private sector lessens the influence that local communities can have over resources and service delivery' (Boston et al., 1996: 180-181). Others said that this outcome was 'a deprecation of the holistic ideal of community and democracy, an ideal considered to be inefficient by Treasury and the Business Round Table'. Local government was therefore becoming more like a 'branch office of central government' than a 'creature of Parliament'. This was a far cry from the devolved system once hoped for by social democrats (Richardson, 1999).

Whether local government is an equal partner under devolution or a branch office agent under decentralization, central government has a duty to ensure that its agencies (both central and local) have the capability for carrying out its national mandate. The Government never, however, accepted fully the role to build capacity in councils for its co-operative resource management mandate. Very strong business-oriented forces in the Government, led by Treasury with support from the Ministry of Commerce (now the Ministry for Economic Development), opposed funding from the centre. They wanted to reduce costs to the Crown, and to reduce as much as possible costs on business. They did not like the RMA either ('the most intrusive instrument after taxation'), so it suited them to argue 'no money' for the agencies responsible for its implementation. Thus, except for roading, within two years of the RMA becoming law, grants to local government were cut to almost nothing (Boston et al., 1996). It was forlornly hoped that, through the structural and managerial reforms of local government, new efficiencies would pay for its effective implementation. Extending the user-pays policy to resource developers so that they might bear the cost of internalizing environmental externalities proved politically unpalatable, because the public and developers did not want to pay the costs of information-gathering and environmental investigations.

Into the bone While councils were able to raise funds through user charges and dividends from their trading in services, in addition to gathering property taxes (rates) and raising loans, any new capacity found through this, and the 1989 amalgamations, was inexorably squeezed by the managerial approach and ratepayers' reluctance to contemplate increases in the rating bill. Restructuring within councils bit deep, and in some cases often, senior staff losses resulted and, with them, local knowledge and planning experience was dissipated. Suddenly, councils seemed to have much more to do, but much less to do it with. Our interviews showed that, in retrospect, the new RMA functions — of both local and national importance — were pushed onto councils by central government, without local government being given the necessary funds for carrying them out or the powers to raise the money to do so (see Chapters 9-12). Indeed, the devolution, both intended and unintended, of non-RMA functions continued throughout the decade, leaving councils to fill the gap with very limited funding from the centre (Local Government New Zealand, 2000).

In summary, governance for resource management quickly became a hybrid of devolution and decentralization that reflected the political compromises struck in its development by liberal and social democrats. The policy shift by the National Government was not 'upfront'; it was done by 'not acting' and 'not funding' activities that were important to the development of the hierarchy of integrated resource management clearly articulated in the RMA. The effect this had on local government and iwi partners is taken up in Chapters 4 and 5 respectively. The effect it had on its own central agencies is detailed below.

Resources for Implementation

Our concern here is to assess the Government's commitment to resourcing its environmental agencies to ensure that their actions led to the successful implementation of the environmental mandate in local government. In particular, we examine in some detail the financial resources provided by central government over the past decade to the Ministry for the Environment and Department of Conservation to facilitate plan-making. The analysis is not about the commitment of staff within these agencies. Suffice it to say, they worked hard in very difficult circumstances and provided the best they could.

Ministry for the Environment

As noted in the Introduction, the Ministry for the Environment (MfE) functions under the *Environment Act* (1986) and is responsible for overseeing implementation of not only the RMA, but other statutes, including *Agricultural Pest Destruction Act* (1967); *Noxious Plants Act* (1978); *Soil and Rivers Control Act* (1941); *Harbours Act* (1950); *Marine Pollution Act* (1974); *Transit New Zealand Act* (1989); *Transport Services Licensing Act* (1989); and *Hazardous Substances and New Organisms Act* (1996). MfE therefore has much to deal with, such as policy development and advice on climate change and ozone depletion, international environmental treaties, Agenda 21, wastes, hazardous substances, new organisms, information and education, sustainable land management, biosecurity and environmental indicators. It provides advice to regional councils, local authorities and resource users; establishes environmental standards when necessary; deals with water conservation orders (a residual from the previous regime); and administers the Sustainable Management Fund (see Chapter 4).

From the outset there prevailed in the MfE a philosophy about its activities that was influenced by two interrelated factors. First, the political environment dictated that central planning was to be eschewed in favour of local actions. Second, the Government wanted to avoid having its new environmental policy ministry evolve into a large operational bureaucracy. In its formative years the MfE leadership therefore proffered the view that it was a 'neutral policy agency' — a 'ministry in the middle', a 'ministry of balance' — rather than a hands-on operational agency (Ministry for the Environment, 1987). The paradox in this is that, while MfE had a wide mandate to implement, being portrayed as a small policy advisor meant it could be held to low levels of funding by the Government — and it was. A comparative analysis in 1993 showed that the cost of policy advice from MfE to the Government was the lowest, being less than 30 per cent of such agencies as Treasury and the Ministry of Commerce (Morris, 1994, in Boston et al., 1996: 130-131).

Funding trends   Trends in MfE expenditure are shown in Figure 3.1. It excludes grants for catchment works and resource management subsidies, which fell from around $50 million[2] in 1989 (soon after MfE took over managing them from the disestablished Ministry of Works and Development) to $9.5 million by 1995.

After MfE was established in 1987, annual expenditure increased to around $11 million by 1990, but then sagged until it returned to that level in 1995. This four year period was a crucial time for implementing the RMA, given the two year statutory deadlines for producing regional policy statements and coastal plans. Funding started to increase in 1995 (to $12 million, excluding grants) following the

---

[2] NZ$ values, unadjusted for inflation, are used throughout the book. For the 1990s, the average value of NZ$1 was equivalent to UK£0.30 and US$0.50.

re-appointment of Simon Upton as Minister for the Environment in 1993 (having been replaced by Robert Storey in 1991). He argued for more funds resulting in the *Green Package* of 1997 (Government of New Zealand, 1997). It provided $51.6 million for:

- establishing a new Environmental Risk Management Authority;
- promoting good practice under the RMA;
- assisting adoption of sustainable land management practices;
- supporting the development of climate change policy; and
- promoting improved waste management.

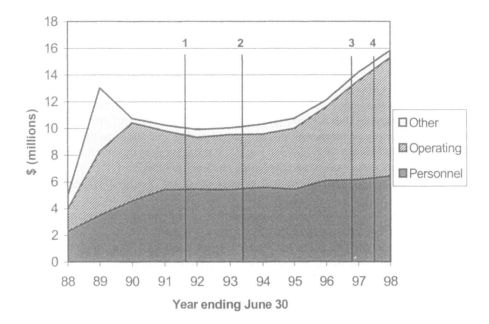

Figure 3.1 Ministry for the Environment: trends in expenditure for personnel, operating and other items between mid-1988 and mid-1998 (The period 1991 to 1996 was critical for policy- and plan-making, as indicated by: 1 RMA passed into law; 2 mandated due date for notified regional policy statements, regional coastal plans, and New Zealand coastal policy statement; 3 mandated due date for initial notified district plans; and 4 Green Package funding)

Source: Data extracted from Annual Reports to Parliament of the Ministry for the Environment

By 1998, the MfE's expenditure had reached nearly $16 million. (These dollar values are not inflation-adjusted.)

While increased 'provision for good practices under the RMA' was welcome, much of it was diverted to other activities in the MfE to meet more politically urgent needs. Subsequent increases in funding of MfE for good planning practice have, at times, had the same fate.

You want more? The early rise in funding in 1989 reflected two main activities: the transition arrangements for the grants and subsidies transferred to MfE from the former Ministry of Works and Development; and the resource management law reform process (1988-90), which Ministry staff indicated cost around $3 million.[3] To facilitate implementation of the RMA starting in 1991, the Ministry 'sought from Government, a further $2.2 million' for its proposed *Resource Law Transition Plan*, with a range of activities to ensure the efficient and effective transition from old to new resource planning regimes (Ministry for the Environment, 1991b). These included, for example:

- broadening the principles and practices of project-oriented environmental impact assessment — formerly administered by the Commission for the Environment — into a wider framework for assessing environmental effects;
- dealing with new plans with planning aspects that were outside the *Town and Country Planning Act* (1977), such as those under the *Water and Soil Conservation Act* (1968) and the *Conservation Act* (1987);
- bringing the subdivision regime across from the *Local Government Act* (1974);
- improving local government understanding of the purpose and principles of the new RMA as outlined in Part II, sections 5-8; and
- providing public education for a new way of thinking about urban and rural planning that was based on environmental effects, not land use activities.

The proposal was rejected by Treasury because the money already earmarked for developing the RMA had run out and there was a view at the Cabinet of Government ministers (who met weekly) that the Act was done and it was now up to local government to implement it. Indeed, the Government's commitment to funding its overall environmental programme in the early 1990s was so poor that a senior MfE manager described its budget as 'quite simply anorexic'.

In the critical five years for plan-making in councils (1991-96), we estimate that the MfE was allocated about $9.5 million, an average of $1.9 million a year, for implementing the RMA. Most of this work was carried out by its Resource Management Directorate (RMD) and regional offices, but a good deal was also done in its Māori Secretariat (Maruwhenua) and Legal Unit.

---

[3] According to Sir Geoffrey Palmer, architect of the resource management law reform, the overall cost to enactment of the RMA was around $8 million (Palmer, 2003).

Staffing Trends

By the mid-1990s, MfE had the dubious distinction of hovering around the lowest salary structure in central government. This affected staff quality and numbers, and hence activities. Figure 3.2 shows a downward trend for staff numbers throughout the 1990s, especially in the RMD.

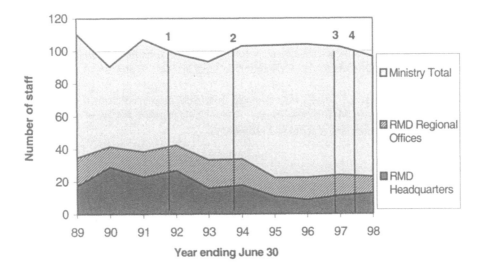

Figure 3.2 Ministry for the Environment: staffing trends for RMA implementation (head and regional offices) compared with total ministry staffing (The period 1991 to 1996 was critical for policy- and plan-making, see Figure 3.1 for points 1-4)

Source: Data extracted from Annual Reports to Parliament of the Ministry for the Environment

In 1991, the year the RMA came into being, overall MfE staff declined by 8 per cent (107 to 98), although the RMD staff (head office and regional offices) temporarily increased by a similar percentage (39 to 42). The RMD staff numbers then fell nearly 24 per cent the next year to 33, and 33 per cent the following year to a low of 22 in 1995, where it has remained — in spite of the *Green Package* of 1997. Staff numbers in the Māori and legal units, which also had a role in reviewing the policies and plans of councils, averaged around four and two respectively throughout the decade.

The RMD had only $0.9 million for its work in 1991 and, while this had quadrupled by 1999, it is still a puny sum, and must have had serious effects on retaining quality staff. As elsewhere in the MfE, there was a constant struggle to maintain critical mass. For example, 'where one person handled several key issues and left the agency, there was no guarantee that the work would be continued in the same shape or continued at all'.

Paradoxically, while total funds for the MfE increased after 1995, total staff numbers did not rise correspondingly (see Figures 3.1 and 3.2). There are two reasons for this apparent gap:

1. to improve staff retention after 1995, the MfE employed fewer more senior staff and paid them higher salaries; and
2. the shamrock approach was used for employing a core of high-quality permanent staff, retaining a pool of semi-permanent staff to call on when needed, and contracting external consultants for specialized tasks. As well, some secondments to the MfE aimed to help provide expertise.

Given the shortage of funds, this strategy probably increased both the efficiency and effectiveness of MfE's operations. Nevertheless, staff turnover continued to climb (see MfE's annual reports to Parliament).

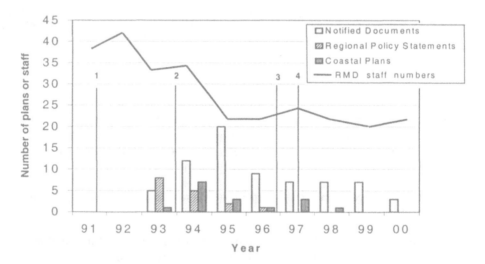

Figure 3.3 Ministry for the Environment: comparison of trends in number of notified documents (regional policy statements, regional coastal plans, and district plans) with total number of staff in the Resource Management Directorate (RMD head and regional offices), 1991-2000 (see Figure 3.1 for points 1-4)

Source: Data extracted from Annual Reports to Parliament of the Ministry for the Environment, 1998-2001 and annual local government surveys, 1993-2000

Bleak house! The significance of the Government's tight-fisted staffing is best seen by relating the annual expenditure and staffing trends for implementing the RMA to the annual number of policies and plans that had to be assessed. These trends are shown in Figure 3.3 (above). Clearly, on this one dimension alone, it can be seen that the burden of work was dramatically increasing while staff numbers were falling. For example, the workload created by the preparation and notification of one plan alone could span several years.

We can only concur with Dormer (1994) that, while the Government had made the MfE its main agency for implementing the RMA, it was unwilling to commit to it adequate resources for performing its functions. This lack of commitment had several important consequences. Staff stress and burn-out, already high by 1991, increased. Staff turnover further increased due to low salaries and stress, and was still 24 per cent of the total in 1998 (MfE, annual reports to Parliament). Relatively junior staff had to act where the involvement of more senior and mature staff was desirable. Very few early recruits to the MfE had expertise in planning and/or experience in local government, most being scientists or policy analysts, but salaries were so low that the MfE could not compete in the market-place for experienced planners.

Department of Conservation

For the Department of Conservation (DoC), the RMA added responsibility for the coastal marine environment, and by implication advocacy through local government planning, to its existing responsibilities of looking after the Crown estate and advocating conservation values generally. In addition to preparing the *New Zealand Coastal Policy Statement* (NZCPS), DoC's main responsibilities under the RMA included ministerial approval of the mandatory regional coastal plans produced by councils, controlling activities in the coastal marine area in conjunction with regional councils, and monitoring policy outcomes in these areas.

Outside the coastal marine area, DoC did not have a direct statutory role under the RMA other than being a required consultee for policies and plans prepared by councils (see next section). However, as the nation's advocate for conservation values under the *Conservation Act* (1987), it had to ensure that regional and local councils effectively implemented methods in policy statements and plans for conserving and protecting natural and historic resources of national importance in their areas. (See Appendix 1, sections 6-7.) This meant not only providing information and advice on these matters to councils, but also advocating a conservation viewpoint on them.

Funding trends While budgetary pressures were keenly felt in DoC, many difficulties it experienced early in the RMA's implementation resulted from repeated restructuring and down-sizing in the four years to 1993 — critical years for plan-making. In that period, the $119 million budget was cut 10 per cent (Figure 3.4). By 1995, a regional conservator lamented that a 'poverty mentality' had emerged in the conservancy that adversely affected operations, while another said that 'vehicles were put up on blocks owing to lack of funds for running them'.

In the same year, a poorly constructed viewing platform at Cave Creek near Punakaiki on the South Island's West Coast collapsed, killing 14 people. A public inquiry into the Cave Creek disaster highlighted DoC's resource-poor condition and recommended changes. 'No government organization can do its job without adequate resourcing. Here, the evidence is clear that the Department of Conservation lacked and continues to lack those resources' (Commission of Inquiry, 1995). DoC was re-organized and funding increased from $100 million to $131 million in 1996 following a review by the State Services Commission (1995).

Given its statutory role on the coast (including preparation of the NZCPS) and advocacy for environmental planning, DoC tried to ensure that, in spite of financial stringency and budget cuts in the four years to 1996, activities for these were, relative to elsewhere in the department, adequately funded. In addition, the annual purchasing agreements for RMA-related activities tended to be more flexible, which helped deal with crises as they emerged. It was also possible to predict reasonably well the outputs from councils that would have to be dealt with, and plan accordingly.

Nevertheless, funding shortages were keenly felt, and many working in conservancies reported that advocacy planning under the RMA suffered at the expense of coastal obligations, and that conservation in general suffered because of the demands of the RMA. In the longer run, DoC suffered because its advocacy role at regional and local council hearings was perceived by many as too often heavy-handed (see Part IV).

Figure 3.4 shows expenditure trends for activities under the RMA in relation to DoC as a whole. Against the overall trend, funds for these activities doubled from 1990 to 1991 to $7.5 million. Funding then fell steadily to about $5.4 million in 1996, when nearly all regional policies and coastal plans had been notified, then fell steeply following the new round of restructuring in 1996, and a partial retreat from advocacy planning to reach $3.3 million in 1999.

Staffing trends Between 1991 and 1993, there was a 20 per cent reduction in DoC staff as budget cuts bit deep. The consequences were most keenly felt in the regional conservancies. For example, in Canterbury Conservancy, the total number of fulltime equivalent (FTE) staff professionals in 1994 was 44, of whom only 3.6 worked on RMA matters. Of these, 1.4 FTE worked on coastal issues and 2.2 on conservation advocacy. While being somewhat better resourced for the coastal work, staff reported that they 'were struggling to keep up with the workload'. Because the 'coast was emphasized in the RMA and needed urgent attention, other conservation work was suffering'. Staff in other conservancies expressed similar concerns. For example, the number of coastal staff in the Otago Conservancy was halved to 0.5 FTE in 1993 (DoC staff interviews, 1994).

Unlike earlier restructuring, that of 1996 did not result in staff losses overall. Rather, it down-sized head office, redistributed resources to regional conservancies and field offices for pest control and recreational activities, and increased accountability through tighter line management. Early restructurings aside, staff turnover in DoC was relatively low, even though salaries were low in comparison with many other units of government.

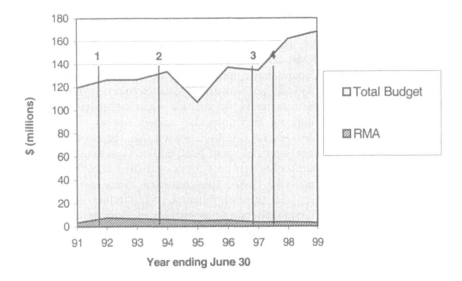

Figure 3.4 Department of Conservation: expenditure on RMA implementation compared with total expenditure mid-1990 to mid-1999 (The period 1991 to 1996 was critical for policy- and plan-making. See Figure 3.1 for points 1 to 4)

Source: data extracted from Annual Reports to Parliament of the Department of Conservation

Parliamentary Commissioner for the Environment

The third agency in the environmental administration is the Office of the Parliamentary Commissioner for the Environment — Parliament's environmental guardian and ombudsman.   Its focused work resulted in publications of considerable value in guiding councils on organizational processes and various environmental issues, some of relevance to plan-making (Parliamentary Commissioner for the Environment, 1992, 1998 and 1999). Its budget of $0.6 million in 1994 had risen to $1.45 million by 1998 and supported 12.9 professional FTE (Parliamentary Commissioner for the Environment, Annual Reports to Parliament, 1993-1998).

Implications for Local Planning

Overall, compared with other units of government, such as those devoted to business interests, the three environmental agencies were, throughout the 1990s, poorly funded — especially the MfE with its responsibility for overseeing implementation of the RMA, including plan-making. DoC started its work on the

RMA with a budget four times greater than that for MfE. Indeed, the $7.5 million DoC spent on the RMA in 1991 was about three-quarters of the MfE's total expenditure for all of its operations.

In the seven years to 1998, the three environmental agencies had spent around $60 million facilitating the RMA (DoC, $39 million; MfE, $15 million; and the Office of the Parliamentary Commissioner, nearly $6 million). We estimate that as much as half of this amount ($30 million) was directly related to plan-making.

These seven year totals were really very small when compared with the overall Government expenditure on environmental matters (policy, operations, management and planning). McDermott (2000) estimated that, for the single year 1996/97, nine central government agencies spent $250 million (less than 1 per cent of total government expenditure), of which only 5 per cent ($13 million) went to MfE. As for implementation of the RMA, Government's three environmental agencies received only 3 per cent ($8 million) of the total $250 million in 1996/97. Of this amount, only part went into facilitating plan-making in councils.

As already noted, funding for implementing the RMA improved after the reappointment of Simon Upton as Minister for the Environment in 1993. His keynote address to planners at the New Zealand Planning Institute conference in 1994 acknowledged the need for more resources to improve the capacity of local government to respond, including resources for MfE to monitor how well councils were doing (Upton, 1994). His aim to improve funding was difficult to achieve in a climate of fiscal restraint and managerial changes, and, while Figures 3.1 and 3.4 suggest he had some successes, the financial consequences for local government of not doing better were considerable.

By 1998, analyses showed that the financial burden of implementing the national mandate for the sustainable management of natural and physical resources had fallen heavily on local government (McDermott, 2000). By then, the mean cost to councils for plan preparation was: $0.525 million for regional policy statements; $0.894 million for regional plans; $1.495 million for district plans; and $1.477 million for combined plans (Ministry for the Environment, 1999). Overall, we estimate that plan-making alone had cost local government around $120 million by 1998, being perhaps about 10 per cent of its total cost for implementing Government's environmental policy, planning, management and operations (McDermott, 2000). Costs to non-government groups and individuals are not known, but must have been high, as case studies in Part 4 indicate.

In summary, the National government adopted the ambitious and rather sophisticated planning mandate proposed by its predecessor Labour government. With little moral commitment to the mandate's provisions the Government was, however, unwilling to provide the level of funds necessary to build sufficient capability within its agencies for helping local government to implement it.

Just what constitutes adequate funding is difficult to ascertain. But when activities needed for the transition are cut entirely, when funding each year inexorably falls while activities increase, when staff turnover is high due to overwork and stress, when salaries are so low that attracting quality staff is difficult — then clearly central government is not providing a satisfactory level of funding to its agencies.

Actions of Government Agencies

Having reviewed the effects of government policy swings on, and low levels of commitment to, the environmental planning mandate, we now consider the roles of central government agencies in implementing the RMA process for making new plans.

Although co-operative plan-making in local government was not new, the environmental effects-based approach and statutory processes were. As well, there was now a much greater need for rigour, both in policy evaluation and monitoring processes and outcomes (sections 32 and 35, respectively). This demanded a higher level of performance from councils than hitherto. How well did central government agencies deal with regional and local councils when formulating new policies and plans, and how squarely did their behaviour fit with our characterization of the RMA as a co-operative mandate?

Statutory Consultees

During the preparation of a regional policy statement or regional or district plan, the RMA requires that councils consult with ministers of the Crown whose interests as 'statutory consultees' may be affected by their implementation (Clause 3, First Schedule). Staff in relevant government agencies review local government policies and plans on behalf of their minister and make comments to councils as appropriate. After public notification of the policy or plan, they may make submissions at council hearings to ensure their point of view is addressed. Thus, a council has to synthesize different views from different government agencies on the same issues when revising its plan.

This process was fundamentally different from that under the previous planning regime. The Town and Country Planning Division in the Ministry of Works and Development took the lead role in co-ordinating into a unified Crown view the responses of different government agencies to regional and district schemes (i.e. plans). Local councils therefore had a 'united face' of government and a 'point person' to deal with. Building a partnership between Crown and local government under this system would seem to be much easier than doing so with a multiplicity of central agencies under the RMA.

In the 1990s, there were over 40 government departments (including ministries) and state-owned enterprises, most of which did not have an interest in commenting on the policies and plans of local government. For those that did, their roles ranged from resource development in the interests of business through to the preservation of natural resources in the interests of conservation. These two could be seen as the extremes on a continuum of roles and interests.

With different interests to protect, the agencies either supported or opposed objectives in plans, and favoured differing methods for achieving them. For example, at one end of the continuum, DoC tended to support the use of rules in plans that would compulsorily protect nationally significant natural environments. And, being concerned with human welfare, the Ministry of Health, responsible for dealing with the effects of air pollution, also supported regulatory controls.

At the other extreme, the Ministry of Commerce (now the Ministry of Economic Development) was charged with looking after the economic and commercial interests of the nation, being the advocate for industries such as tourism, energy and mining. Keeping costs down for business was its main concern. The RMA was seen as impinging on economic development by intervening in the market's allocation of natural and physical resources. Compulsory regulations were therefore eschewed in favour of economic instruments and voluntary restraints in dealing with the externalities associated with resource development.

Some agencies, like the Ministry of Forestry and Ministry of Agriculture and Fisheries (now Ministry of Agriculture and Forestry and Ministry of Fisheries), had both production and conservation functions, and therefore lay more towards the centre of the continuum. The responses of these ministries to plans varied, but tended to encourage councils to adopt alternatives to regulatory rules in the interests of sustainable resource development. Agencies that developed infrastructures with an impact on resources and the environment, like Transit New Zealand and Transpower, tended to focus on reducing regulatory variation across jurisdictions on their works. In this way, they were like large private companies (forestry, energy and utilities) that prefer uniformity and certainty in plans across the country, rather than the variation, and therefore the uncertainties, that community-based plans encourage.

How much effort each agency expended on policies and plans depended on their level and range of interest. Clearly, pro-business agencies posed a considerable challenge to councils that held faith with the environmental effects-based approach. As such, they could not be regarded as capability-builders for plan-making in local government, except indirectly. Rather, they responded to plans simply because they perceived that their interests would be adversely affected. In this, they found considerable support from pro-development groups within the regions and districts.

Capability-Builders

Ministry for the Environment To help get policy and advice to local government, businesses and other resource users, the MfE was organized into a head office and three regional offices — northern (Auckland), central (Wellington) and southern (Christchurch and Dunedin). This structure endured until 1994 when, in response to cost-cutting, the central region was absorbed into head office (Wellington), while the Dunedin office in the southern region, originally an independent unit, closed in 1997. At head office, the RMD was responsible for advising on and monitoring implementation of the RMA, including regulations and amendments.

Consistent with the philosophy of minimalist policy advice noted earlier, the MfE's activities were influenced by four main factors. Staff recognized that a lot of plan-making expertise already resided in local government, and this reduced the need for MfE help. History suggested that councils did not appreciate being told what to do by central government with state-subsidized projects, thus reinforcing its minimalist approach. District schemes (plans) prepared by former councils under the *Town and Country Planning Act* (1977) served as transitional plans for the new councils, and most had up to five years before new district plans had to be produced. Finally, some MfE help could piggy-back on the activities of professional organizations, such as the New Zealand Planning Institute, Local Government New Zealand and the Resource Management Law Reform Association.

In these conditions, according to MfE staff interviews in 1992 (May et al., 1996) the MfE set out to encourage and facilitate, rather than impose and coerce, in the belief that innovative planning might emerge within councils. It adopted the view that the 'best way to learn is by doing' rather than being told what to do — reinforced by the reality of the 1991 budget cuts. Learning would be 'facilitated by exchanging information and being participants in discussions with soft quality control'. Providing examples to councils for meeting given ends was thought to be useful, especially for small and poorly resourced councils. Thus, a 'somewhat flexible approach' was deemed appropriate. Only when councils 'failed to respond in ways that would meet statutory obligations would the Ministry use stronger means — going to court being a last resort' (MfE staff interviews, December 1992). For example, four regional councils missed the two-year deadline for notifying their regional policy statement. Through the regional offices they were asked for explanations (typically lack of resources for extra duties and functions), and only after several months had passed did head office become involved. The only sanction made was against Gisborne unitary authority, when the ministry withheld its grants for one year.

The MfE's approach was driven by these practical realities rather than by any explicit view of the RMA as a co-operative mandate. Nevertheless, the approach is consistent with our characterization of the RMA. It was when moving from philosophy to practice that the approach got compromised, because the MfE's transition plan was declined by the Government. Thus, from the outset, 'the Ministry had to pursue low-cost methods for reaching clients', including councils, because that is all that the minister would, or could, purchase on behalf of the Government (MfE staff interviews, 1997).

THE ACTIONS: Ministry activities for implementing the RMA can be grouped into 13 somewhat overlapping types, as listed in Figure 3.5. The strength of activity over time is shown by the density of shading, and where appropriate the number of items. The range of actions seems impressive. They were, however, too often carried out in a reactive rather than proactive manner, and aimed more at issues of internal consistency and legal certainty in draft and notified plans than with matters of national importance. What is more, the lack of funds affected what could be delivered, and how.

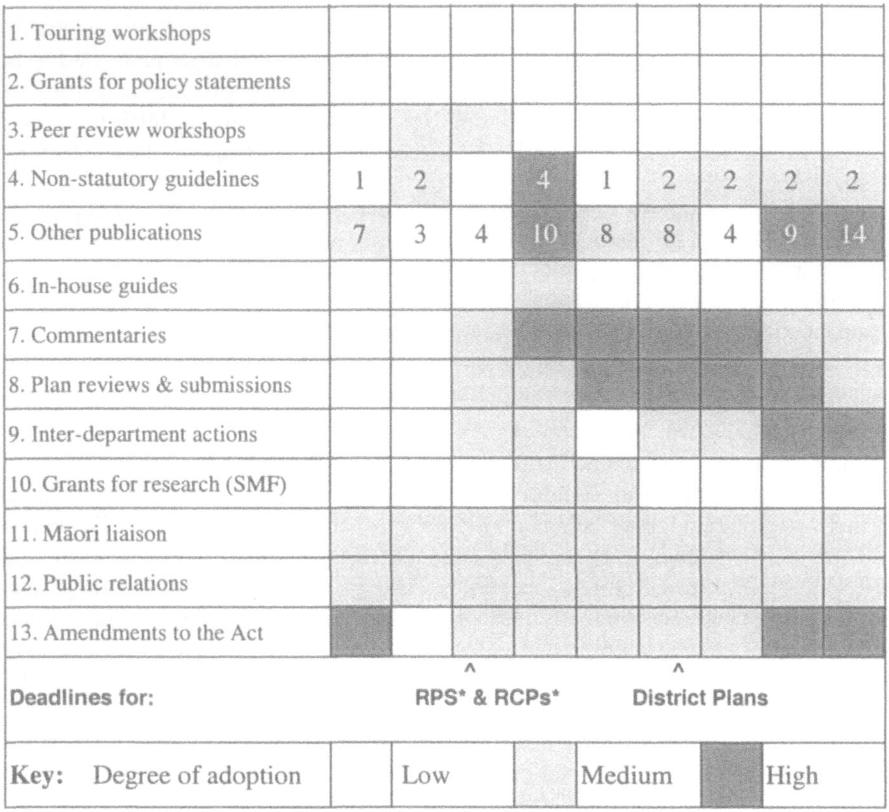

1. Touring workshops									
2. Grants for policy statements									
3. Peer review workshops									
4. Non-statutory guidelines	1	2		4	1	2	2	2	2
5. Other publications	7	3	4	10	8	8	4	9	14
6. In-house guides									
7. Commentaries									
8. Plan reviews & submissions									
9. Inter-department actions									
10. Grants for research (SMF)									
11. Māori liaison									
12. Public relations									
13. Amendments to the Act									

Deadlines for: RPS* & RCPs*   District Plans

Key: Degree of adoption | Low | Medium | High

* RPS = regional policy statements; RCP = regional coastal plans

Figure 3.5 Actions undertaken by Ministry for the Environment, 1991-99

Source: Ministry for the Environment staff and Ministry for the Environment 1988b, 1991c-i, 1992a-c, 1993a-d, 1994a-k, 1995a-h, 1996a-i, 1997a-d

Thus, for example, while 'publications' explaining the new planning process and how it would affect councils and iwi appeared early, the 'touring workshops' arrived at regional councils too late to help them complete their compulsory policy statements and coastal plans. Earlier, in 1992, it was the regional councils rather than the MfE that initiated the successful *peer review workshops*. Their extension to local councils was stifled by lack of resources. Likewise, the *grants* to regional councils for preparing policy statements were not extended to local councils for district plans. And, as is detailed in Chapter 4, the 'grants for research' fell mostly to agencies outside local government. Likewise, 'non-statutory guidelines', some of which contained 'standards' relevant to plan-making, waned in the critical years 1991-94 precisely when they were most needed by local government. Guidelines for other aspects of the mandate, such as pollution, were produced in good number by 1996 however. Thus, many local councils struggled for years to find clarity in the mandate and how best to develop environmental effects-based plans, even though the MfE worked on 'amending' the RMA every year to 1999.

An obvious action missing from the list in Figure 3.5 is the provision of a 'model plan' to show councils how to prepare and present the new environmental effects-based plans. It was believed that too many councils would adopt a model plan without due consideration of local circumstances — as happened in the early years of the previous planning regime. This would stifle the hoped-for innovation and creativity, given that many planners were used to a more prescriptive approach. Further, (consistent with a co-operative mandate) it was expected that, as negotiated documents, regional policy statements and regional and district plans would vary around the country in response to local environmental issues (Interviews, 1992). There was also concern that a model plan would be too generalized, providing a high-level document that may well have misled councils (Letter, MfE Deputy Secretary, 1995). Even had it wanted to, however, the MfE had no money and little expertise for developing a model plan. Its early attempt at providing focused case studies as simulations was terminated when the staff responsible left (Ministry for the Environment, 1992c). Also largely absent from its activities was the provision of guidance and information to councils for dealing with issues of national importance (sections 6-8 in Part II of the RMA).

THE REVIEWS: While the MfE lagged behind in helping explain what councils had to do, it was soon faced with reviewing what they were doing. At the coalface were its regional office staff advising councils on plan preparation, including 'commentaries' on their public discussion documents, like those dealing with local issues and objectives. But this interaction depended a lot on council staff approaching the MfE as plan-makers called for help in, for example, evaluating alternative policies or implementing the cascade of issues, objectives, policies and methods (see Figure 3.3).

Consequently, the MfE held 'workshops' that produced 'good practice guides' for plan-makers to follow (Ministry for the Environment, 1993c; 1994h). But this happened after the deadline for regional councils to publicly notify their compulsory policies and plans had passed, some local councils had already notified their district plans, and council planners had requested help from the MfE's Auckland regional office. Reflecting on this situation six years later, a plan-maker (whose proposed district plan scored well) stated: 'That cascade guideline was a god-send. It demystified what needed to be done'.

Once policies and plans were notified, MfE head office staff had to 'review' them, and then, if need be, prepare a *submission* for the minister to make at the relevant council hearing as part of the public process. Although helpful, these reviews were mostly legalistic and often seen as somewhat negative by council staff and councillors.

In retrospect, some MfE staff viewed their overall actions as having been 'sub-optimal for plan preparation in councils' (MfE staff interviews, 1996 and 1999). The many activities they engaged in were not necessarily the most useful ones. It is easy to be wise after the event, but one could ask the obvious: how much more useful would it have been to the plan-makers had the MfE surveyed them early on to find out the nature of help needed, the priority in which it should be given, and how best it might be delivered? But, politically, was that a purchasable item? On the other hand, would those in councils have known what they really needed for making good plans in the opening years?

Department of Conservation   DoC developed a very formal, hands-on approach for its work on the coast, aimed at gaining consistency in conservation issues covered in regional policies and plans. Responsibility for developing the NZCPS fell to head office staff who also drove the process for assessing policies and plans that emerged from councils (Department of Conservation, 1990, 1992, 1994a, 1994b, 1995). DoC's fourteen regional conservancies had the lead role in reviewing the policies and plans for councils in their areas, and for preparing submissions on them where necessary (DoC staff interviews, 1999).

In the first two years (1992-93), DoC concentrated on developing the coastal policy and associated guidelines for use by regional councils, and was then able to focus more fully on its advocacy role as a statutory consultee. Unlike the coast, activities for conservation advocacy elsewhere were more diffuse, and responsibility for dealing with them resided mainly in the fourteen regional conservancies. As advocates for conservation values, staff sought to persuade councils to a point of view by highlighting issues for the region or district that planners needed to consider. Because issues varied around the country, a uniform or blueprint approach was not used, and the level of help offered depended on the type of council and its ability to deal with the issues. DoC tended to have more contact with smaller councils struggling to come to grips with locally contentious matters of national importance specified in the RMA.

The main advocacy activities of conservancy staff here included, variously: discussions with council staff, field inspections, reports, field surveys to identify important species and habitats, help with developing appropriate plan provisions, reviews of draft policies and plans, meetings with councils, and formal submissions on policies and plans.

In summary, given the Government's lack of commitment to adequately funding its environmental mandate, the MfE and DoC 'chose a series of low-cost activities for reaching clients', including plan-makers in councils. These were the activities that the ministers allowed through their annual purchase agreements with their agencies. While each agency offered a reasonably wide range of activities, despite statutory deadlines and dramatic budget cuts at critical times, in hindsight, staff believed they were less than optimal for facilitating plan-making in councils. The implications of this for councils trying to meet the provisions of the Government's planning mandate are now explored.

How Well Did Central Government Do?

Our interviews with staff in the environmental agencies indicated that even small increases in the annual budget would have gained a large increment in dealing with local plan-making. Staff found not having the funds frustrating. They focused mainly on ensuring certainty, internal consistency and legality when advising councils on their plans. But what of the bigger issues, like clarity of mandate, matters of national importance, and compensation to councils and property owners complying with the RMA? What difference would it have made here had the Government not only provided the $2.2 million the MfE sought for its transition plan in 1991, but also a stable budget costing only, say, another $2.2 million for the next five years? In other words, what important matters in the RMA ought Government have been funding its agencies to deal with, to ensure that councils were properly implementing its mandate? We review performance on: 1) clarifying the mandate; 2) matters of national importance; and 3) policy learning.

Clarifying the Mandate

Our surveys indicated that at least half of the staff in regional and local councils responsible for preparing policy statements or plans found five of six key provisions in the RMA to be unclear (as detailed in Part III of this book). Although the RMA was amended almost every year before our survey in 1997, these key provisions remained unchanged except for section 31, where the relationship between regional and local councils on hazards was clarified — a resolution forced by costly court action (see Chapter 4). Yet, early meetings, workshops and conferences with local government signalled strongly to the MfE that widespread confusion existed over the purpose and principles of the Act and its meaning for plan-making.

The purpose of the RMA as laid down in section 5 allows for diverse interpretations, especially when read in conjunction with the meaning of environment, which includes people and their socio-economic aspects (section 2). Thus, the extent to which social and economic issues, in addition to environmental ones, could be dealt with in plans varied from none or some to a lot, as could the degree of regulations. These interpretations reflected the ideological spectrum (Nixon, 1997: 13-14). This flexibility fitted a devolved mandate well because it provided for local choice.

Not wanting to intervene either by developing national policy statements or by making section 5 of the RMA more clearly environmental, the Minister for the Environment emphasized effects-based planning and alternatives to regulatory rules in his frequent public addresses (e.g. Upton, 1991; 1994; 1995a; 1995b; 1997a; 1997b; 1998; 1999). The inclinations of many others, including many plan-makers, lay elsewhere in the range of possibilities. The MfE was obliged to toe the political line in this, but saw council reactions to it as simply carry-over thinking from land-use activities-based regulatory plans under the old regime. The resulting conflicts generated uncertainties for planning as decision-makers grappled with what environmental effects-based planning really meant without reference to the land-use activities that generated them. Little wonder that MfE staff, with little planning and/or local government experience, too often found themselves talking past, rather than with, the plan-makers in councils. In many councils, too, political interpretations of section 5 were polarized.

As the discontent among councils and resource users grew ever louder, something had to be done (Pavletich and McShane, 1997; Upton, 1998). In 1998, the Minister called for advice on this and other matters affecting the effective implementation of the RMA. He employed a dialectical approach by commissioning a consultant identified with the New Right to 'analyze the extent to which particular sections of the Act are contributing to regulation which is unnecessary in achieving (its) purpose (section 5) ...' (McShane, 1998, Terms of Reference), and then having three expert planners, seemingly from across the political spectrum, offer a critique of the recommended solutions.

In the report that followed, there was agreement that the purpose (section 5) and principles of the RMA (especially sections 6 and 7) lacked clarity. All four reviewers agreed that the definition of the 'environment' (section 2) should be narrowed to focus more firmly on the natural and physical environment — for example, by excluding people and communities from being part of ecosystems. This was because when section 2 (environment) and section 5 (purpose) were considered together they could be seen as providing support for social and economic planning. The intent had been that councils should consider the socio-economic 'effects' (positive and negative) only when controlling the environmental effects of using and developing resources (Nixon in McShane, 1998: 10). One reviewer pointed out that councils could accomplish social and economic planning under other statutes (see Figure 1.2), and that anyway, since the RMA was passed in 1991, they had come to focus more extensively on developing strategic plans, annual plans and long-term financial strategies (Tremaine in McShane, 1998: 8).

The definitions of 'environment' and 'amenity' were narrowed in the *Resource Management Amendment Bill* (1999), but sections 5-8 (Part II) remained unchanged, including matters of national importance (section 6).

Matters of National Importance

Dimensions of the Problem

Matters of national importance are specified as principles in section 6 of the RMA (see Appendix 1). According to MfE staff, two aspects were seen as critical to plan-making for matters of national importance (interviews and correspondence, 1999). First, 'it is nationally important for councils to address issues of national importance in their plans'. Most of the MfE's effort, they say, was put into this aspect. For example, where it had the skills or knowledge on such issues and these were missing from a proposed policy or plan, it made submissions on them. This was, however, a reactive rather than proactive approach.

Second, 'there are nationally important aspects to those issues that you would expect government agencies to feed to local government where they had the responsibility, although there were no formal arrangements for this to be done'. For example, MfE thought that DoC had responsibility (under the *Conservation Act*) for nationally important landscapes, heritage, coastal matters and so on, although there were no formal protocols to ensure this. As we shall see later, DoC had a different view.

This very loose arrangement — in an era of managerial protocols and contracts — suggests there was a lack of strategic leadership from the MfE in identifying and clarifying lines of responsibility and accountability with other departments and councils from the outset. This was an interdepartmental problem, which was also identified in a review of environmental administration by the Organization for Economic Co-operation and Development (1996). Senior MfE staff indicated to us in 1999 that more protocols might well be needed in future.

Providing policy direction and data   Our interest here is in the second aspect — the timely and appropriate feeding of information on national issues by central agencies to local government for plan-making. Two aspects of this service, we believe, are critical to making good plans. First, central government must have a clear vision of the issues of national importance and what they mean for plan-making. The best means provided in the RMA for doing this is the preparation of 'national policy statements'. Second, adequate and timely 'practical information' (e.g. methods and data) must be supplied by its agencies to local government if local plans are to be satisfactorily developed. Indeed, the provision of good information by the Government would seem to be essential for the development of innovative plans. To what extent did central government provide these resources for local government planning?

NATIONAL POLICY STATEMENTS: Only one national policy statement has been produced, that which the RMA made compulsory for the coastal marine area. Although preparation of regional coastal plans was already well under way when the *New Zealand Coastal Policy Statement* (Department of Conservation, 1994b) emerged, it has nevertheless been helpful to both regional and local councils, by doing what one would expect in a hierarchy of policies and plans. It enables national goals to be interpreted and applied down the hierarchy, so that local plan-makers gain direction and guidance in aligning their objectives and policies with those higher in the hierarchy. (This is discussed further in Chapters 4 and 11.)

What, then, are the implications of not having national policy statements and related guides and data to help facilitate local plan-making on matters of national importance specified in the RMA? To illustrate, we have selected from the list of principles the requirements for councils to recognize and provide protection for 'outstanding natural features and landscapes' and 'significant indigenous vegetation and habitats for fauna' (i.e. significant natural areas) when managing the use, development and protection of natural and physical resources in their areas (sections 6(b) and 6(c), respectively). Further illustration is provided in case studies (Chapters 9 and 10).

NATIONAL GUIDELINES AND DATA: As Part III will show, only a third of councils considered the two lead government agencies, the MfE and DoC to be useful sources of information. The following review helps to explain why. Before that, however, it is important to acknowledge that by 1997 the MfE had produced many non-statutory guidelines, especially in relation to the pollution of water and land, and that some of these had the potential to become national standards, although none had been produced a decade after the Act had become law (see Figure 3.5 and Ministry for the Environment in the references cited).

Some four years after the RMA became law, serious conflicts were emerging in some local councils, such as over seeking to recognize and protect significant indigenous flora and fauna and outstanding landscapes in new plans. In part this was due to four interrelated problems. First, there was an inadequate appreciation of what and how much to protect, and why, as there were no national policy statements to guide councils in their thinking about these matters, just phrases in the Act. Second, the methods by which natural areas should be identified for protection were flawed. No specific methods were provided by central government to guide local councils. Instead, councils searched the literature or employed consultants to devise their own, or pleaded with the under-resourced DoC for help. Sometimes others told them that they had it wrong when their proposed plans were reviewed. Third, there were limited options for protecting these important areas. A regulatory approach combined with the lack of funds for adequate research and consultation too often resulted in a backlash from property-owners to the notified plans because they would be carrying the costs (this is illustrated in the case studies in Chapters 9-12). Fourth, there was no case law for guiding actions.

Councils, not unreasonably, assumed that DoC had the primary responsibility to help them on natural heritage matters, but it had no statutory responsibility to do so under the RMA, other than to be a consultee for reviewing proposed policies and plans (like any other government agency affected by them). Nor did its purchase agreements call for building capability in councils beyond the coastal marine area. DoC staff explained that it was MfE's role to provide appropriate policies and guidelines to local government and to build their capabilities for planning under the Act (DoC interviews, 1999).

Nevertheless, DoC's conservancies did try to help councils beyond coastal matters as far as limited resources would allow, if only to ensure that the values they advocated under the *Conservation Act* were included in regional and local plans. This led to different approaches being adopted in different regions, and a mixture of good and bad experiences. For example, the experience between the Waitakere City Council and DoC in identifying significant natural areas was positive because they shared in aerial photographic surveys. An aerial survey was not done by DoC for the Far North District Council and this contributed to poor outcomes (Chapter 9).

Such was the confusion caused by DoC's behaviour that it eventually worked with Local Government New Zealand to develop a protocol or contract spelling out the various responsibilities of each party, especially on matters of national importance. This included limits on DoC's ability to provide information on, for example, significant natural areas and outstanding landscape values (Local

Government New Zealand and Department of Conservation, 1996). The agreement was signed in late 1996 — five years after plan preparation had started under the RMA.

This did not, of course, solve the issue of adequate national policies, guidelines, and data for local plan-makers. It only clarified responsibilities for who did what, and why. Thus, for example, a workshop on significant natural areas was convened in late 1997, not by central government, but Local Government New Zealand (Local Government New Zealand, 1997). Both MfE and DoC had a presence at the workshop, and MfE commissioned an expert to present a paper on approaches for councils identifying areas of significant flora and fauna (Froude, 1997). But this was now six years after the Act had become law, 58 out of 74 local plans had already been publicly notified, and public controversy still raged around some of them.

By then, with adverse public reaction to DoC's advocacy in plan preparation mounting, the Minister of Conservation requested staff not to become involved in helping councils on outstanding landscapes, and to draw back from the submission process. And a year later he was signalling not only a review of DoC's role on the coast, with the prospect of reducing it to being a statutory consultee only, but also better ways for working with communities on national conservation issues (Department of Conservation, 1998a, 1998b).

A useful activity to emerge from the confusion and consternation was the development of a national strategy for biodiversity, with prospects for supporting legislation to complement that on biosecurity (*Biosecurity Act*, 1993; *New Zealand's Biodiversity Strategy*, Ministry for the Environment, 1998). It has been proposed that this lead to a 'national policy statement' on biodiversity and that the Government develop 'accords' with landowners and iwi over management of biodiversity on private land (Ministry for the Environment, 2000: 54-59, 46-53).

The absence of formal planning in central government for the provision of national policies, guidelines, methods and data on issues of national importance in the preparation of plans by councils is a major failing by the Government. The reasons for this failure are several and interrelated, including the lack of agency funding and political leadership at critical times, the Government's minimalist approach and desire not to interfere, especially if costs to it could be substantial — as they would be, and equivocation among councils of the need for central direction (Hutchings, 1999). In short, central government was both a poor employer and a poor partner. It starved its agencies, and then failed to provide adequate and timely information on issues of national importance, leaving councils to flounder. Thus, it did not adequately support the implementation of its own mandate.

Policy Learning

When the Government introduced its ambitiously reformist mandate for environmental planning in local government in 1991, it did not seem to fully appreciate that its own agencies would be on a steep learning curve themselves while also trying to implement its mandate within very tight statutory timeframes. As well, at least one of the creators of the framework for the RMA did not envisage the resulting plans to be as long or complex or to take as much time to prepare as they did (MfE interview, 1999).

One problem was that, while the MfE wanted to encourage innovation in councils (consistent with a co-operative mandate), it seems to have given too little direction through its technical and financial assistance programmes (which a co-operative mandate requires). The dilemma facing councils — that they needed direction but were expected to be innovative — was exacerbated by staff in government agencies having themselves to learn what environmental effects-based planning really meant and how best it might be applied in local government. But it was due also to Government's desire for quick nationwide implementation of the mandate, when some focused experimentation might have been better as a way of fostering much-needed policy learning at all levels. For regional councils, the statutory timeframes were, however, simply too short to allow significant experimentation. Instead, the MfE could have worked with a few representative regional councils to help them develop the basis for regional policy statements and selected regional plans, while at the same time working with representative local councils within these regions to frame their district plans.

The irony in all this is that a co-operative mandate recognizes that there will be variation across the nation not only in the abilities of councils to comply, but also in how they seek to solve their environmental problems. Freed from substantive prescriptions, local councils in particular were left by the Government to experiment, but too often they fell back on what they knew best (land-use activities-based planning), or waited until others had undertaken the costly exercise of working out environmental effects-based planning before applying it to their own plans. In the meantime, the Minister cajoled planners about the need to prepare environmental effects-based plans, but was unable to adequately fund the MfE to provide the technical support needed for helping to produce them (his absence from the job from 1991 to 1993 notwithstanding). Also, where some councils dared to be innovative, they paid dearly as communities rebelled against their effects-based plans and ministers did not rally in their defence (see Part IV).

Innovation in this context applies to methods for achieving common goals in plans. Two points are germane here. First, being innovative in methodology is difficult when years later the community is still arguing over the interpretation of purposes, such as sustainable management of significant natural areas or landscapes of outstanding value, in the absence of policy direction from central government. Second, the Minister for the Environment spoke often of the need for methodological innovation, but his examples were seen as New Right rhetoric: against compulsory rules, in favour of economic instruments and voluntary restraints. Local government had no experience of these, and many no enthusiasm.

Much learning has occurred over the decade since. Until 1997, the MfE was deeply immersed in the certainty, internal consistency and legality of policies and plans. In that year, it gained a modest increase in funding and shifted its focus to more technical matters. Thus, in early 1998 it proposed a new three-year work programme to support its involvement in the formal statutory process, development of regulations and amendments to the RMA, and improvement in implementing the Act (Ministry for the Environment, 1998a).

For 'improvement in implementing the RMA' year one would focus on getting councils to speed-up their resource consenting processes; year two would focus on getting councils to improve the quality of planning; and year three would focus on whether plans were achieving environmental outcomes. By 2000: resource consents were being processed more quickly by councils (although the quality of their implementation was questionable); a Quality Plans Project was underway that included the development of a web site containing good practice examples; but work on whether sound environmental outcomes were being achieved under the RMA had not started. As in earlier years, often funds gained for this work programme got diverted into other areas to meet political needs.

Good progress was, however, made on other elements of the 1998 work programme: national environmental indicators were under development; guidelines (such as on pre-hearings, auditing assessments of environmental effects, and interactions between iwi and councils) were developed; and annual surveys of councils to monitor performance on a range of indicators were being carried out (Ministry for the Environment, 1998b).

Conclusion

The Government (as distinct from staff in its environmental agencies) did very poorly in implementing its environmental planning mandate. Trapped between competing political ideologies (social and liberal democracies), the implementation process was shot through with mixed messages and contradictions, not only in the environmental mandate, but also the expected cost of its implementation. A devolved mandate suddenly became more like a decentralized one; managerialism stopped reforms for power- and revenue-sharing with local government; environmental agencies were starved of resources for giving leadership and direction to councils on matters of national importance; costly plan-making mistakes were replicated around the country; and local government was left to carry the political, financial and emotional consequences. Thus, the hoped-for policy coherency of the late 1980s degenerated into a series of hybrids in resource management, environmental planning and local government. More 'planning' by central and local government was being done than ever before (financial, transport, biosecurity), but coherency was taking a long time to achieve because planning at the national level had been almost non-existent.

In providing for a devolved environmental mandate that is implemented through a managerialist regime, the Government set up tensions that got played out at the local level. Economic efficiency from privatizing services, like water supplies, impacted on principles of democracy because these services then fell outside community control through elected councils. There was a fundamental contradiction between the diversity, variation and flexibility allowed for at local level through local government and resource law reforms, and the national consistency and uniformity demanded by resource users that are extra-local. This, of course, could be removed only by providing either a series of national policy statements and standards on matters of national significance or a prescriptively coercive Act that ran counter to local people determining their futures. This provided for yet another dynamic tension: seeking local variation in methods while at the same time achieving national goals, while the corporate entities (like Telecom) sought to standardize methods as a means of getting a clear plan.

Another contradiction in the age of economic efficiency and managerialism was to have governments embarking on major reforms and new statutory goals without thinking through the strategic and cost implications of implementing them. This in itself required careful research, which was not done for the RMA. Instead, the Labour Government set out in the belief and hope that implementing the mandate would not cost it much, and that amalgamations and managerialism in local government would create the capacity to pay for it. Thus, even though the incoming National Government had adopted a co-operative mandate for resource management but under a decentralized system of governance, subsequent actions demonstrated that it did not accept it had a major role in sharing the burden of cost. A more cynical view is that, being luke-warm towards a mandate that intruded so much on its liberal values, the system was quite simply set up to fail.

Learning by doing is a legitimate approach, as is establishing a blueprint for action at the outset. But both approaches need adequate funding to be effective. What happened was inefficient and costly and does not sit squarely with the rhetoric of liberal managerialism, where freedom of choice requires — indeed demands — adequate information for effective decisions to be made. The Government's insistence on cutting fat from an already anorexic environmental administration charged with implementing its sophisticated and costly mandate was irresponsible and counter-productive.

Like everyone else affected by the RMA, government agencies were on a steep learning curve. Deadlines within the Act made it difficult for policy learning to take effect in reasonable time. Consequently, two main trade-offs had to be made. First, the Government traded off directing councils in how to do their plan-making for facilitating spontaneous innovation. Second, the Government traded off measured experimentation for quick national implementation. Clearly, innovation and experimentation go hand in hand, and both require ample time and resources for learning to occur.

To conclude our analysis of the Government's role in implementing its co-operative planning mandate, we give two political anecdotes — one from near the start of the process, the other from near the end.

At a head office pre-Christmas function in 1988, the Minister for the Environment (Geoffrey Palmer, also Deputy Prime Minister) profusely and most sincerely thanked his staff for having achieved so much, in such a short time and with such limited resources. Key activities had included the countrywide assessment of climate change impacts, the demanding resource management law reform consultations, and some preparation for the *Resource Management Bill*. This recognition for having achieved much more than expected with much less than was needed was, unfortunately, a harbinger of things to come.

By 1997, the Minister for the Environment (Simon Upton) had publicly acknowledged for some time that implementation of the RMA had been under-funded. To his credit, he had successfully argued with his colleagues for the Green Package and, in his defence, he was not Minister in the critical years from 1991 to 1993. Nevertheless, at a New Zealand Planning Institute conference in April 1997, he said that 'success or failure' in implementing the Act was 'largely' in the hands of 'practitioners and councillors'. In stressing the point he used this analogy:

> .... what we have at present can be likened to a high-performance motorcar coughing and spluttering its way along the road. The Government is the designer of this car but the driver is the local councillor. The co-driver — who advises the driver which way to go, where the next pot-hole or pitfall might be — is the planner.
>
> Any faltering in performance tends to be blamed either on the designer of the vehicle or the designer's failure to provide a comprehensive operator's manual. Passengers (resource users?) claiming that they are heading in the wrong direction are informed that this is the way they've always gone and no road map has been supplied to show an alternative route. With cries of 'gross inefficiency' those who are asked to fill up the tank (tax-payers?) of this car are constantly asking me to pull it off the road for a major overhaul.
>
> I've tended to suggest that, subject to some fine tuning, the vehicle is fine but that the driver and/or co-driver may need a defensive driving course and more time to become familiar with the new fangled machine. An oversimplification perhaps (Upton, 1997: 2).

Not only an oversimplification, but also, our evidence suggests, misleading. To be blunt, the Government did not adequately design, fuel or service its car. The user-guides for councils to drive it came far too late and are still incomplete. Indeed, the Government did not bother to see whether councils were properly licensed to drive the new effects-based model in the first place. Little wonder that so many councils had, by 1997, publicly notified poor-quality policies and plans. In acknowledging the degree of poor practice at a conference in 1999, however, the Minister did concede that:

> As with most areas of the Act our biggest gains (for assessing environmental effects) are likely to come from improvements in practice. ... Central government has been slow in its provision of guidelines to assist councils, staff, applicants (for resource consents), consultants ... in determining the type of information required, how much is needed and how it should be evaluated... Ministry projects to improve practice by training, guidelines and a bit of 'peer comparison' are well advanced (Upton, 1999).

We concur with his observation of the problem and solution to it, but it had taken nearly a decade to walk the talk and for the necessity of this to sink into the political psyche of the Government. It seems 'learning by doing' does work — but then so does a Model-T Ford.

Chapter 4

Regional Government: A Non-Partner

Worldwide, the trend towards decentralization included creating an intermediate tier of regional government. Regional governance is being explored in many democracies as a pivotal means of achieving decentralization. Historically, the centre-periphery model of statehood entrenched the notion of a pyramid model of hierarchical government, each tier being subordinate to the one above. More recent conceptions portray intergovernmental relations in an equilateral triangle, where local government has equal footing with regional government in carrying out central government policy goals. Thus, each partner has special powers, competencies and tasks for working co-operatively in the same territory. More than one government exercising power over the same territory within a country is now commonplace (Elazar, 1991; Goldsmith, 1993), although in practice the model often skews towards regional government as the senior partner.

The rationale for regional government is that it is of sufficient size or scale for addressing problems and/or mitigating undesirable effects where local jurisdictions are too small to do so. Regions can develop expertise beyond the day-to-day needs of individual local jurisdictions, and mediate between jurisdictions to enhance their communication and service provision (FACIR, 1991: xvii-xviii). They can function as single purpose entities, such as for forests, land use or transport, or as multiple purpose entities, such as for comprehensive development planning involving social, economic and environmental issues (e.g. Burby and May et al., 1996; Michaels, 1996; Mitchell, 1990).

Press (1995) notes that regionally based environmental governance, such as was adopted in New Zealand for implementing the RMA, attempts to bridge: 1) political boundaries of local jurisdictions; 2) different media, including water, land and air; and 3) management functions, including planning, enforcement and monitoring. The intent is to enhance policy implementation and increase effectiveness by consolidating environmental management functions within one regional entity.

Here we examine the role of regional government in integrated resource management in New Zealand. In particular, we assess the capability of regional councils for carrying out their functions while also acting as partners in providing guidance to districts within their boundaries. We start by reviewing briefly some historical antecedents to the RMA.

Evolution of Regional Governance in New Zealand

Since provincial government in New Zealand was abandoned in favour of central-local governance in 1876, many kinds of regional entity have evolved. The Local Government Commission (1973) mapped 40 overlapping types, such as for education, employment, hospital, police, civil defence, transport, harbour, land, water and pest destruction.

Integrated Management Prior to RMA

By the time of reform in the 1980s, there were several forms of regional governance for resources and environmental management: 18 united councils (with regional planning functions); three metropolitan regional councils (with metropolitan planning functions); and 20 regional water boards and catchment authorities (with water resource, flood control and soil conservation functions).

The concept of integrated resource management in regional entities is not therefore new to New Zealand. It lay behind the comprehensive catchment-wide planning underpinned by the *Soil Conservation and Rivers Control Act* (1941), which enabled local communities to form regional catchment boards. It was extended in 1967 when the *Water and Soil Conservation Act* required twenty regional water boards to be established in association with the 17 catchment boards then in existence (Ericksen, 1986). As well, both comprehensive and integrated planning were hallmarks of regional planning under the *Town and Country Planning Act* (1977), if not the earlier 1953 Act.

Comprehensive and integrated water and soil planning in the combined catchment/water boards did not progress all that well. This was because each had a rather narrow view of its mandate, the catchment component dominated the staffing profile, physical works (like land drainage and flood control) attracted the highest grants from central government, and highly parochial local jurisdictions (municipalities, boroughs and counties) feared loss of control and increased costs (Ericksen, 1990). For example, by 1967 20 per cent of the country was still not covered by a catchment board, even though legislation allowing for their establishment had existed for more than a quarter of a century (Ericksen, 1986). On the other hand, good progress had been made in improving databases for water resource planning.

Efforts to push local authorities into organizing for regional planning (e.g. regional and united councils) through the planning Acts (1953, 1977) were spectacularly unsuccessful — as were efforts to require local councils to plan for their own integrated water and land management. As with the catchment/water boards, only more so, the ability of regional/united councils to function fully depended on the attitudes of landowners in local communities and hence the politicians in local councils. The latter successfully argued with central government for control over both representation and financial contributions. Given the negative attitudes in local councils, few regions had been formed before an amendment to the *Local Government Act* (1974) made them compulsory by 1979.

By the time of the local government reforms in 1989, some regional councils had done very well, especially in metropolitan areas, but most had not. For example, the Auckland Regional Authority (with its own enabling legislation since 1963) developed plans based on detailed investigations that placed water and land matters in a wider ecological and cultural context (e.g. Auckland Regional Authority, 1983). In contrast, even though central government had large-scale energy projects under way in the rural Waikato region in the 1970s and 1980s, local councils starved the Waikato United Council of funds, and only 1.5 staff could be employed in regional planning.

Linking land and water planning and management responsibilities of local councils to those of the catchment boards through amendments to the *Town and Country Planning Act* (1977) and *Local Government Act* (1974) did provide some progress in integrated resource planning and management in some areas. By the 1980s, however, a complicated system of administration and rules had emerged.

Thus, by 1985, a new *Water and Soil Bill* had been drafted to consolidate existing legislation. Key elements were the preparation of mandatory Water and Soil Policy Statements (as part of mandatory regional planning schemes) and of mandatory Water and Soil Management Plans (to deal with the details of management in the regions), and a greater devolution of functions from central government in the form of the National Water and Soil Conservation Authority (NWASCA) to regional catchment/water boards. Clearly, the confusion of existing regional entities would remain, as would functions for regional councils under the *Town and Country Planning Act* (1977). It is not surprising, therefore, that these proposals were soon overwhelmed by the resource management law and local government reforms of 1988-90. They did, however, provide a good basis for that comprehensive review (Ericksen, 1990: 78-84).

Over a period of nearly 50 years, a mix of regional entities emerged in New Zealand with varying responsibilities for comprehensive and integrated resource management. The catchment/water boards were strongly controlled by central government and tended to focus on planning for environmental effects, while the regional councils, and especially the united councils, were strongly controlled by local councils and tended to focus on resource management in the context of socio-economic development, and therefore activities-based planning. The activities of all regional entities were, however, subject to the requirements of public participation, including public notification of proposals and judicial hearings and appeals. What had emerged was a complex set of institutional arrangements and statutes that impeded, rather than facilitated, integrated regional resource management. There was a need for 'system change'.

Reshaping Regions for the RMA

The integrated management of natural and physical resources under the RMA was of special significance for the new regional councils created in 1989. As is clear from Chapter 3, the resource management reforms were buffeted by competing political ideologies. At one extreme were social democrats who had historically been largely responsible for advancing comprehensive, integrated resource management through various regional entities. At the other extreme were liberal democrats who had always preferred a minimalist approach, wherein the resource management functions in central government agencies needed to be curtailed in favour of local government, albeit with firm central controls. For them regional government could stay, but only if focused on the integrated management of biophysical resources, to the exclusion of comprehensive planning that included economic and social development. In other words, liberal democrats in particular opposed the comprehensive activities-based planning of the *Town and Country Planning Act* 1953 and 1977.

As we saw in the previous chapter, key provisions in the RMA were sufficiently confused and unclear (especially sections 2 and 5, environment and purpose respectively) as to enable regional (and local) councils with a predilection for social and economic planning to continue that theme (McBride, 1990). An early political response to this predilection was to curtail it by reviewing not only the functions of regional councils, but whether or not regional councils were needed at all. The new Minister of Local Government (Warren Cooper) therefore put regional councils under great pressure at a time when they were struggling to come to grips with their new mandate under the RMA. He visited regional councils around the country and challenged them to justify their existence, arguing that their number should be reduced by up to half, that dual-function unitary authorities be made easier to create, and that socio-economic elements be pruned from the list of regional functions.

The outcome of his review was an amendment to the LGA in 1992, which limited the functions of regional councils so that they dealt primarily with the natural and physical environment (*Local Government Amendment Act*, 1992), but relevant sections of the RMA remained unchanged. The amendment also allowed for the creation of unitary authorities where 10 per cent or more of ratepayers were in favour of doing so. In consequence, there was a flood of applications to the Local Government Commission, but only one regional council (Marlborough-Nelson) was broken into dual-function unitary authorities (Tasman District, Nelson City and Marlborough District). When added to the original unitary authority (Gisborne), this resulted in four unitary authorities and 12 regional councils — that is, 16 councils with regional functions.

Role of Regional Councils

In essence, the 16 regional councils (including the four unitary authorities) have several duties and functions under section 30 of the RMA, and specified instruments for giving effect to them. Each must establish, implement and review objectives, policies and methods for achieving the purpose of the Act — the sustainable management of natural and physical resources. Regulatory functions include the control of land uses for soil and water conservation, risk avoidance for natural hazards and hazardous substances, control of the coastal marine area, quality of water use, and the control of discharges of contaminants to air, land or water. (See section 30 of RMA in Appendix 1.)

To give effect to these functions, regional councils had to produce within two years of the Act being passed a *regional policy statement* — a strategic planning document for identifying issues, objectives, policies, methods, rationale and environmental outcomes for achieving integrated resource management within the region (sections 59-62). The regional policy statement was to provide a framework for the development of the mandatory *coastal plan* (also required within two years of the Act being passed) and optional *regional plans*, as well as mandatory *district plans* prepared by local councils (sections 63-67).

Finally, regional councils were required to develop methods to achieve the objectives and policies in their regional plans (sections 68-71), provide guidance for integrating the resources for which local councils have responsibility, and establish a comprehensive basis for assessing the environmental effects of public and private developments through the resource consents process of both regional and local councils.

Joint Functions with Local Councils

The strength of the RMA is that regions and districts have to follow the same processes when developing policies and administering plans. For hazard management and controlling the effects of land use on soil and water, there are overlapping functions. Further, the presumptions made for regional and district plans differ. For regional plans, effects must be expressly allowed, whereas for district plans, activities are permitted unless otherwise specified.

Where there were overlapping jurisdictions and partnership difficulties, clear separation of roles and responsibilities was not easily achieved between the regional and district councils. With time, however, the process could enable problems to be resolved as co-operation between councils improved and monitoring in all councils provided better data. This could result in more delegations and transfers of power between jurisdictions allowed by the RMA. Because the regional councils had a more focused environmental mandate than the multi-functioned district councils, they could be expected to take the lead in building co-operation and partnership by providing advice and technical assistance.

Some argued that one way of achieving greater integration of the sustainable management of resources was by adopting the dual function unitary authority structure. This had the potential to reduce the sense of hierarchy and allow greater interaction between regional and district staff when dealing with resource management problems. As we shall see later in Chapter 12, others argued that the checks and balances of having separate regional and district councils could be compromised in unitary authorities, blurring functions. Another concern was that calls for unitary authorities tended to have arisen in order to cut costs to ratepayers, which could lead to under-funding for their dual functions.

Challenges for Regional Councils

Regional councils have a role consistent with the devolved co-operative mandate described earlier. This, however, provided several important challenges to their success. Although supposedly equal partners with local councils in achieving the purpose of the RMA, regional councils had to produce policy instruments that fell within a hierarchy of policies and plans.

In spite of the review of regional councils' functions in 1992, the RMA still contained ambiguities over how far social and economic development could be dealt with in policies and plans because the most relevant provisions remained unchanged. The Act was also still unclear about the nature of overlapping responsibilities between regional and local councils. And it did not contain an explicit statement that regional councils were to help build the capabilities of local councils in their areas, just a strong implication that they should when need be.

Co-operation between regional and local councils depended on overcoming the notions of territoriality and turf protection evident in the previous planning regime. Building a partnership with local councils in these circumstances would depend a great deal on the capabilities of regional councils, and hence the extent of support from central government for doing so.

These challenges prompted two questions for our analysis. Did the newly created regional councils have the capability to fulfil their responsibilities, as well as meeting the need to support local councils within their territories? How well were regional councils able to perform their role as partners under a co-operative mandate?

Resourcing Regional Councils

Here, we first examine the efforts by central government to build up the capabilities of the new regional councils, and then its efforts to resource them for carrying out their responsibilities, including technical assistance to local councils.

Government Capability-building in Regional Councils

Was the amalgamation of the various regional entities into the new regional councils sufficient to provide the capability (i.e. commitment and capacity) to succeed? The main problem from the start was to convince ratepayers of their value, and then, as we have seen, convince the incoming National Government in late 1990 that they were essential for integrated regional management.

Effects of Amalgamation   Early work in some regional councils was hampered by in-fighting between former managers of the united councils (planners) and catchment boards (scientists and engineers) for control. This spilled over into interpretations of the RMA as being either an essentially comprehensive planning mandate or an environmental management mandate fostered by the unclear section 5 of the Act.

The amalgamations were also accompanied by a loss of central government services through the administrative reforms. For example, the systems for obtaining data on which regional councils could base state of the environment reporting (so as to help local councils predict and mitigate adverse environmental effects of development) were fragile, especially in financially strapped councils. This was exacerbated by the loss of data-gathering systems and databases after key central agencies were disestablished in 1988 (Hughes, 1991). The viability of these systems under the care of regional councils depended on their level of resources. Where this was minimal, as in West Coast Regional Council, the number of data-gathering gauges, such as for rainfall and flooding, was worryingly reduced.

Clearly, the material wealth of a region had much to do with its capabilities for carrying out its role, while the character of its natural and physical resources determined how difficult this role would be. Thus, a sparsely populated region with relatively little wealth, but possessing within its boundaries natural and/or physical resources that are nationally or regionally important, would be faced with a much greater challenge than a metropolitan region to provide technical services for developing effects-based policies and plans.

Capabilities: perceived and actual   Perceived commitment and capacity can serve as indicators of the extent to which regional councils came to terms with the new responsibilities assigned to them by the RMA. In 1994, just over two years after the Act was passed, the study by May et al. (1996: 115-16) found that the perceived commitment and capacity of planning staff in regional councils and unitary authorities for carrying out environmental management functions were higher than expected — given the system change in local government. Using a seven-point scale, they found mean scores for commitment to be over 5, and for capacity over 4.5. Given that the existence of the regional councils had been threatened by central government in 1992, this good result may, however, have reflected staff wanting others to believe that they were more capable than they really were.

The 1994 surveys showed that the regional councils had, as expected, many more (over seven times) environmental science staff than did local councils, and much greater familiarity with computer modelling of environmental elements and their effects. This would seem to position them reasonably well as technical advisors on environmental matters to their local councils. On the other hand, the 1994 surveys also found that regional councils had fewer planners and policy analysts for preparing policies and plans than did unitary authorities and especially local councils, and it is likely that this impaired their ability to prepare good-quality policies and plans. Certainly, early reviews of regional policy statements led to concerns about their quality (Berke, 1995; Gow, 1994; Hutchings, 1994; May et al., 1996). Our own research not only clearly linked the poor quality of regional policy statements to the low capability for preparing them, but also to unrealistic time constraints by central government (see Chapter 8).

Overall, these findings suggest that, while regional councils were in a precarious state in 1992, by 1994 they had 'at least the minimum capabilities for the roles assigned to them' (May et al., 1996: 115; Michaels, 1996: 20-1), although not much of this capability seems to have been devoted to the preparation of regional policy statements (see Chapter 6). Little was done by central government to help. Indeed, the managerial reforms and cost-cutting often led to the best planners in local government becoming managers, as their salaries were much higher. In regional councils that had a limited rating base, the stresses and strains caused by managerialism and in-house rivalries were acute. Not only was there a paucity of funds, but employees were too readily shuffled into jobs including plan preparation, for which they had no training. Some of our interviews extended in time well beyond their need, as stressed-out staff seemed to use them as a therapeutic opportunity.

Some eight years after regional councils were created, and after they had spent nearly six years trying to implement the RMA, we found from our surveys in 1997 that staff capacity had a direct influence on the quality of regional policy statements and plans, and that many regional councils did not have sufficient resources to support staff because the areas they serviced were not wealthy (see Chapter 8). Given the limited funding of the mandate detailed in Chapter 3, what had the Government been willing to commit to helping its new regional councils carry out their mandated actions?

Government Inducements for Regional Councils to Respond

Concomitant with capability-building are inducements, the purpose of which is to transfer resources to agencies in exchange for particular actions (McDonnell and Elmore, 1987). Two sources of funds from central government materially assisted regional councils, although their sufficiency remains in doubt: 1) a Resource Management Subsidy programme that provided needs-based grants for specified mandated activities; and 2) a Sustainable Management Fund that provided contestable grants for research projects on how best to implement the mandate.

Needs-based grants   The MfE provided financial grants to assist regional councils to meet obligations under the RMA, but only in the first two years. They were aimed at four specified mandated activities: 1) preparing policies and plans; 2) environmental quality standards for inclusion in policies and plans; 3) assessment of options for risk management associated with contaminated sites, natural hazards and waste management; and 4) assessing risks of adjusting to natural hazards. For the first, the 'eligible activities for new and devolved resource management functions under the *Resource Management Act*' were:

> regional policy statements, coastal plans, iwi input into regional policy, including iwi management plans, the transfer and performing of functions for air pollution control, the granting of resource consents for lake and riverbed areas, and mining (particularly rehabilitation), geothermal management, and allocation of coastal space including marine farming (Ministry for the Environment, 1992: 1, 2.1).

All applications for grants had to show they were the 'most effective measure for achieving defined outcomes' within given time-frames, and be 'subject to the annual plan process' or 'other acceptable council processes' (Ministry for the Environment, 1992: 1, 2.1).

The availability of these grants recognized not only that regional councils were mandated to produce overarching regional policy statements and coastal plans within two years of the RMA becoming law, but also that some councils had much less capability for producing them than others. It also recognized that some councils needed help in carrying out other important functions. Consequently, the MfE varied the level of grant according to the wealth or need of each regional council and the activity's importance.

These rates are shown in Figure 4.1 for the thirteen regional councils and one unitary authority then in existence across the four activities noted above. For example, all councils received financial support for developing their mandated regional policy statements and coastal plans. The wealthiest regional councils, Auckland and Wellington, could claim a 10 per cent grant, while the least wealthy, West Coast Regional Council and East Coast Unitary Authority, could claim 60 per cent.

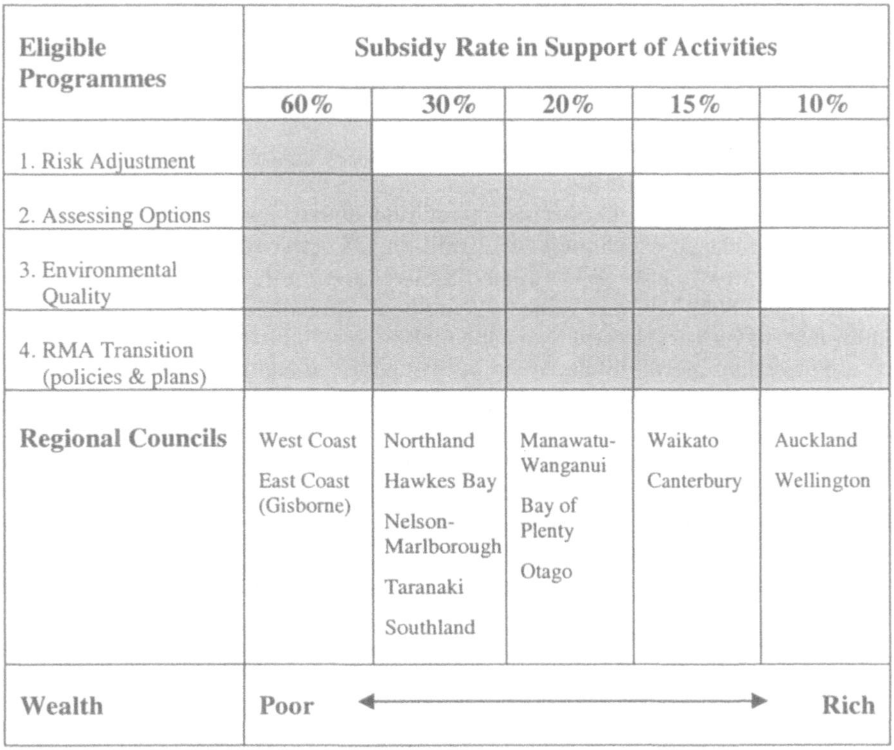

Eligible Programmes	Subsidy Rate in Support of Activities				
	60%	**30%**	**20%**	**15%**	**10%**
1. Risk Adjustment					
2. Assessing Options					
3. Environmental Quality					
4. RMA Transition (policies & plans)					
Regional Councils	West Coast East Coast (Gisborne)	Northland Hawkes Bay Nelson-Marlborough Taranaki Southland	Manawatu-Wanganui Bay of Plenty Otago	Waikato Canterbury	Auckland Wellington
Wealth	Poor ⟵			⟶	Rich

Figure 4.1   Government subsidy rates for activities supported in regional councils, 1992-93

Source: Adapted from Ministry for the Environment, 1992: 5

The total of funds allocated for the RMA Transition (policies and plans) is not known, nor is the total for the overall scheme, which is likely to have been several million dollars.[1] Figure 4.1 shows that fewer regional councils gained grants for the three other eligible programmes, progressing to the point where only the two least wealthy councils were allowed to apply for assistance for adjusting to natural hazards. The grants under this scheme ceased in 1994 — just as pressure was mounting both for the preparation of other regional plans and for assisting local councils with preparing their district plans. Significantly, most staff we interviewed in 1997 could not recall the grants even having being made.

As noted earlier, central government did not provide subsidies to local councils for developing district plans. Instead, money from the Resource Management Subsidy programme was diverted into a contestable Sustainable Management Fund for studies aimed at maximising environmental benefits in line with the Government's *Environmental 2010 Strategy* (Ministry for the Environment, 1994a).[2]

Contestable grants   The Minister for the Environment (the reinstated Simon Upton) established the non-departmental Sustainable Management Fund (SMF) in 1994 to develop methods for facilitating the implementation of the RMA, including regional and district plans. The fund was made open to all parties, including councils, on a competitive (and therefore more accountable) basis, and covers the categories shown in Figure 4.2.

Over the six years to 1999, $25.2 million dollars was provided for studies through this fund, of which only $6.72 million (25 per cent) went to regional and local councils (two-thirds of that in the first two years of the fund's operation). Of this amount, about half was gained by regional councils, 32 per cent by unitary authorities and just 18 per cent by local councils (see Figure 4.3).

Funds went mostly to the larger and wealthier councils, where the capability for carrying out research was greater, but the smallest and largest grants ($7,650 and $215,977) went to the relatively poor Northland Regional Council (Figure 4.4). The high level of funding to Tasman District Council suggests fertile staff minds at work and/or political interest in unitary authorities succeeding as a regional entity. Its research projects included development of voluntary environmental controls in the private forestry industry, and use of economic instruments, like tradable permits, as alternatives to regulatory rules in plans — all of which fitted the New Right perspective of the Government well.

Apart from 1995, the local government share of Sustainable Management Fund grants decreased, shrinking to less than 10 per cent in 1999 (Figure 4.3).

---

[1] We estimate from MfE Annual Surveys of Local Authorities and the level of subsidies in Figure 4.1 that around $5.5 million in needs-based grants were provided by the Government to regional councils for their mandatory regional policy statements and coastal plans.

[2] In 1994/95, MfE administered $9.56 million of Resource Management Subsidies and Catchment Works Grants (letter from Lindsay Gow, MfE Deputy Secretary, 3 April 1995). The apportionment between subsidies and grants is unknown.

The critical years   Clearly, 1991-93 was a critical period for preparing mandated documents in regional councils. While by the end of 1994 they had publicly notified both regional policy statements and coastal plans, they were still deeply involved in making these operational and in developing optional regional plans (water, soil, air, etc.).   They were also developing systems for state of the environment reporting and monitoring. At the same time, they had to deal with the burgeoning number of publicly notified district plans from local councils.

It is difficult to obtain good data on the costs of implementing the RMA, but an estimate of the total cost borne by regional councils in the three years to 1995 was $100 million, nearly half of which was for preparing policies and plans (Scarlet and Matthews, 1995). The rest was for monitoring and enforcement, and dealing with resource consent (permit) applications. The proportion of central government funds (i.e. through needs-based subsidies and contestable grants) contributing to the $50 million cost of mandated policy and plan preparation in regional councils to 1995 is not known for sure, but was probably under $10 million. Most costs were borne by local ratepayers (property taxes), resource consent applicants and people making submissions on regional policies and plans.

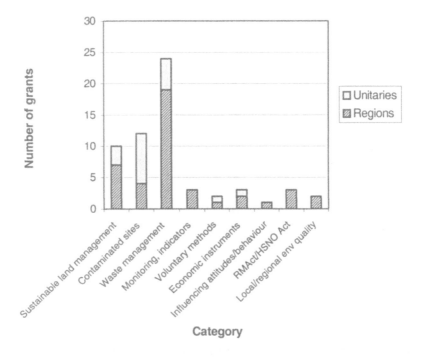

**Figure 4.2   Sustainable Management Fund (SMF) grants to regional entities
by category of interest, 1994-99**

Source: Compiled from Ministry for the Environment, SMF annual reports

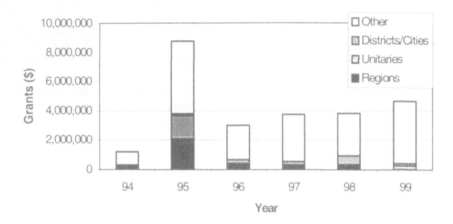

Figure 4.3   Sustainable Management Fund (SMF) grants to regional and local councils and other entities by amounts per year, 1994-99

Source: Compiled from Ministry for the Environment, SMF annual reports

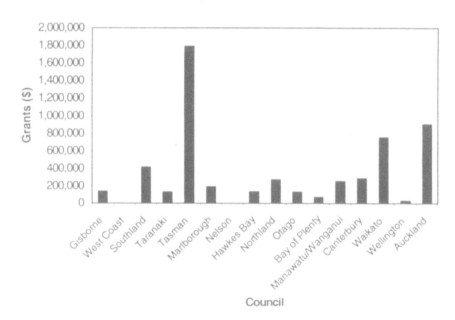

Figure 4.4   Sustainable Management Fund (SMF) grants to each regional council and unitary authority, 1994-99

Source: Complied from Ministry for the Environment, SMF annual reports

Co-operation and Partnership

Sitting between central government and district (local) councils, regional councils ought to be important communicators up and down the intergovernmental system. They should provide the inter-agency integration needed for sub-national and extra-local functions, and facilitate resource management activities in district councils through technical assistance and advice. They should therefore help to achieve integration in local planning across the region (Michaels, 1996). How successfully was this model of a co-operative partnership between regional and district councils realized?

Sharing Among Regional Councils

Before considering relationships between different parts of local government (i.e. regional and district councils), we consider how well the regional councils related to each other.

The siege-like state within which the new regional councils were born did much to enhance their commitment to support each other and to learn from shared experiences, in order to demonstrate their value nationally and locally. There are many examples of this sharing ethos. Uncertain as to how to proceed with the new mandate, the Waikato Regional Council organized a seminar with adjacent regional councils to discuss it. More generally, after an initial period of intense competition among councils to produce the first regional policy statement, many then entered a period of co-operation. The peer review of draft regional policy statements through a buddy system was an early example. Another was when one of the least-resourced members (West Coast) set out to prove its worth by being the first to notify a regional policy statement, but got into difficulties and had help from neighbouring councils (Canterbury and Otago) to bring it back on track. As well, some regional councils grouped together to deal with common problems. When notified regional policy statements and regional plans were found wanting, five regional councils worked together to produce a guide to improving their preparation (Hawke's Bay, Taranaki, Manawatu-Wanganui, Otago and Southland regional councils, 1997). Others looked at ways in which the often-large number of regional plans on various environmental topics could be amalgamated into fewer plans, or even a single plan, to improve the integrated management of a region's resources.

This self-help ethos was also evident among technical staff. They frequently communicated with one another about common problems across regional councils. Typically, informal communications lead to more formal ones, such as workshops on waste treatment, natural hazards, coastal environments, improving cost-reduction and the like. At the political level, there were frequent meetings and annual conferences about important issues in promoting integrated and sustainable resource management.

Skewing the Local Partnership

Although the Local Government Commission (1988) had clearly signalled to regional and local councils that they were to operate in an equal partnership, many regional councils failed to acknowledge that the hierarchy of policies and plans in the RMA was not intended to mean that local councils were subordinate to them. Many operated as if they were the superior entity, and in consequence many local councils became antagonistic. Thus, the equilateral triangle model of local government partnership in reality became quickly skewed.

This had important consequences for the development of a co-operative partnership by heightening the turf-protection mentality evident historically in most local councils. The degree of antagonism can be seen in the fact that, within three years of the 1992 amendment to the LGA, half of the regional councils had faced challenges from local councils seeking dual-function unitary authority status (Local Government Commission, 1995). Interviews with staff and politicians in local councils in 1994 revealed a surprising degree of hostility towards their regional partner. There was considerable discord in West Coast and Northland regions. Elsewhere, support from local councils for the partnership was varied, as in Canterbury and Waikato. Relations in the large metropolitan regions were mixed, being amicable in Wellington but discordant in Auckland. In Otago and Canterbury, the primary cities saw little need for the services of a regional council (May et al., 1996).

Building Capability for Local Councils

Clearly, supporting antagonistic local councils would prove difficult. In the early years, too, regional councils were in a precarious state politically and financially. Central government had introduced massive budget cuts in 1991-92, provided only limited inducements to regional councils up to 1994, and largely withheld inducements from local councils for carrying out their RMA responsibilities. Thus, it was not surprising to find in our 1997 interviews that staff in local councils reported that co-operation with, and facilitation by, regional councils had not worked well, due to issues of political control, territoriality, overlapping mandated functions, poor resources and other distractions — like the regions struggling to meet their own requirements within the two-year Government deadline.

As explained later in Chapter 6, we used three measures to gauge how useful regional councils had been in providing technical advice and assistance to local councils preparing district plans. On a five-point scale, the lead plan-writer was asked to indicate the *usefulness* of the regional council in: 1) providing relevant information; 2) participating in consultation during the early stages; and 3) providing guidance on the interpretation of RMA provisions.

We recast the results of our survey into the form given in Figure 4.5. This shows that the first and second activities proved more useful than the last. Overall, the results are skewed too far towards being not useful to suggest that regional councils had, by 1997, matured into agencies capable of carrying out what was expected of them under a co-operative mandate.

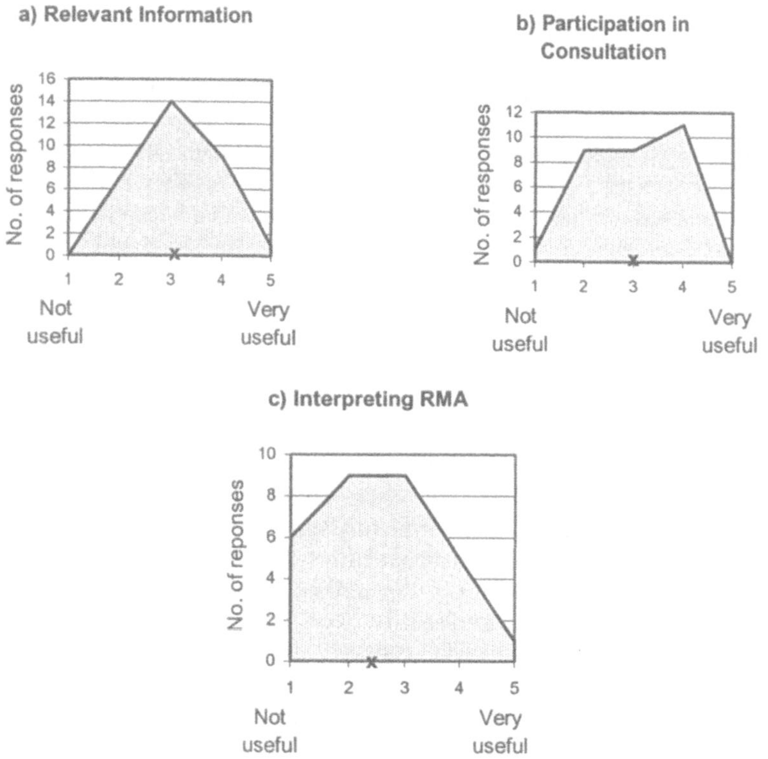

a) Relevant Information

b) Participation in Consultation

c) Interpreting RMA

Figure 4.5   **Local councils' ratings of the usefulness of regional councils in providing technical assistance and advice to local councils for plan-making** (a) relevant information; b) participation in consultation; and c) guidance on interpreting the RMA. The x indicates the mean score. N=30 local councils with notified plans (excluding unitary authorities))

Source: questionnaires to district councils, 1996

Building Local Government Partnerships

One might expect that, as common crises were dealt with, more enduring partnerships between regions and districts would form. It is clear that decisions of the Local Government Commission helped cement the role of regional councils, as only local councils in the Marlborough-Nelson Region succeeded in gaining unitary authority status after 1992. Declaratory judgements and case law also helped to clarify important provisions in the RMA, such as overlapping responsibilities for controlling land use effects, or whether local councils were subordinate to regional councils (e.g. Court of Appeal 99/95 Judgement, 4 July 1995).

Nevertheless, by 1997 an assessment of features in the RMA that might have been expected to foster co-operation showed that little progress had been made. For example, most regional councils had made only token efforts to deal with cross-boundary issues, although Nelson and Tasman districts (as unitary authorities) worked through them systematically. Another example is that only four regions (Auckland, Hawke's Bay, Canterbury and Otago) had *delegated* functions to a district, mainly for matters under the *Building Act* (1991) or noise control in coastal marine areas. A fifth, Wellington, had delegated functions to iwi for resource consents in Māori precincts. Only Auckland Regional Council was at that time discussing further delegations. Delegation does not confer on the recipient agency ultimate responsibility and power for such functions. That remains with the delegating authority. Regional councils can, however, *transfer* functions and thereby authority in certain circumstances, but none had occurred by the time of our surveys, although three were made in the following year (Ministry for the Environment, 1998: 40).

Reasons given by respondents for lack of power-sharing varied. For transfers, it was in part because of the complexity of the process and lack of clarity over what would happen if the recipient local council performed poorly. For delegations, it was because regional councillors were reluctant to share power with local councils, and to pay them for the service. One respondent explained that, although the RMA had been in place for over five years, the massive changes in local government helped engender caution over the delegation or transfer of functions to or from local councils, but once the effects of system change settled down, and trust between councils grew, greater power-sharing would occur. Since 1996 MfE has monitored the performance of local government, including the number of delegations and transfers of functions, through its annual surveys of councils. As foreseen, the number of delegations is increasing slowly.

Our research into regional and local councils occurred when relationships for many of them were strained. Since 1997, there has been a period of settling down that might enable more enduring partnerships to emerge. In the longer run, too, crises can help force co-operation. For example, faced with growing infrastructure problems (transport, sewerage, and water supply) within metropolitan Auckland, hitherto wayward local councils were induced into a more co-operative partnership with the regional council to help solve them. This was encouraged by threats of intervention by central government, and more recently by its financial inducements. Another example was a grant from the Sustainable Management Fund to Waikato Regional Council for developing an integrated monitoring strategy between region and districts (Beanland and Huser, 1999).

Managing Natural Hazards

It is not clear how far the overlapping functions in the RMA were intentional. One would expect that partnership between the regional council and local councils within the region would form best where the RMA indicated overlapping functions. One instance of this is natural hazards, which was also one of the two special topics we identified for our evaluations of plan quality (see Part III). For natural hazards and hazardous substances, the intention of the MfE policy-makers was, through the RMA, to force regional and local councils to sort out a sensible allocation of functions for themselves. This induced co-operation makes hazards a good case for illustrating development of a partnership as hazard-creating effects are often extra-local and reducing them requires the efforts of both regional and local councils. How effective has co-operation been in avoiding and mitigating natural hazards through policies, plans and other supportive methods?

Seeking Mandate Clarity

Staff we interviewed in most councils found the overlap in various functions confusing, including natural hazards. Section 30(1)(iv) of the RMA says that regional councils shall have 'control of the use of land for the purpose of avoiding or mitigating natural hazards', whereas section 31(b) says that local councils shall have the 'control of any actual or potential effects of the use, development, or protection of land, including for the purpose of avoidance or mitigation of any natural hazards'. It was not clear whether these were fundamentally similar or different statements, and whether regional councils could have land use rules that controlled what local councils could do.

In these circumstances, it would require much goodwill to develop co-operative arrangements, and goodwill was at a premium. Thus, the RMA was amended in 1993 to ameliorate the problem. By 1994, however, Canterbury Regional Council wanted even greater clarity, because the confusion caused by overlapping functions extended beyond natural hazards to related issues. It sought declaratory judgements from the Planning Tribunal (now Environment Court): one concerning the control of land uses for managing natural hazards, and the other the control of land uses for managing soil and water. The Planning Tribunal decisions did not favour Canterbury Regional Council, which took them beyond the High Court to the Court of Appeal, where they were overturned (Ericksen, Dixon and Berke, 2000; Court of Appeal 99/95 Judgement, 1995a).

In spite of this lengthy and very costly process, the 1995 declaratory judgements do not seem to have lifted the veil of confusion. In a case where local councils sought to overturn policies of the Auckland Regional Council for controlling urban growth in their areas, a Court of Appeal decision supported the regional council, suggesting that it was superior to local councils (Court of Appeal 29/95 Judgement, 4 July 1995b). It is little wonder that, more than a year later, staff interviewed in regional and local councils still found sections 30 and 31 to be unclear (see Chapters 6 and 7).

Allocating Responsibilities

The *Resource Management Amendment Act* (1993) authorized each regional council to include in its regional policy statement which local councils would have responsibilities for what natural hazards through land use rules in a district plan.

Otherwise, the regional council retained responsibility for the hazard and dealt with it through rules in its regional plan (section 62(1)[h(a)] of the RMA).

Analyses of regional policy statements found considerable variation in allocation of responsibilities. One did not even include natural hazards as an issue (Griffiths and Ross, 1997). Two had no division of responsibility between regional and local councils, so that these regional councils retained primary responsibility for managing natural hazards. Two others had a division of responsibility, but gave no direction as to how these were to be achieved. And eight had a division of responsibility, as well as specific responsibilities and methods assigned to local councils under the policies and methods sections of the regional policy statement (Hinton and Hutchings, 1994).

Generally, 'a co-operative approach [was] proposed within regional policy statements', with regional councils focusing on threats of regional significance (coastal erosion, drought and catchment-wide flooding) and local councils focusing on localized threats (flooding, erosion, land slip and subsidence) (Hinton and Hutchings, 1994: 4). For example, of eleven natural hazards identified within the Waikato Regional Policy Statement, drought was not allocated to local councils, because it is normally a widespread and pervasive phenomenon. On the other hand, flooding, which is often quite localized, was the only hazard common to the six district plans reviewed.

Some planners acknowledge that using 'regional significance' as a threshold for allocating responsibility for natural hazards is fraught with difficulty over political control. Thus, for example, in the Canterbury Region the councils have simply agreed that *which* authority manages *which* aspects of *which* natural hazards will be negotiated should any change from the agreed *status quo* be contemplated by any council (Ericksen, Dixon, and Berke, 2000).

Hierarchical Consistency in Policy and Planning

The hierarchy of policy instruments identified in the RMA must surely mean that local plans gain guidance from those above. Mostly, the hierarchy begins at the regional tier with the regional policy statement, but for the coast it begins with the mandated national policy statement. We examined these hierarchies to see how beneficial they are to planning in local councils, first for regional policy statements, then for the coastal hierarchy.

Regional policy statements There is considerable variation across regional policy statements not only in the resolution of natural hazards (Griffiths and Ross, 1997), but also between the planning instruments in regional policy statements and those in the district plans. For example, for natural hazards, the Waikato Regional Council identifies its roles in conjunction with local councils as being:

- identifying areas at risk and developing a centralized database;
- preparing hazard maps, policies, procedures and plans for high-risk areas; and
- discouraging development in areas where there is a high hazard risk.

While there is general alignment between the six district plans we reviewed and these policy directions, half of the plans did not identify at least one of these three policies. More successful in Waikato, were instances where the regional council and a local council agreed to produce a joint plan for specific, more localized hazards, like riverine and coastal flooding in the town of Thames (May et al., 1996).

The imperfect congruence between regional policy statement and district plans ought not to be too surprising, because the RMA simply requires that they be not *inconsistent* with one another. Since local councils do not have to strive for consistency, variation within regions seems assured. By creating more generic and less directive regional policy statements, it may well be that regional councils helped to avoid increasing confrontation with disaffected local councils in the throes of preparing their own district plans.

Conclusion

Regional councils are agents of central government in helping to ensure that districts within their boundaries have an integrated perspective on land use management with regard to regional responsibilities. Regional councils are also to work as partners with district councils in achieving government's goal of sustainably managing the regions' natural and physical resources.

In essence, we found limited capability-building for regional councils by central government, and mixed results in the relationships between regional and local councils. Expecting them to operate in an equal partnership was naïve, for various reasons. The hierarchy of policies and plans in the RMA suggested a hierarchy in the levels of governance. Overlapping functions in the RMA between regional and local councils were very unclear. Creating new entities and functions would not likely dissolve a history of discord between local councils and regional entities. Each new council was struggling to come to grips with the massive changes introduced by the new mandate, with limited resources at their disposal. And too many partnerships had to be worked out at once.

With the passage of time, some of these impediments have lessened, but there are still many ways in which councils could act in a more integrated manner for sharing technical resources. Taken far enough, however, this might well challenge the very rationale for having separate regional and local councils.

Chapter 5

Māori Interests: Elusive Partnership

A feature of the new RMA mandate was the special provision for Māori — the indigenous people of New Zealand — in local government decision-making. This resulted from increasing recognition of Māori rights by the courts and Waitangi Tribunal over the previous 20 years, along with the commitment of a reforming government to the principles of the *Treaty of Waitangi* (1840). It was also a response to growing Māori demands for greater recognition of their rights as Treaty partners with the Crown. The Government's efforts in this regard have drawn worldwide attention.

For implementation of the RMA to be satisfactory, it was essential that the roles of central and local government in relation to Māori and the Crown were clear; otherwise the legislation would founder on issues of poor governance. In this chapter we explore how well the Crown and sub-national government met the challenges created by new provisions for Māori in the RMA.

Diminishment and Resurrection of Māori Rights

When James Cook became the first European to set foot in New Zealand in 1769, the tangata whenua (people of the land, later known as Māori) had been in Aotearoa for up to 1000 years, having migrated south from what is now French Polynesia, probably via the Cook Islands. (See Figure I.1 in the Introduction.) By 1800, estimates of their population range between 100,000 and 500,000, mostly in the warmer northern half of the country, clustered around 36 tribes (iwi) and sub-tribes (hapū). (See Glossary of Māori Terms.)

In succession, whalers, sealers, traders, missionaries and settlers soon followed (King, 1997). Like indigenous people elsewhere in the world, Māori suffered greatly in the process of British colonization.

By 1900 they numbered around 45,000 and owned only about 10 per cent of the total land area. Introduced diseases, devastating intertribal wars facilitated by the acquisition of guns, loss of resources through dubious land deals, government confiscations following the 1860s land wars — all contributed to this parlous state. By the 1950s, significant numbers of Māori had moved to towns and cities in search of work, which hastened the Government's programme of assimilation and acculturation that had been under way for decades.

Treaty, Sovereignty and Governance

The taking of land and erosion of culture happened even though most tribal chiefs had signed the *Treaty of Waitangi* in 1840 with the British Crown (see Annex 5.1 of this chapter).   The English version of the Treaty states that, in return for complete governance over their lands (Article I), Māori would still enjoy the 'full exclusive and undisturbed possession of their lands and estates, forests, fisheries and other properties' (Article II). The failure to uphold Articles I and II is all the more damning since Article III granted Māori 'all the rights and privileges of British citizens'.

For Māori it was much worse because, while the Māori version of the Treaty did cede to the Crown the right to govern (kawanatanga), it did not cede sovereignty (rangatiratanga) (McDowell and Webb, 1998; Boston et al., 1996; Boast, 1989; Parliamentary Commissioner for the Environment, 1998). Clearly, in spite of the Treaty, Māori lost most of their resources and their rights were substantially discounted (Baragwanath, 1997).

Māori Renaissance

Civil rights movements after the Second World War influenced Māori, and several activist groups emerged in the early 1970s arguing for recognition of rights under the Treaty and a better deal. As well, various international agreements on human rights and indigenous peoples brought pressure on countries, such as New Zealand, to improve their treatment of indigenous people (Ministry of Justice, 1999).

While a good deal of progress had been made since then, the economic reforms and restructuring of major government agencies in the 1980s had an especially severe effect on Māori employment and income. Māori made up only 15 per cent of the total population of 3.7 million people in 1998, but their unemployment rate was 18 per cent compared with 6 per cent for non-Māori, although the situation has improved since then (Te Puni Kōkiri, 2000). What is more, statistics for Māori ill-health, poverty, and crime were significantly worse than for non-Māori. Moreover, proportionately more Māori households had extended families than non-Māori, and household income was required to cover more people. As well, proportionately more Māori than non-Māori lived in state-subsidized housing, and were adversely affected when government began charging commercial rents in the mid-1990s as part of its user-pays, market-led policies (Statistics New Zealand, 1998; Te Puni Kōkiri, 1998).

Political Ideology and Māori Rights

How the rights of indigenous peoples are construed by the dominant culture today depends largely on the political ideology underpinning modern democratic governments. Left-of-centre social democracy is more likely to recognize indigenous people as a distinct group that requires special treatment in matters of governance, including collective rights. In contrast, right-of-centre liberal democracy is more likely to argue for democratic equality, uniform citizen rights and individual choice (Boston et al., 1996: 152).

In New Zealand, political power had resided with the centre-right for 38 of the 50 years to 2000. During that time, Māori allegiance had lain mostly with the centre-left, so that in all but three of those 50 years candidates from the Labour Party occupied the reserved Māori seats in Parliament. Four Māori seats in Parliament were established in 1867, but increased to five in 1996 and to six in 1999. Māori have, however, been able to stand for election in any other electorate. In 1999, there were sixteen Māori members of Parliament out of a total of 120.

Until the 1980s support for Māori was from a paternalistic government channelling resources through a large Department of Māori Affairs in a manner that fostered dependency within the mainstream rather than cultural independence and self-reliance.

Achieving self-reliance required settling past grievances over loss of land and giving much greater power to Māori for managing their own affairs, including the use and development of their renewable resources. It meant honouring the Treaty partnership. This was slowly being achieved through political change, and through the courts — which have been crucial in clarifying what the Treaty means in practice.

Towards Devolution and Partnership

When the reforming Labour Government came to power in 1984, it quickly built on work started when Labour was briefly in office in the 1970s (1972-75). As we saw in Chapter 3, it favoured a model of governance that devolved decisions away from the centre and out to the periphery where their effects were most keenly felt. In addition to holding summits in Parliament to gain public help in strategizing for improved economic and environmental performances during its first year in office, the new Labour Government held one on Māori issues.

Hui Taumata, the Māori economic summit meeting, signalled the commencement of the Decade of Māori Development. Six themes emerged, covering the *Treaty of Waitangi*, tino rangatiratanga (Māori self-determination), iwi development, economic self-reliance, social equity and cultural advancement. Māori were to be advanced through such measures as the settlement of land claims, clarification of Māori-Crown relationships, development of a sound economic base, elimination of social disparities, use of iwi social service delivery systems, promotion of the Māori language and Māori educational systems (Durie, 1998).

Māori solutions to Māori problems The new Māori call was for 'Māori solutions to Māori problems' (Durie, 1998: 8). There was recognition that Māori could achieve greater economic self-sufficiency and less dependency on the state if they took advantage of their own institutions and tribal resources. There was growing support among Māori and social democrats in the Government, and beyond, for devolving central services and funding away from the paternalistic and somewhat unresponsive Department of Māori Affairs and into iwi authorities (Boston et al., 1996: 147; Ministry of Māori Affairs, 1988a).

Devolution requires some form of partnership between the Government and Māori. In 1988, the Government's policy for devolution to Māori, Tukua Te Rangatiratanga, was released in discussion papers highlighting the nature of the partnership (Minister of Māori Affairs, 1988a, 1988b). First, the Department of Māori Affairs would be replaced by a much smaller policy ministry (Manatu Māori), and many of its responsibilities mainstreamed into other hitherto-unresponsive government departments, like Social Welfare, Health, and Labour. Departments would be made accountable for results through purchase agreements negotiated with iwi for the provision of services. Second, an Iwi Transition Agency would be created to help transfer land development and other programmes to iwi over a period of five years. These initiatives would require, however, clarification of the legal identity of iwi in order for them to work with government agencies (Durie, 1998: 224).

The *Rūnanga Iwi Act* (1990) was passed to provide this clarity, but not without opposition from Māori groups concerned that it could undermine traditional tribal autonomy, as it would identify iwi organizations and require them to be registered with the Māori Land Court. The new Act included, at its very end, a provision for the preparation of iwi management plans, which were to provide a 'resource management planning overview of those matters that are of significance for the organisation and development of iwi' (section 77).

Considered alongside the devolution-based reforms of local government and resource management laws, there was considerable potential for the Government to develop a clear co-operative partnership with iwi, as well as with local government.

Issues of concern   While Māori strongly supported the emerging policy on devolution and partnership as it involved their interests, they were suspicious that it was more likely to result in administrative delegation than a genuine transfer of power, especially given the plans for mainstreaming Māori responsibilities across government departments (Boston et al., 1996: 148). More important, they saw the policy as the Government affirming governorship (kawanatanga) over them, rather than providing for real tribal control over their own resources (rangatiratanga).

Further, as with local government, iwi members suspected that devolution would be accompanied by cuts in public expenditure to their programmes through the general managerial and financial reforms that were poised to strike all government agencies. Finally, they saw the creation of rūnanga iwi (tribal councils) as another top-down construct imposed by the Government, which wanted to simplify the highly differentiated whānau, hapū and iwi (extended families, sub-tribes and tribes) into a few Māori regional institutions. 'Before long, Māori development was seen by many Māori as little more than a restatement of mainstream preoccupation with economic engineering at the expense of collective state responsibility' (Durie, 1998: 11).

The extent to which these concerns over partnership under devolution were more real than imagined was never tested. Within seven months of the *Rūnanga Iwi Bill* (1990) being passed into law, it was repealed when government changed hands and the liberal democrats took hold for the rest of the decade. By then, several important decisions had been made in the courts that not only recognized Māori interests in environmental planning, but also their partnership with the Crown under the Treaty.

Rights Through the Courts

Legislation has always played a dominant role in suppressing or protecting Māori interests in New Zealand (Williams, 1997: 33). The establishment of the Waitangi Tribunal in 1975 as a forum in which disputes between the Crown and Māori could be addressed was a major acknowledgement of Māori grievances. Shortly after, section 3(1)(g) of the *Town and Country Planning Act* (1977) recognized Māori relationships with their ancestral lands as a matter of national importance, a considerable step forward when there had been no prior recognition. However, not until 1987, when a conservative interpretation of this provision was overturned by a High Court decision, were significant gains made. This had the effect of broadening the scope of environmental planning to cover any Māori land and resources, irrespective of ownership (Matunga, 2000). Similarly, in the same year, another High Court decision determined that Māori spiritual and cultural values could be taken into account in the administration of water rights by regional water boards (Boast and Edmunds, 1994).

Treaty gains a legal toe-hold   After the mid-1980s a raft of legislative changes made some reference to the *Treaty of Waitangi* or Māori interests. The reforming Labour Government had signalled that the Treaty was to be a priority (Hayward, 1999), although as we have seen this was to be short-lived (Rikys, 1999). The Treaty was recognized by statute for the first time in the 1980s through the *State-Owned Enterprises Act* (1986), *Environment Act* (1986), *Conservation Act* (1987), *Crown Forests Assets Act* (1989), *Māori Fisheries Act* (1989) and *Education Act* (1989). This made legal appeals about Treaty rights a realistic proposition, even though the Treaty had not yet been fully incorporated into statute law (McDowell and Webb, 1998).

Māori rights under the Treaty gained a major boost through a 1987 decision of the Court of Appeal (New Zealand Māori Council v Attorney General [1987] 1 NZLR 641). This case produced the most significant decision reflecting the 'new' attitude towards the Treaty and Māori interests (McDowell and Webb, 1998). To corporatize Crown assets, the Government proposed in the *State-Owned Enterprises Bill* (1986) to allow the transfer of large tracts of Crown land (to which Māori laid claim) to newly created and commercially-operated state-owned enterprises.

In response to concerns by Māori and the Waitangi Tribunal, the Government inserted two additional provisions requiring that the Crown did not act inconsistently with the principles of the *Treaty of Waitangi* and set out a procedure for dealing with pending and future Tribunal claims (sections 9 and 27 of the *State-Owned Enterprises Act* (1986), respectively). However, the provision omitted to deal with land that was subject to a Māori claim after the Act was passed, but which had been sold in the interim to third parties. A court action was brought by the New Zealand Māori Council to stop the transfer of Crown assets. Because of its importance, the High Court sent the case to the Court of Appeal, which found that the Crown had breached section 9 of the *State Owned Enterprises Act* (1986) and found in favour of the New Zealand Māori Council.

Five guiding principles   The Appeal Court defined five principles that derive from the *Treaty of Waitangi*, and noted that they must be capable of adaptation to new and changing circumstances. The five principles are summarized below (McDowell and Webb, 1998; Williams, 1997).

- Kawanatanga: the Crown has the right to make laws, but is obliged to accord Māori the interests specified in the second Article of the Treaty.
- Rangatiratanga: the second Article of the Treaty guarantees to Māori the control and enjoyment of those taonga (valued resources) they wish to retain. This includes tribal self-regulation in accordance with customary preferences.
- Partnership: the principles of the Treaty require the Treaty partners to act towards each other reasonably and in good faith. This includes effective, early and meaningful consultation by both partners.
- Active protection: the guarantee of rangatiratanga is consistent with an obligation on the Crown to actively protect Māori interests and values in their lands, water and other taonga, including spiritual values and beliefs.
- Hapū/iwi development: the third article of the Treaty confers the rights of citizenship on Māori. Thus it accords the rights for hapū/iwi to develop resources in accordance with their needs and aspirations.

Linking Māori rights to law reform   At this time, the Labour Government was developing not only the *Rūnanga Iwi Bill* (1990), but also the *Resource Management Bill* (1989) and reform of local government (see Chapter 3). Considerable attention focused on the need to recognize the *Treaty of Waitangi* in legislation and provide for Māori autonomy in decision-making (Ministry for the Environment, December 1988b; Officials Co-ordinating Committee on Local Government, February 1988). Prime Minister David Lange released a statement of the principles of the Treaty on which the Crown proposed to act. Thus, there was considerable optimism among Māori that they would be given a decisive voice in resource management matters (Hayward, 1999).

In January 1989, a report of the Māori consultative group on local government reform argued that local and regional government needed clear statutory guidelines outlining their Treaty obligations, and that there should be statutory representation for tangata whenua on councils (Matunga, 1989). In October that year, close to the triennial council elections, a draft Bill (*Local Government Amendment Bill No. 8*) to establish Māori advisory committees was circulated for comment by the Minister of Local Government, though it did not proceed further.

Thus, the newly created regional and district councils were elected under the reformed *Local Government Act* (1974), which did not require councils to recognize the *Treaty of Waitangi* (Chen and Palmer, 1999; Hayward, 1999; Palmer, 1993), as the Cabinet had refused to include a Treaty provision (Rikys, 1999). By now, the Government had also made it clear that resource *ownership* issues of importance for Māori would not be considered in the forthcoming *Resource Management Bill* (1989). Rather, the legislation would focus on resource management.

Under increasing political pressure, not only in the political opposition, but also from the liberal democrats in its own ranks, the Labour Government was in retreat from its short-lived Treaty policies and planning to return Treaty issues to the political arena (Durie, 1998; Kelsey, 1990).

Advance and Retreat

When the new National Government repealed the *Rūnanga Iwi Act* (1990), this was consistent with its political philosophy of not giving preferential treatment to any group on the basis of 'racial origins or status as tangata whenua' (Boston et al., 1996: 152). Further, when the new Minister of Māori Affairs (Winston Peters, a Māori) came back to his colleagues in 1991 after very extensive consultation with iwi, with a proposed policy that highlighted a unique and distinct status for Māori *(Ka Awatea: It Is Day)*, excuses were found to sack him. While some parts of the policy were implemented, the Government then by and large adopted its predecessor's proposals for mainstreaming public services through central agencies and allowing them to enter into contracts with competent, legally formed Māori organizations — preferably through competitive tendering.

The Government also restructured, again, the main agency for Māori services, Manatu Māori (Ministry for Māori Affairs), and renamed it Te Puni Kōkiri (Ministry for Māori Development). Overall, its policies aimed at closing the socio-economic gap between Māori and non-Māori (Boston et al., 1996: 154).

Ironically, *Ka Awatea* returned as the blueprint of Māori policy in a National-led coalition government with New Zealand First (led by Winston Peters) six years later (Durie, 1998).

Settling land claims   Having disposed of devolution, embraced mainstreaming and reformed the Māori Ministry, the National Government then focused its attention on reaching full and final settlement of Treaty claims. In 1994 it set a fiscal cap of NZ$1 billion (less $170 million from the Sealord deal, which provided Māori rights to fishing quotas in return for dropping further claims on that resource) as the limit within which all claims would have to be made — a prospect strongly rejected by Māori. The cap was subsequently removed in 1997 (McDowell and Webb, 1998).

By 1999, when the National Government was voted out of office, there had been eleven settlements totalling $530 million and 700 unsettled claims lodged with the Waitangi Tribunal, although many overlapped (Ministry of Justice, 1999).

However commendable the Government's settling of past grievances may be judged, these actions had limits in helping to advance the rights of Māori. The National Government did not clarify partnership rights and responsibilities in its legislative agenda. Rather, whatever advances were made throughout the 1990s came through Māori challenges in the Waitangi Tribunal and the courts.

Thus, by drawing back from devolution in 1991, the Government failed to clarify the nature of the partnership between Crown and Māori and between Crown and local government in relation to Māori interests. This in turn had serious implications for environmental planning under the RMA, because it was unclear whether councils are agents of the Crown with respect to its partnership with Māori under the Treaty. Without policy guidance from the Government on this matter throughout the 1990s, councils were left to work out relationships for themselves, with wide variations resulting.

With the change to a social democratic Labour-led Government in 1999, there were renewed moves to clarify governance and partnership issues, especially with respect to Māori and local government through review of the *Local Government Act* (1974) (e.g. Clark, 2000; James, 2000; Lee, 2000; Ministry of Justice, 1999; and Rosson, 2000).

Māori Rights and the Resource Management Act

Despite the repeal of the *Iwi Rūnanga Act* (1990) in March 1991, reference to 'iwi authorities' and 'iwi management plans' remained in the *Resource Management Bill* (1989) when it became law late in 1991, as did other provisions for Māori.

The RMA contains five substantive provisions involving Māori rights that must be addressed in plans:

- plans 'shall recognize and provide for' the relationship of Māori and their culture and traditions with ancestral lands, water, sites, wāhi tapu and other taonga (treasures) as matters of national importance (section 6(e));
- plans are to 'have particular regard to' local Māori responsibility in the guardianship and stewardship of the land and resources (section 7(a));
- principles of the *Treaty of Waitangi* shall be 'taken into account' (section 8);

- councils shall have regard to any 'relevant planning document recognized by an iwi authority' affected by a regional policy statement, regional or district plan (sections 61, 66 and 74); and
- councils shall consult with original people of the land (or tangata whenua) during plan preparation (Clause 3, First Schedule).

Whether and to what extent councils took notice of the responses of Māori was at their discretion (Boast and Edmonds, 1994).

Obstacles to Effective Implementation

While these five provisions gave considerable opportunity for local government to enter into partnerships with Māori for resource management, there were, and still are, many obstacles to effective implementation.

There was scepticism owing to the imprecise wording of each of the above provisions. How a local plan was to recognize, or account for, or have regard to, a particular provision (sections 6, 7 and 8 respectively) was not clear (Nuttall and Ritchie, 1995: 7). Problems with the clarity of the mandate in these provisions are described in Chapters 6 to 8, where council staff expressed difficulties in interpreting key sections of the RMA. Consequently, we found, as have others, considerable potential for widespread non-compliance, especially where the commitment of councils was weak (Matunga, 2000; Nuttall and Ritchie, 1995; Parliamentary Commissioner for the Environment, 1998).

Also, the RMA does not offer clear guidance for local government on how to incorporate Māori environmental concepts into plans. For example, the RMA refers to kaitiakitanga (ethic of guardianship), but is unclear whether this concept applies only to Māori. Instead, the Act indicates that it is an ethic for which 'all persons exercising powers and functions under the Act shall have regard'.

In addition, because central government did not provide clear policy guidance for local government on how to address Māori interests, affected parties had to rely on decisions of the courts and reports of the Waitangi Tribunal (Hayward, 1999). Further since 'ownership' of Māori resources was not addressed in the RMA because the Government specifically decided against doing so (Hayward, 1999), matters relating to this, such as loss of land, water, geothermal and mineral resources, were being dealt with separately by the Waitangi Tribunal and through settlements negotiated between the Crown and Māori claimants.

Finally, the inconsistencies between boundaries of the 36 iwi and the regional and district boundaries generated extraordinary demands on Māori (see Figure 5.1). As mentioned in Chapter 4, the boundaries of regional councils were based on major river watersheds modified by communities of interest. These non-Māori interests also helped delineate district boundaries. Consequently, the tribal and administrative units do not compare well for planning.

All of these obstacles contributed to councils struggling in plans to deal adequately with Māori interests. Difficulties with interpretation of the mandate, lack of guidance for councils, the separation of ownership and management issues, along with the heavy demands placed on whānau, hapū and iwi by the RMA, combined to undermine the implementation of potentially empowering provisions for Māori.

Figure 5.1   Map comparing iwi and regional/district council boundaries

Source: Kelly and Marshall, 1996; reproduced from Berke et al., 2002: 119

Iwi Management Plans

A major tool for the inclusion of iwi interests in local government plans was the requirement for councils to consider iwi management plans prepared by iwi authorities. The Ministry for the Environment (1988: 32-4) foreshadowed the potential for this tool in December 1988, by which time, a few iwi were already preparing plans. It recognized that such plans would fit with the Government's wider strategy for devolution and provide iwi with a mechanism for developing proactive strategies for iwi development. The plans could articulate 'tribal sentiment, resource information, environmental quality standards, strategies for conflict resolution and other Māori expectations of resource management' (ibid: 33). They could also be a means of expressing tino rangatiratanga (Māori self-determination) as outlined in Article II of the Treaty.

Government ambivalence   The Government deferred any decision on a formal requirement for iwi management plans until the relationship between iwi authorities and government agencies was further developed, but noted the desirability of formal recognition of iwi management plans by resource management agencies. This was a weak commitment. However, the provision for iwi management plans was, in fact, implicit acknowledgment of the existence of another cultural planning tradition outside the framework provided for in the RMA (Matunga, 2000).

Given this government ambivalence, it should be no surprise that no accurate record could be found of how many iwi plans had been prepared. There was no statutory requirement for iwi to formally notify plans (as there was for council plans under the RMA) or lodge plans with any government agency. However, despite the lack of formal integration into the planning system, the number of plans produced by iwi was increasing. For example, iwi groups in the Bay of Plenty Region had prepared nine plans by 2000, while at least six plans existed in the Auckland Region.

Variability of iwi plans   Because of the lack of guidance, there was considerable variability in the plans that had been produced, and hence a very loose definition of what actually comprised an iwi management plan. As the Parliamentary Commissioner for the Environment (1998) noted, hapū or iwi plans could include statements of tribal identity, rangatiratanga and rights; and requirements for management of particular resources and areas, and for consultation and involvement of tangata whenua. Sunde, Taiepa and Horsley (1999) identified three types of iwi plans:

- comprehensive plans (e.g. *Nga Tikanga o Ngati Te Ata: Tribal Policy Statement 1991* prepared by Ngati Te Ata);
- resource-specific plans (e.g. resource management plans prepared for the Manukau Harbour and Waikato River by Huakina Development Trust); and

- broader resource management plans (e.g. *Otaki River and Catchment Iwi Management Plan 2000* prepared by Ngati Raukawa).

While the structure and content of these plans remained highly fluid, new plans were becoming more sophisticated expressions of hapū and iwi interests, embracing notions of place and community sustainability.

Challenges for iwi planning   The preparation of resource management plans set significant challenges for iwi.   On the one hand, plans were seen by iwi as a constructive and important exercise in moving towards better environmental outcomes, and as a means of communicating their interests to government agencies.   On the other hand, while some groups wished to maintain their independence and fund the work themselves, others did not have the resources to do so.

Even when iwi plans had been prepared, an important issue was how much weight councils gave them. The RMA requirement to 'have regard to' did not place an obligation on councils to formally adopt the concerns or priorities expressed in iwi plans. The Waitangi Tribunal has recommended that the RMA be amended to ensure that hapū and iwi management plans are accorded an appropriate weight by councils (Parliamentary Commissioner for the Environment, 1998: 78-81).

We did not find many references to iwi management plans in regional policy statements and district plans. Some councils reported that they had supported their development, others indicated they could not endorse their content.   The discretionary judgment that councils could exercise over iwi management plans greatly weakened their potential for influencing council policy-making.

Another challenge to Māori created by the RMA was the extraordinary diversity of policies and plans for them to consider. The complexity of the planning process is illustrated in Figure 5.2, which shows a hypothetical portion of Māori land in cross-section, and the other policies and plans relating to resource management under the RMA and other statutes (such as *Conservation Act* 1987, *Biosecurity Act* 1993, *Fisheries Act* 1996, and *Historic Places Act* 1993).

The complexity of plans is overwhelming, highlighting the fact that the *ad-hoc* incremental approach so roundly criticized under the old planning regime remains, and indeed has been encouraged by new statutes enacted since the RMA in 1991.

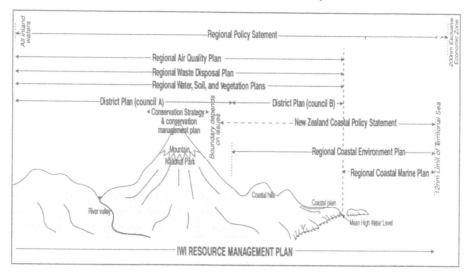

Figure 5.2   A plethora of policies and plans for resource management to which iwi management plans need to relate (The complexity for co-operative action increases for an iwi if there is more than the one regional council and two district councils noted in the diagram. It is further complicated if policies and plans under other statutes are included, such as land transport, civil defence and biosecurity)

Implementing the Bicultural Mandate

Having acknowledged principles of the *Treaty of Waitangi* in the RMA and created considerable expectations among Māori for their involvement in local decision-making, what did central government and councils do, in turn, to support Māori? We first review the actions taken by several central government agencies to assist councils in meeting these obligations, and then the efforts made by councils to incorporate Māori interests in their plans.

Capability-Building by Central Government

Te Puni Kōkiri   The old multi-functioned Department of Māori Affairs had a budget of $264 million supporting 958 staff in its last year of operation (1988/89) (Boston et al., 1996: 156). The new policy Ministry (Manatu Māori) had a similar budget ($270 million), but this fell dramatically to $35 million and 200 staff when the National Government replaced it with Te Puni Kōkiri (Ministry of Māori Development) in 1992. By 1996, the budget of Te Puni Kōkiri had risen to $48 million supporting 306 staff. Nevertheless, the new Ministry still had difficulties in recruiting and holding senior policy analysts.

In response to claims under the *Treaty of Waitangi* and requirements in the RMA, Te Puni Kōkiri published a guide (*Mauriora Ki Te Ao*, 1993) to help Māori prepare environmental and resource management plans so that hapū and iwi could articulate their aspirations for environmental management. The guide outlined some preliminary steps for preparing plans, including making an inventory of resources and setting goals and strategies. In 1996, the Ministry published a guide on the protection of sites significant for Māori. By and large, however, operational aspects of resource management were not a priority for Te Puni Kōkiri, whose main role was to advise the Government on the Crown's relationship with hapū and iwi on key government policies as they affect Māori.

Ministry for the Environment   The MfE had the main capability-building role for iwi resource management. It provided resources to Māori in two main ways. Soon after the mainstreaming policy on Māori development was implemented in 1989, it established the Māori Secretariat (Maruwhenua) to drive Māori issues.[1]   In part this included outlining the responsibilities of Māori (and councils) under the RMA. This included numerous meetings and field visits, as well as the production of pamphlets, reports and guidelines (Ministry for the Environment, 1991, 1992a, 1992b, 1993a, 1993b, 1993c, 1998, 1999, 2000a, 2000b; Ministry for the Environment and Te Puni Kōkiri, 1996). Maruwhenua also had a significant input into Māori Treaty claim settlements, many of which could have had an important, but unexpected, impact on local government and their resource management plans. An example is the *Ngai Tahu Settlement Claims Act* (1998), which requires councils to have special reference to statutory acknowledgments in relation to areas of particular association for Ngai Tahu as described in the Act.

    In addition, iwi and/or iwi in co-operation with others applied for grants from the Sustainable Management Fund for developing hapū and iwi management plans. By 1999, ten grants totalling $864,000 had been allocated specifically for plan development, but this was only 3.4 per cent of the total funding allocated since the fund was established in late 1994 (see Figure 4.3). However, a total of 16 grants were awarded for the maintenance and enhancement of the exercise of kaitiakitanga.

Parliamentary Commissioner for the Environment   The Office of the Parliamentary Commissioner for the Environment (PCE) produced some significant publications on Māori involvement in environmental management. On very limited staff resources, nine reports about Māori and the *Treaty of Waitangi* were produced between 1987 and 1996 (Wickliffe, 1997). In response to a request from local government, the Office produced the *Proposed Guidelines for Local Authority Consultation with Tangata Whenua* in June 1992. This report was reviewed six years later and a further influential publication produced in June 1998 (*Kaitiakitanga and Local Government: Tangata Whenua Participation in Environmental Management*).

---

[1] Maruwhenua's staff fluctuated between three and six throughout the 1990s.

How well did central agencies do? The early part of the 1990s, so crucial for council plan-making, was characterized by limited capacity-building for implementing statutory provisions addressing Māori interests. MfE as the lead implementer was poorly resourced and got off to a slow start. The PCE contributed effectively although was not responsible for implementation of the RMA. Te Puni Kōkiri was only marginally involved, as environmental planning was not part of its core business.

The ambivalent provision for iwi management plans in the RMA was an example of 'unfinished business' resulting from the halt to the devolution policies of the Labour Government by the new National Government after 1990. With little support from the Government, most iwi did not have the capacity or expertise to contribute to policy development by councils. Yet the RMA required their active participation if the provisions in Part II were to be fulfilled by councils. It created expectations that iwi could not possibly meet, and thus they were largely set up to fail in terms of the mainstream planning mandate.

More fundamentally, the issue of iwi management plans underscored several important and interrelated issues that impeded councils in their ability to get on with the job. First, central government did not clarify its role in respect of Māori, and the obligations of local government in respect of the Treaty. Second, this in turn created a national policy vacuum which councils struggled to fill. No national policy statement or guidance on Treaty rights and obligations existed. Third, an opportunity to integrate the Treaty and environmental planning was lost. Finally, attempts to achieve 'system change' were compromised by changing policies of successive governments, which led to confused signals for Māori. Weak 'inducements' and 'capacity-building' did not help.

Recent efforts by central government Since 1997, many publications from central government agencies and the quality of the advice offered to both the councils and Māori demonstrate that all participants achieved significant learning.

Publications by the Parliamentary Commissioner for the Environment (1998) and the Ministry for the Environment (2000a) provided guidelines for good practice in improving iwi participation in local government decision-making. The MfE's report on iwi and local government interaction (2000b) represented the first output of its new iwi-local government programme. Further outputs were planned, including an iwi management plan toolkit, training of iwi-based resource managers and the development of a kaitiakitanga website to assist hapū and iwi resource management practitioners.

With this uneven record of central government activity throughout the 1990s, how well could councils be expected to do?

Capability-Building by Councils

Council efforts to engage with Māori to address their obligations under the RMA focused primarily on building iwi capacity to support council plan-making, rather than long-term building of iwi capacity for resource management. Councils took a wide range of approaches in attempting to implement provisions in the Act relating to partnership under the *Treaty of Waitangi* (sections 6(e), 7(a) and 8). We report on these efforts, and then discuss several factors that influenced the degree to which plans supported these provisions. More specific analysis of plan quality as it relates to iwi interests is provided in Part III.

Council-iwi interactions   Our interviews with well over 100 council respondents revealed a range of council-iwi interactions similar to those described in other reports (Local Government New Zealand, 1997; Parliamentary Commissioner for the Environment, 1992, 1998). Resources did not enable us to extend interviewing to hapū/iwi representatives, so we have relied on case studies and secondary sources.

Various models involving iwi in plan-making processes were established by regional and district councils, including political representation on council committees, appointment of iwi liaison staff within councils, payment for advice from iwi by means of consultancies, and service contracts with iwi.

Where working arrangements between iwi and councils were well-established, as with Ngai Tahu in the South Island, relationships were constructive and productive. In other cases, where new relationships were developed, outcomes were also positive. For example, Hurunui District Council negotiated a protocol with iwi that provided a process for meetings and material for the plan. Two iwi set up their own resource management groups as a result. Queenstown Lakes District Council paid for iwi to travel to Queenstown to attend meetings with them as part of establishing contact with a group in its community that had been previously ignored, and engaged iwi to draft sections of the plan. Several regional and district councils reported that their experiences of working with iwi were very positive and had fostered good working relationships beyond the plan-making process.

For some councils, partnership-building was more problematic. Several councils experienced difficulties where disputes over which group had tangata whenua status impeded consultation, or disrupted formal arrangements that had already been established, such as standing committees. Some councils found that not all iwi groups in their district wished to be part of an umbrella group for consultation, and they needed to meet separately with groups. Not all councils wanted to build relationships with iwi.

In other cases, representation issues delayed plan preparation at critical stages. For example, Horowhenua District Council needed to consult with thirteen iwi and hapū in developing its plan. Disputes between iwi over tangata whenua status delayed writing parts of the plan, and the relevant chapter was finally notified as a variation to the plan. Similarly, while Manukau City Council negotiated a service agreement with one group to assist with plan-writing, disputes over representation resulted in further rounds of consultation and a rewrite of a chapter of the plan.

Recent activities of councils Since our council interviews in 1997, some progress in improving capability for council-iwi interactions has been made. However, fundamental issues of governance and power-sharing still needed to be addressed.

By decades end, at least one regional council was developing templates to assist Māori in preparing iwi management plans. District councils were producing guides, such as Manukau City Council's (2000) *Treaty of Waitangi Toolbox*, designed to assist staff to improve their council's delivery on its obligations as a Treaty partner. As we found, and others have noted (Hewison, 1997), many councils were entering into formal protocol agreements with iwi to manage resource consent processes. The Bay of Plenty Regional Council even proposed legislation giving Māori statutory representation on its council. The *Annual Survey of Local Authorities* released by the MfE in 2000 states that 63 per cent of local authorities had made a formal budgetary commitment to Māori/iwi participation in RMA processes, an increase of 5 per cent on the previous survey. It also indicated that 50 per cent of councils committed themselves to providing staff training on Māori issues in 1998-99.

At a more fundamental level, and acknowledging growing concerns about legal and political issues, Local Government New Zealand published a review (Chen and Palmer, 1999) addressing the relationship of local government to the *Treaty of Waitangi*. The purpose was to clarify the implications of the Treaty for local government. The report suggested several options for moving forward, including incorporation of the Treaty within the *Local Government Act* (1974) and giving local government a statutory duty to consult with Māori.

Conclusions

Overall, in the first decade under RMA, a great deal of progress was made regarding the recognition of Māori interests in environmental planning and governance — and beyond. However, the tension between competing political ideologies of left and right in central government towards dealing with the indigenous people of Aotearoa was reflected in the range of responses to iwi interests in local government.

In most councils staff worked hard to meet the intentions of the RMA, and in many instances iwi consultation included Māori consultants being employed to help develop the plan. Too often, however, these activities were undercut by councillors, especially in district councils, either out of ignorance of their responsibilities under the Act, sympathy with the ideological position of universal rights, or simply deep-seated racial prejudice (Chapter 12). Where governance was poor, progress was slow, and may remain so until Government gives better direction. In this, we would agree with The Parliamentary Commissioner for the Environment (1998), who called for a national policy statement to provide clarity and direction for councils dealing with the relationship of Māori and their culture and traditions, kaitiakitanga and the principles of the *Treaty of Waitangi*. The Waitangi Tribunal also called for the strengthening of provisions in the RMA.

While most hapū and iwi wanted to participate in the RMA processes, they were hindered by limited capacity and overwhelming demands on their resources. This made it difficult for councils that may have been so inclined to transfer powers and money to hapū and iwi so that they could take responsibility for self-regulation of resources and engage in collaborative and co-management initiatives with the Crown and local government (e.g. Sunde, Taiepa, and Horsley, 1999; Taiepa, 1999; Moller et al., 2000; Rennie, Thomson and Tutua-Nathan, 2000). Calls for the restoration of rangatiratanga through the control and management of resources allocated to Māori are therefore unlikely to be met soon.

As argued by Matunga (2000: 45), the inability of councils to share even limited decision-making with Māori resulted in the 'Māori Treaty partner on the outside, looking in on a passing parade of environmental decision and policy processes controlled by the other'. This can only change with clarification of governance issues, commitment by central government to transfer resources to local government and Māori, and willingness by both central and local government to share power with their Treaty partners.

Annex 5.1

The Treaty of Waitangi 1840

Translation of Māori Version

The First Article The Chiefs of the Confederation and all the Chiefs who have not joined that Confederation give absolutely to the Queen of England for ever the complete government over their land.

The Second Article The Queen of England agrees to protect the Chiefs, the Subtribes and all the people of New Zealand in the unqualified exercise of their chieftainship over their lands, villages and all their treasures. But on the other hand, the Chiefs of the Confederation and all the Chiefs will sell land to the Queen at a price agreed to by the person owning it and by the person buying it (the latter being) appointed by the Queen as her agent.

The Third Article For this agreed arrangement therefore concerning the Government of the Queen, the Queen of England will protect all the ordinary people of New Zealand and will give them the same rights and duties of citizenship as the people of England.

English Version

The First Article The Chiefs of the Confederation of the United Tribes of New Zealand and the separate and independent Chiefs who have not become members of the Confederation cede to Her Majesty the Queen of England absolutely and without reservation all the rights and powers of Sovereignty which the said Confederation or Individual Chiefs respectively exercise or possess, or may be supposed to exercise or to possess, over their respective Territories as the sole Sovereigns thereof.

The Second Article Her Majesty the Queen of England confirms and guarantees to the Chiefs and Tribes of New Zealand and to the respective families and individuals thereof the full exclusive and undisturbed possession of their Lands and Estates Forests Fisheries and other properties which they may collectively or individually possess as long as it is their wish and desire to retain the same in their possession; but the chiefs of the United Tribes and the individual Chiefs yield to Her Majesty the exclusive right of pre-emption over such lands as the proprietors thereof may be disposed to alienate — at such prices as may be agreed between the respective Proprietors and persons appointed by Her Majesty to treat with them in that behalf.

The Third Article In consideration thereof Her Majesty the Queen of England, extends to the Natives of New Zealand Her royal protection and imparts to them all the Rights and Privileges of British Subjects.

PART 3
PLAN QUALITY AND
CAPABILITY UNDER THE RMA

PART 3

PLAN QUALITY AND
CAPABILITY UNDER THE RMA

Chapter 6

Regional Councils: Lightweight Policy Statements and Limited Capability

The quality of regional policy statements is crucial for effective integrated resource management. They provide guidance to policy-makers in regional councils when preparing regulatory regional plans, and serve as a framework for co-ordinating the individual district plans that are prepared by each district council within each region. As noted in Chapter 2, plan quality is defined through eight core criteria, which were also used to evaluate the 16 regional policy statements.

The Quality of Regional Policy Statements

Our analysis showed that regional policy statements were of only fair to poor quality. The mean scores for each of the eight plan-quality criteria (Figure 6.1), and the overall ranking of each council in relation to these (Figure 6.2), are examined in turn below. (Derivation of an index score for each of the eight criteria is explained in Annex 6.1.)

Mean Scores for the Eight Plan-Quality Criteria

Results of the mean scores reveal a fair to poor quality rating for each criterion. All mean scores are 6.88 or lower out of a possible score of 10. Half of the criteria scored below the half-way mark of 5.0 for regional policy statements (Figure 6.1).

Content analysis revealed that most policy statements were descriptive and superficial. The 'fact base' criterion scored lowest. Almost all policy statements were quite weak in using the results of systematic analysis to support identification and prioritization of issues, and thereby the selection of objectives, policies and environmental outcomes. The 'monitoring' criterion scored second lowest, as most policy statements did not contain indicators for tracking the performance of policies within them. The next lowest score was for 'organization/presentation', as most policy statements did not include supporting evidence in the form of documents or databases (e.g. mapped information, tabular summaries of data) to help understand the rationales for issue identification or policy selection. In addition, almost all regional policy statements did not contain a users' guide and a detailed index to enhance their user-friendliness.

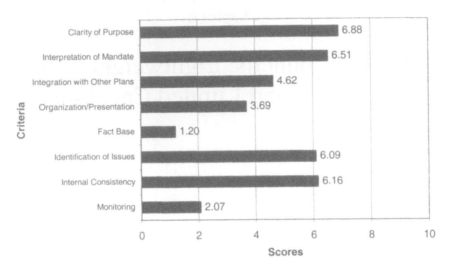

Figure 6.1   Mean scores of regional policy statements by quality criteria
(maximum possible score is 10)

The four strongest criteria had only a fair score (between 6.09 and 6.88). They were 'clarity of purpose', 'clear interpretation of key provisions' in the RMA mandate, 'internal consistency' (indicating that issues, objectives, policies, anticipated results and indicators for monitoring were clearly linked), and clear 'identification of issues'.

The 'integration' criterion showed a poor score (4.62). Most regional policy statements acknowledged other relevant plans and statements, but did not clearly explain the relationship between these and the regional policy statement.

Clearly, overall, the fact base and monitoring aspects were poorly done. Thus, any council that made a reasonable job of one or the other, or both, shot up the plan score ladder.

Ranked Regional Policy Statement Scores

From our coding of regional policy statements, we also created a single index that reflects their overall quality. (The index scores for the eight criteria were summed.) The ranking of each regional policy statement is shown in Figure 6.2. The mean score of 37.4 out of a possible 80, or 47 per cent, provides further evidence of the generally lacklustre quality of regional policy statements, with about two-thirds scoring substantially below half of the maximum score. The worst two (Marlborough and Northland) scored around 22 out of 80, or 27 per cent.

The mediocre quality of regional policy statements was confirmed in interviews. One manager acknowledged that there is a 'very uneven quality of regional policy statements and this could have been avoided'. One council chairperson observed that 'due to the [very limited] timeframe for preparing [it], the RPS is so generic that it could apply anywhere in New Zealand. There is deficient issue identification and hence the policy framework is inadequate'. This was explained in part by a planner who commented that the regional council 'didn't identify things early enough in the process to see the need for research', which accounts for the poor fact base.

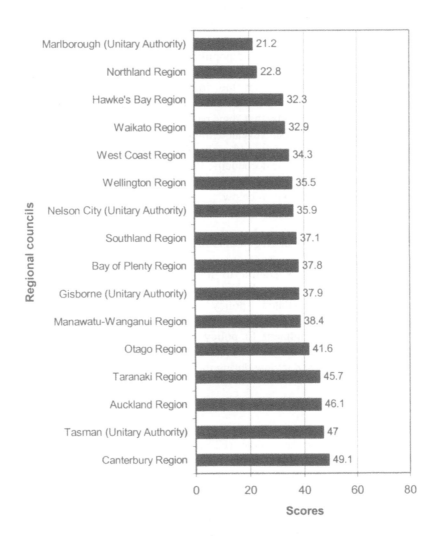

Figure 6.2 Ranking of overall regional policy statement scores (maximum possible score is 80)

On the other hand, several regional policy statements scored comparatively well and contained some commendable components, such as those produced by the Canterbury, Tasman, Auckland and Taranaki regional councils. The *Canterbury Regional Policy Statement*, for example, had the highest score of 49.1 out of 80, or 61.5 per cent, although that score is still only mediocre. It did better than others because it was well-organized and cross-referenced, with a particularly good tangata whenua section. It also had strong integration of cross-boundary issues. The *Taranaki Regional Policy Statement* contained a particularly strong fact base for the hazard mitigation section (one of the topics we selected for evaluation) and included a comprehensive assessment of economically productive lands (i.e. agriculture and forests), buildings and population at risk from all major hazards in the region (floods, volcanic eruptions, earthquakes, and so on).

It is important to note that the four dual-function unitary authorities (Marlborough, Nelson, Gisborne and Tasman) were located evenly from bottom to top of the rankings, suggesting that they performed no better or worse than the regional councils. Our analysis revealed three main reasons why the scores for regional policy statements were of only fair to poor quality: the limited organizational capability of most councils; the clarity of provisions in the RMA; and the usefulness of information provided by central government agencies. These three reasons are dealt with in turn below.

Organizational Capability of Regional Councils

Organizational capability is central to the planning programmes of regional councils. It provides the means through which resource management issues are defined, objectives formulated, policy solutions determined, and policy performance monitored. In particular, it involves the region's technical capacity to plan, and to build commitment to policy statements, plans and plan-making. Specific elements of capacity include staffing levels and procedural actions taken to improve consultation and review during the preparation of policy statements.

Staff Size

The level of staff capacity is an important factor. The discussion here focuses on the number of full-time equivalent (FTE) staff members in the core group responsible for writing the regional policy statement.

As indicated in Figure 6.3, there was considerable diversity in the size of the core group across regional councils, with 44 per cent of all councils allocating fewer than two FTE staff compared with 44 per cent allocating over four. These staffing levels seemed low, given the scale and importance assigned to planning and policy-making by the RMA and the short timeframe required for producing the regional policy statements. With such low staffing numbers, regional issues cannot be fully identified, evaluated and prioritized, and the likelihood of generating viable policy solutions decreases.

Several regional council managers commented on the lack of policy writing skills of staff. One noted simply a 'lack of ability to write'. Another commented that their council staff did 'not have the skills required', adding that he 'would have liked staff with initiative, attitude and pizzazz, as well as public policy analysis'. On the other hand, one senior manager commented that the 'hallmark of our approach is our skilled staff'. Interviews suggested that writing skills were highly variable across councils.

As noted in Chapter 4, too often the amalgamation of catchment boards and other entities into regional councils led to staff being moved from their specialized jobs into ones for which they had little or no training. This is likely to have affected policy and plan production in some councils as well.

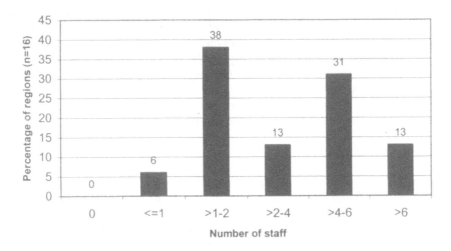

Figure 6.3   Number of staff in core group responsible for writing RPS

Peer and Legal Reviews

A second aspect of organizational capability is the facilitating actions taken by councils to enable critical review of regional policy statements, particularly by peer groups and lawyers. Peer and legal reviews can be an effective means of raising the quality of regional policy statements.

Peer reviews offer opportunities for councils to benefit from the insights and experiences of staff in other regional councils. Survey results suggest that regional councils took advantage of peer reviews, with 75 per cent of all councils using them. One regional manager acknowledged that the initiative taken by the Ministry for the Environment (MfE) in setting up peer review was good. Another commented that it was an 'important process in helping our thinking of what an RPS should be'. Despite some collaboration between regional councils in developing the structure of regional policy statements, one senior manager said that he would like 'to have seen more collaboration among councils to achieve more consistency of wording'. One council chairperson also observed that 'we put all policy/plans to audit at draft stage and again on the final document. It's an essential part of the process'.

Reviews by lawyers can be crucial in improving the legal viability of regional policy statements. There is limited precedent and considerable opportunity for legal challenge of policy statements developed under the RMA. The effects-based approach was a radical departure from the well-established activity-based planning. The new regime posed a major challenge by requiring councils to focus on outcomes, and hence greater use of non-regulatory methods. It is thus somewhat surprising that regional councils use this facilitating activity only partially, as indicated by the fact that slightly over half (55 per cent) retained a lawyer for legal review.

There may be two reasons for this finding. First, unlike district and regional plans, regional policy statements do not include regulatory rules. They set a policy framework, but do not have as much impact on resource consent applicants as rules in plans. The need for legal review may not be critical. Second, while district councils were accustomed to working with lawyers in preparing plans, regional councils at that time were not. Historically, regions under the old catchment board regimes used by-laws and water discharge permits to implement water resource management policies, and the planning role was new to many.

Enhancing Consultation

The third aspect is how councils enhanced consultation during preparation of the regional policy statement. The extent to which councils were capable of generating community involvement was indicated by the actions they took in this regard. In turn, the response of the community should be reflected in the relevance of issue identification and policy solutions adopted in regional policy statements.

An evaluation of the effectiveness of six common types of action is presented in Figure 6.4. The most effective was to establish early and ongoing consultation with interest groups. Nearly three-quarters of respondents rated this action as effective (rating of 4 or 5 on a 5-point scale, with 1 = not effective and 5 = very effective). An important advantage of interest group involvement during the early stages of planning is that awareness and knowledge about key issues and potential solutions can be heightened, and thus forestall or prevent decisions that could constrain viable courses of action in the future. Another advantage is that it involves interest groups at the outset, when they are most likely to be open to suggestions.

The next most effective actions were production of discussion papers, establishment of focus groups, and involvement of councillors early on. About two-thirds of all respondents rated these three as effective. Just over half considered establishment of iwi liaison committees to be effective. The least effective consultation action was to hold public education workshops, with only 25 per cent rating this action as effective.

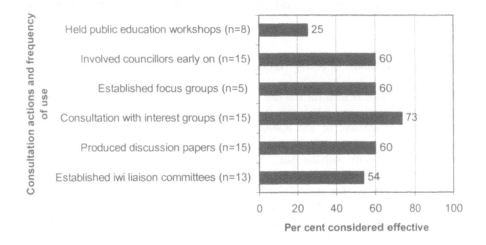

Figure 6.4   Actions considered effective in enhancing consultation during preparation of regional policy statements

Also shown in Figure 6.4 is the frequency of use for each type of consultation action. Interestingly, the pattern of frequency of use generally follows the pattern of effectiveness rating. In other words, those actions rated the most effective were used most frequently. The one exception to this pattern is focus groups: while 60 per cent of regional councils rated this action as effective, only five of the 16 regions and unitary authorities used it.

The consultation carried out by regional councils in preparing regional policy statements was not as extensive as that carried out by district councils in preparing plans. This reflects the different nature of the two planning instruments, as well as much less experience with community consultation processes within the regional councils. Furthermore, the timeframe was too short for extended consultation, given the two-year deadline for regional councils to prepare regional policy statements.

One senior planner noted that 'groups had a high expectation of what council might do, or what they might be able to do'. Others commented on the difficulties of engaging interest in the wider community. One manager commented that it was 'very hard to get interest in the RPS because the RPS was so early [in the process] and so vague'. This was confirmed by a planner who observed that his council 'heard mostly from interest groups'. 'Most people do not understand what the RPS is or what RMA is about'. This interviewee went on to note 'a degree of confusion about regional functions'. The means of engaging with communities were not always sufficiently extensive. One planner commented 'consultation was limited to being presented with something to react to'. One councillor acknowledged that, while in the submission process 'councillors were tested in terms of their understanding, [they] did change a lot of things based on submissions', and a planner suggested that 'the most influential people [in the process] were the councillors'.

Clarity of Provisions in RMA Mandate

The second major reason for the fair to poor quality of regional policy statements was the lack of clarity in key provisions in the RMA mandate. Clarity about the substantive provisions of the RMA can reduce the degree of uncertainty and confusion in preparing regional policy statements. It also enhances the likelihood that they will comply with the Act.

Ratings of six key provisions (sections 5, 6, 7, 8, 30 and 31) are indicated in Figure 6.5. Each section is described in Appendix 1 of the book. By far, section 6 (matters of national importance) was seen as the clearest section, with 74 per cent of respondents rating it as clear (ratings of 4 and 5 on a 5-point scale, with 1 = not useful and 5 = very useful). This section is clearly written and is similar to the provisions of the earlier *Town and Country Planning Act* (1977). Regional councils thus had some precedent for interpreting and translating the provisions for matters of national importance in the RMA.

Most of the responding councils did not consider the remaining five sections to be clear. In section 8, dealing with *Treaty of Waitangi* issues, the language in the Act is clear, but the means to operationalize the intent of Parliament is not. Answers to the question of how regions are to implement the bicultural mandate have not been clearly articulated in regional policy statements. Section 7 requires a regional assessment of what is important, and therefore relies on local knowledge, values and research. Again, the Act's language is clear, but applying the intent requires judgement, and little is provided on which to base such judgement. Sections 30 and 31 are at the heart of the respective roles of districts and regions. They were not clearly drafted, providing for overlaps in functions, and confusion in responsibilities for managing land uses by regional councils (and the effects of land uses by district councils) that took both court decisions and amending the Act to resolve. Councils have had to work out what these overlaps mean and how to deal with them in practice.

Section 5 (the purpose and principles of the RMA) received by far the lowest rating among all sections. This is likely due to the confusion surrounding the Act's definition of sustainable management of natural and physical resources, together with the difficulty of conceptualizing what this means for a particular region. The RMA contains some fine rhetoric about sustainable management, but individual councils have thus far struggled with translating this vague definition of sustainable management into practice.

Interviews yielded some insights into how the regional councils developed their understanding of the mandate. Councillors and planners acknowledged that there was a lot of uncertainty at the beginning in deciding how to proceed with policies and plans. One council chairperson observed that, because 'interpretation of RMA has been left to individual councils, retrospectively, we can get into trouble'. Another manager noted that the consideration of key sections of the Act led to deep philosophical discussions within his council. 'The Act contemplates what should be in the RPS, but this is not clear. Should an RPS address section 6? What does section 31(a) mean? How far should we go in respect of the built environment?' This person also observed that the 'government's decision to back away from environmental policy to the extent they have is wrong'.

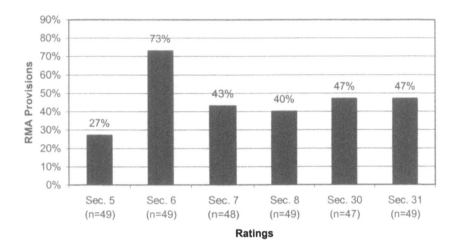

Figure 6.5   Sections 5, 6, 7, 8, 30 and 31 of RMA rated as clear by regional council staff

On the other hand, two planners commented that the 'RPS has served a good function in developing thinking under the Act' and that 'the plan preparation process educated the council about the RMA'. In at least two councils organizational restructuring had been informed by the RMA's requirements. Another planner commented that there had been 'a learning curve for everyone at the beginning' but that 'time constraints meant that issues were dealt with in a traditional manner rather than reconfigured in more modern environmental thinking'.

Usefulness of Central Government Agencies

The third main reason for the fair to poor quality of regional policy statements concerns the usefulness of information provided by central government agencies. Regional council planning and policy-making initiatives are embedded in a larger policy-making context and, therefore, must receive and maintain educational and technical support to be effective. Thus, the usefulness of technical and educational information for preparing regional policy statements can have an important effect on their quality.

As Figure 6.6 shows, regional councils in general did not give high ratings to central government organizations in the provision of information. Specifically, only about a third of all respondents considered the Department of Conservation (DoC) and the MfE to be useful (ratings of 4 or 5 on a 5-point scale, with 1=not useful and 5=very useful), about a quarter rated the New Zealand Historic Places Trust as useful, while the Institute of Geological and Nuclear Sciences and Transit New Zealand were considered useful only by less than a tenth of the respondents. A likely explanation for these low ratings is that these organizations were not geared up to fill the role of information providers in time for contributing to the preparation of regional policy statements. Because almost all regions did not have the expertise in-house, except for the most wealthy regions (Auckland, Wellington and Canterbury), the absence of effective external assistance led to most regions operating in an information vacuum.

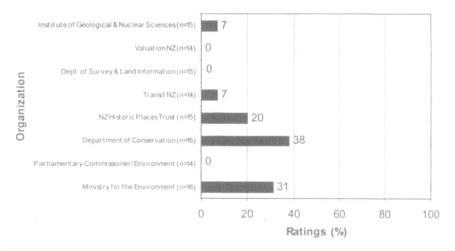

Figure 6.6   Usefulness ratings of central government organizations in providing relevant information for RPS preparation

Further interpretation of these findings for each organization is necessary. The low rating for the Parliamentary Commissioner for the Environment was not unexpected, since this body's primary role is to act as an environmental ombudsman and not as an information provider for regional (and district) planning activities. Valuation New Zealand provides individual property valuations nationwide, which are used to set local property taxes, but this has more relevance for district councils, which collect rates (taxes) on behalf of the regions. However, low ratings of the other organizations indicated that regional councils were not taking advantage of potentially useful information sources. The Department of Survey and Land Information was an important source of information about maps and geographic information systems. Districts have traditionally used this information to produce zoning maps, but regions do not appear to have considered it useful in producing regional policy statements. The New Zealand Historic Places Trust gathers data on archaeological sites and historic buildings, and has been responsible for financing conservation. Regional councils were uncertain about their role in historic places issues, and thus did not use the Trust's potentially valuable information-base for identifying and conserving regionally important sites, while the Trust itself set out to target district councils. The regional councils' inaction may be due, in part, to uncertainty over whether they would be held responsible for financing the conservation of historic places. Transit New Zealand has been active in assisting the preparation of regional land transport strategies throughout the nation, while the Institute of Geological and Nuclear Sciences produces, under contract, information on hazards' location and risks.

DoC was given a low rating by regions for its usefulness as a source of information, even though it had an interest in protecting indigenous flora and fauna, and a key role under the RMA for the coastal zone. Similarly, the MfE had a low rating for providing useful information even though it had the key role in providing advice and technical assistance to help councils prepare their regional policy statements. Its low rating here was likely due to the common perception that it has provided changing and sometimes conflicting advice. This caused the MfE to lose some credibility. Many regional councils thus took a 'go it alone' attitude rather than seeking the MfE's assistance.

There were, however, contrasting views on the role of the MfE and DoC. On the one hand, for example, one council chairperson noted that the 'MfE and DoC had tremendous input on all our plans [they] gave us a certain amount of direction since they were consistent across the country'. Another manager noted that peer review by the MfE was an 'important process in helping our thinking of what an RPS should be' and that the 'MfE desperately longed for everyone to be more rigorous than council was able to do given the resources'. On the other hand, a council planner described his view of assistance from MfE staff as 'not how to do it, but if we don't like it, we will take you to court'. Another commented that the 'MfE was struggling internally with what they had set up [and that there was] no consistent message from them'. This was confirmed by a council manager who said that the 'MfE kept changing their views. District councils did not know what they [MfE] wanted in the RPS and their own plans'. With reference to a council's relationship with DoC, another planner observed that 'there was a significant tension between [their] regional council and DoC because the New Zealand Coastal Policy Statement was not out when the RPS was publicly notified'. In reflecting on the relationship of councils and government agencies, a regional council chairperson offered the insight that 'the RMA to be truly effective needs a real partnership between central, regional and local levels'.

In addition to central government organizations considered useful in providing relevant information for RPS preparation, we asked council staff about the usefulness of adjoining councils in helping prepare policy statements. Answers suggested that collaborative efforts between regional councils varied around the country. Considerable inter-regional competition and a lack of willingness to co-operate, as well as the absence of good data on cross-boundary issues between regions, were the main reasons given. One senior manager said 'we largely ignored other regions. We were too busy'. Another observed, however, that 'networking between regional councils was invaluable as a support mechanism'.

Māori Interests and Regional Policy Statements

From the outset, regions took a wide range of approaches in attempting to fulfil those aspects of the mandate related to partnership under the *Treaty of Waitangi* (sections 6(e), 7(a) and 8 of the RMA), but the success of their actions was not entirely clear. Some councils addressed the bicultural mandate by setting up a Māori Standing Committee (e.g. Manawatu-Wanganui, Hawke's Bay and Taranaki). In Auckland and Wellington, the council chairperson dealt directly with iwi, whereas Marlborough had appointed an iwi representative to the Regional Policy Statement committee. Hawke's Bay had gone further, by having one Māori appointee for each regional plan committee. Some regional councils had appointed an iwi liaison officer, but the extent of their involvement in preparing either the regional policy statement or regional plans was unclear.

Involving Māori in writing the regional policy statement also varied across councils. Hawke's Bay commissioned the local rūnanga (council) to write the relevant chapter — a model of its kind. Canterbury set up a tangata whenua liaison group (three staff) and they did much of what was necessary, aided by a well-resourced and demanding iwi. As unitary authorities, Marlborough, Nelson and Tasman employed an iwi planning consultant to prepare the relevant chapter, whereas Taranaki sent all documentation to all eight iwi with interests in its region.

So how good were the resulting regional policy statements with regard to iwi interests? The mean scores for selected plan-quality criteria (Figure 6.7) and the overall ranking for each council in relation to these for Māori (Figure 6.8), are dealt with in turn below.

We applied four of the eight plan quality criteria to examine how well regional policy statements were addressing Māori interests: clarity of interpretation of provisions of the *Treaty of Waitangi*; issue identification; the fact base of the RPS; and its internal consistency. The mean scores by plan quality criteria and the total of these means are shown in Figure 6.7. (Derivation of an index score for each of the four criteria is explained in Annex 6.1.) Overall, regional policy statements received quite low scores, as the total mean score was just over 50 per cent (20.9 out 40), while the range was from low (0.9 fact base) to high (*Treaty of Waitangi* 6.9 and clarity of issues 6.9). These higher scores did emphasize that in general, regional councils were sensitized early in their operations to the need to deal with iwi as the Treaty partners.

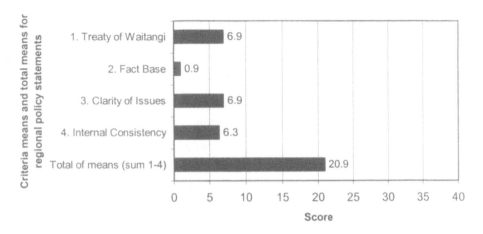

Figure 6.7   Mean for each criterion and total means for regional policy statements regarding Māori interests (maximum score for each criterion [1-4] is 10.0, and total of means is 40.0)

The ranking of total scores is shown in Figure 6.8 (next page). The scores show great variety spreading from a low of 4 (Marlborough) to a high of 32.4 (Canterbury) out of a maximum possible score of 40. (The index scores for the four criteria were summed.) Some councils did not do much in their RPSs for Māori interests, but others did put in a lot of thought and effort. Mostly, those councils that used more robust means of engaging iwi tended to have better scores. Inexplicably, Marlborough did not gain a good score for having done so.

Our research showed that regional councils understood what was required by the directive to take account of the principles of the *Treaty of Waitangi* and that the regional policy statements had better scores for iwi matters than they had overall. However, this did not led to better results for Māori. Councils began well by initiating relationships with iwi, creating opportunities for representation, and in many cases involving them in plan-writing. Once regional policy statements had been produced, however, the relationships became somewhat fractious. Iwi expectations for partnership were not being realized, and councillors were not ready to genuinely share power with iwi. Most iwi were under-resourced, and although some councils provided funds to help build their capabilities, it was not only too little, but aimed at councils' priorities and not those of iwi. Thus, difficulties in implementing the RMA pushed iwi into a reaction mode, exacerbated conflict, reduced integration, created a huge workload (for iwi), and generated a public backlash against perceived preferential treatment.

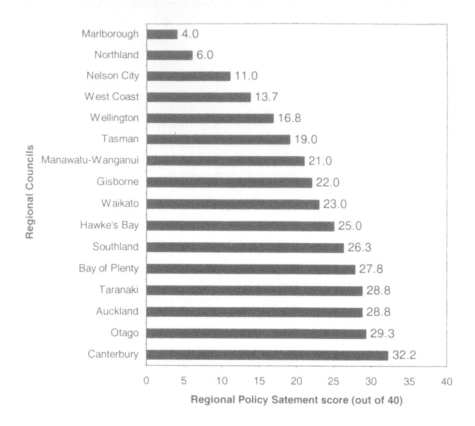

Figure 6.8   Scores of regional policy statements for Māori interests
(maximum score is 40)

Conclusion

We hypothesized that a clear mandate and sound technical assistance from central
government agencies would help regional councils to prepare high quality policy
statements, as would the strength of commitment and capacity within councils.
The major conclusion to be drawn from our evaluation of regional policy
statements is their overall lacklustre quality. A few regional councils produced
good policy statements, but none were good or excellent according to our
evaluation criteria. The overwhelming majority were of only fair to poor quality.

From our analysis of inter- and intra-organizational processes, we can
summarize three possible reasons for this. First, staffing levels were quite low,
given the extensive demands placed on councils to produce regional policy
statements within two years of the RMA being passed. Nearly half of all regional
councils employed only two or fewer FTEs on their core plan-making groups.

Second, the intent of the mandate legislation was not clear. Apart from one key provision (section 6, matters of national importance), staff in the majority of regional councils did not consider the five other key provisions clear. Lack of clarity in the mandate's intentions generates confusion and uncertainty about what to do and how to do it.

Third, regional councils did not consider central government agencies useful in providing information needed for the preparation of regional policy statements. About two-thirds or more of the councils rated eight central government agencies as not helpful. Regional councils thus appeared to have taken a 'go it alone' approach. Prospects for achieving the resource management intentions aimed at in the mandate are likely to be quite low, therefore. Conversely, there was a lack of capability building (see Chapter 3).

However, our findings on the use of peer and legal reviews, as well as consultation techniques, were not consistent with the unimpressive quality of regional policy statements. Three-quarters of all regional councils used peer reviews, and just over half used legal reviews. Further, a majority of regional councils used various consultation techniques (e.g. involved councillors early on, consulted with interest groups, produced discussion papers, established iwi liaison committees), and considered these to be effective in fostering consultation during the preparation of regional policy statements. While these activities likely had a positive impact, it does not appear to have been strong enough to overcome the other obstacles discussed above.

In sum, because of the lacklustre quality of regional policy statements resulting from inadequate staff levels, poor understanding of the mandate and the lack of useful information from central government, regional policy statements are likely to have little influence on resource management policy-making, especially since other plans — regional and district plans — need only be 'not inconsistent' with regional policy statements.

Other studies (Nuttall and Ritchie, 1995) corroborate our findings on iwi interests (see Chapter 8). As well, a study by Berke, Dixon and Ericksen (1997) of natural hazards within nine regional policy statements found that these documents had a weak fact base for natural hazards. As a result, hazard issues were not clearly identified and prioritized, and the rationale for selecting hazard reduction policies was weak. There is considerable evidence, therefore, that regional councils had a limited capacity to prepare regional policy statements, and this resulted in their weak quality.

Annex 6.1

Derivation of indices for plan quality

Variable	Measurement	Source
Clarity of purpose Region; District Range, 0-10; 0-10 Mean = 6.88; 6.76 SD = 3.10; 2.43	2, coherent overview of environmental outcomes; 1, no overview; 0, coherent overview of future development. Score standardized based on a 0-10 scale.	District plan or regional policy statement
Interpretation of mandate Region; District Range, 0-10; 0-10 Mean = 6.51; 3.62 SD = 1.69; 1.21	Index based on the sum of ratings for each key mandate provision: 2, full explanation; 1, vague explanation; 0, no explanation. Provisions include: matters of national importance; *Treaty of Waitangi*; consider alternatives; gather information; monitor and keep records; functions of regions (regions only); functions of districts (districts only). Index score standardized on a 0-10 scale.	District plan or regional policy statement
Integration Region; District Range, 0-10; 0-10 Mean = 4.62; 4.26 SD = 1.80; 2.02	Index based on the sum of ratings for clarity of explanation of how other policy instruments are integrated with district plans or regional policy statements: 2, full explanation; 1, vague explanation; 0, no explanation. 29 policy instruments plus cross–boundary issues were listed. Index score standardized on a 0-10 scores.	District plan or regional policy statement
Organization & presentation Region; District Range, 0-10; 0-10 Mean = 3.69; 4.76 SD = 1.14, 2.02	Index based on the sum of items used for improving accessibility of plan: 1, yes; 0, none. 10 items were listed: table of contents; index; glossary; users' guide; executive summary; cross-referencing; visual illustrations; use of spatial information; identification of individual properties; supportive documentation.	District plan or regional policy statement

Variable	Measurement	Source
Fact base document Region; District Range, 0-10; 0-10 Mean = 1.20; 0.62 SD = 1.02; 0.88	Index based on sum of items used to enhance the fact base: 1, yes; 0, no. 6 items were listed: maps; tabular formats; citation of methods used; issues prioritized; benefit-cost; background information referenced. Index score standardized on a 0-10 scale for Māori and hazard sections.	District plan or regional policy statement
Identification of issues Region; District Range, 0-10; 0-10 Mean = 6.09; 4.63 SD = 3.29; 2.69	Index based on the sum of ratings of clarity of explanation for each issue identified in plan: 2, clear explanation; 1, vague; 0, none. Index score standardized on 0-10 scale for Māori and hazard sections.	District plan or regional policy statement
Internal consistency Region; District Range, 0-10; 0-10 Mean = 6.16; 6.56 SD = 1.44; 1.87	Index based on the sum of ratings of clarity of relationships among 6 items in a cascade of links: 2, clear; 1, moderately; 0, absent. Links are: issues and objectives; objectives and policies; policies and methods; methods and anticipated results; anticipated results and indicators. The most clearly defined issue in two sections of the plan (cultural rights of indigenous people or Māori, and natural hazards) was chosen as the start of each cascade. The mean score of both cascades was determined; index score was standardized on a 1-10 scale.	District plan or regional policy statement
Monitoring Region; District Range, 0-10; 0-10 Mean = 2.07; 3.87 SD = 2.43; 3.49	Index of monitoring based on the sum of ratings of three items: provisions for monitoring policy: 1, yes; 2, no; monitoring outcomes: 2, detailed; 1, vague; 0, absent; identification of agencies responsible for monitoring: 1, yes; 0, no.	District plan or regional policy statement
Overall plan quality Region; District Range, 0-80; 0-80 Mean = 37.22; 35.37 SD = 7.87; 9.32	Index of the sum of scores for each of the 8 plan-quality criteria.	District plan or regional policy statement

Variable	Measurement	Source
Organizational Capacity		
Staff Region; District Range, 2-10; 0-8 Mean = 4.31; 3.27 SD = 2.50; 2.62	Number of staff responsible for writing the plan.	District or regional survey
Mandate		
Clarity of mandate Region; District Range, 1-25; 1-25 Mean = 3.33; 3.43 SD = 0.67; 0.66	Index of clarity of the RMA based on mean respondent rating of 6 items (1-5 scale): section 5 (purpose); section 6 (matters of national importance); section 7 (other matters); section 8 (*Treaty of Waitangi*); section 30 (functions of regions); section 31 (functions of districts).	District or regional survey
Usefulness of central government organizations		
Usefulness of organizations Region; District Range, 1-5; 1-5 Mean = 2.07; 2.68 SD = 0.53; 0.50	Index of usefulness of organizations based on mean of respondent rating of 2 central government organizations (1-5 scale): Ministry for the Environment, Department of Conservation.	District or regional survey
Context		
Population size Region; District Range, 35,380 – 953,980; 7793 – 292,858 Mean = 239,358; 53,004 SD = 238,843; 64,213	1991 population.	Census
Median home value Region; District Range, $71,000 – $225,000; $61,000 - $200,000 Mean = $125,187; $108,485 SD = $35,693; $42,807	1991 median home value.	Valuation New Zealand

Note: the different sources cited are: the sample of plans that we analysed (34 district plans and 16 regional policy statements); the district and regional postal questionnaires that were sent to our sample councils, and the New Zealand Census (Statistics New Zealand, 1997).

Source: from Berke et al., 1999

Chapter 7

District Councils: Mixed Results in Planning and Capability

The questions posed of regional councils apply equally to local or district councils. More specifically, would district councils have the organizational capability to prepare good-quality effects-based district plans?[1] Although reconstituted by the reforms of 1989, their forebears did have a long history of writing local plans. It seemed reasonable to expect that the new district councils would produce good quality plans. The question is pursued using a similar format to Chapter 6.

The Quality of District Plans

Our results show that two-thirds of the 34 district plans we evaluated fell well below the half-way mark for plan quality. This is a very disappointing result, considering that these plans are such an essential component of the inter-governmental planning system operating under the RMA. Unlike regional policy statements, district plans exert a direct influence on land-use decisions through rules controlling the effects of activities. It is important that they are robust.

Mean Scores for the Eight Plan-Quality Criteria

The mean scores for district plans by our plan-quality criteria, as shown in Figure 7.1, reveal just how generally unimpressive the quality of New Zealand's first generation of plans were. (Derivation of an index score for each of the eight criteria is explained in Annex 6.1.) The score for the two criteria that scored highest was still only fair. 'Clarity of purpose' had the highest score (6.76 out of a possible 10.0) among all criteria, but results from the evaluation were mixed. On the one hand, the score was lowered because many plans did not include even a description of purpose. On the other hand, when plans did include a description of purpose, it was correctly oriented towards environmental outcomes rather than focusing on future development patterns. The outcome orientation is consistent with the effects-based approach signalled in the RMA (see Part II).

---

[1] Recall that the 74 local councils include 13 cities, 60 districts and one county, but the plan mandated by the RMA is called a district plan. Local councils are normally called district councils or territorial authorities, but are referred to in this chapter as district councils.

The 'internal consistency' criterion had the next highest score of 6.56. This indicated that the links among issues, objectives, policies, methods, anticipated results and indicators for monitoring were present, but in need of improvement. They were stronger for the first part of the cascade, but much weaker in linking issues to outcomes and monitoring. Several interviewees cited unrealistic political deadlines as forcing notification of plans when another few months would have improved their internal consistency. This was intensely frustrating for planners.

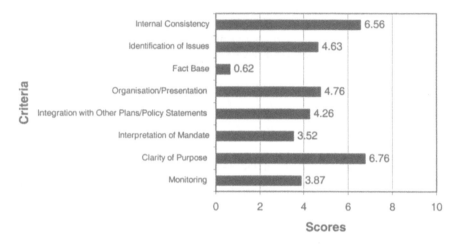

Figure 7.1   Mean scores of district plans by plan quality criteria (maximum possible score is 10)

Five criteria (organization/presentation, identification of issues, integration with other plans, monitoring and interpretation of key mandate provisions) scored poorly (4.76 to 3.52). The 'fact base' criterion received a very poor score of 0.62.

Content analysis provides further explanation of why district plans received generally lacklustre scores for these criteria. Weak fact base scores resulted from the absence of analytical rationales for defining and prioritizing issues, and selecting policy alternatives. The weak 'fact base' also partially explains the poor scores for 'issue identification' and 'monitoring'. Without a strong fact base, it is difficult to define issues clearly and to determine appropriate systems to monitor how far local issues have been resolved.

Interviews revealed that, generally, little significant research was undertaken, which accounts for the weak fact base and consequent poor issue identification. Respondents cited, as contributory factors, constraints on funding and time, lack of staff and insufficient knowledge about which issues warranted further investigation. One consultant stated that 'the budget allocated for the plan meant that a highly analytical approach could not be adopted'. In turn, this impeded the development of monitoring provisions in plans. One planner commented that 'State of the Environment reporting is new and it's difficult to meet statutory requirements. It's like putting the cart before the horse'. One would think, however, that knowing the environment is a prerequisite to planning, to ensure its quality. Several respondents said that monitoring was deliberately left until plans were made operative. The overwhelming view was expressed in interviews that monitoring in plans was universally weak.

The generally poor score for the 'interpretation of the mandate' criterion resulted from the fact that most plans simply quoted or paraphrased RMA provisions. Plans generally did not provide a clear explanation of how the legislation applied to the particular conditions of an individual council, and how such provisions were to be implemented.

Although many other types of plans and policy documents were identified in district plans, the 'integration of plans' criterion did not score highly because such documents were merely mentioned. District plans lacked a clear explanation of how these documents are accounted for in district policies. This finding suggests a lack of inter-organizational co-ordination among the multiple planning efforts operating under the mandate. Further, the timing of notification of plans was often out of sequence. One council planner commented that 'The Regional Policy Statement should lead to the District Plan, but Territorial Authorities [district councils] were often producing plans ahead of the regional documents'. This inhibited integration of regional and district policies. On the other hand, one councillor observed that it was 'hard enough to get the district plan done without trying to get to grips with the RPS'.

The 'organization and presentation' criterion scored fairly because of the mixed results. On the one hand, most plans clearly identified individual properties, provided clear illustrations and included a glossary of terms and detailed table of contents. On the other hand, most did not provide citations for supporting documentation, cross-referencing of issues and a well-developed index. Several interviewees commented on the dilemmas of producing a user-friendly plan which met other requirements as well. One council planner observed that 'it is difficult to make a plan that is legally secure and readable'. One councillor did not think 'that lay people can understand all the complexities of a plan'.

Ranked Scores for District Plans

The overall lacklustre quality of district plans can be seen by producing a single index score for each of the 34 district plans included in the sample. (The index scores for the eight criteria were summed.) The consequent ranking of each district plan is shown in Figure 7.2. The mean score of 35.2 (out of a possible 80.0, or nearly 44 per cent) reflects the overall fair to poor quality of district plans, with two-thirds of the sample scoring well below the half-way mark.

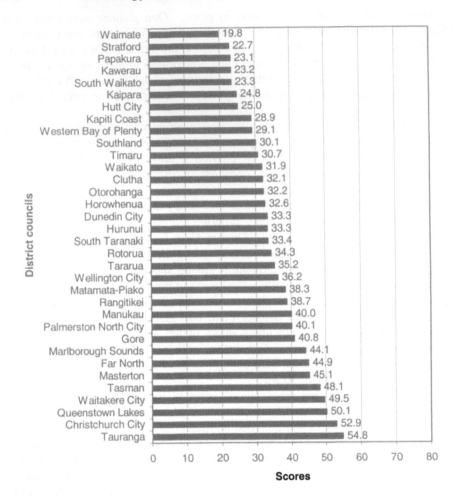

Figure 7.2 Ranking of overall district plan quality scores (maximum possible score is 80)

Several interviewees noted that constraints on budgets affected the overall quality of the plan. One council planner acknowledged that they were 'at the limits of what could be achieved with resources'. Another commented that they 'did not want to be a guinea pig and had limited resources to gamble with'. One council consultant noted 'resentment of councillors at spending dollars on plans'.

Nevertheless, several district plans received good scores, with Tauranga, Christchurch, and Queenstown Lakes scoring above 50 out of 80 (or 64 per cent). These plans contained several particularly commendable features. The Tauranga plan scored highest at nearly 69 per cent (54.8 out of 80) and was well conceptualized and tightly aligned with the RMA. The Christchurch plan gave careful attention to internal consistency. The policies were clearly linked to, and grouped by, the issues, objectives, policies, methods and key indicators for monitoring policy performance. The Queenstown Lakes plan displayed a clear understanding of what sustainable management means for the district and how it might give effect to this.

Organizational Capability of District Councils

Three main factors affected the ability of local councils to produce better plans, of which organizational capability was foremost. As noted in the previous chapter, organizational capability refers to a council's capacity to create a technically sound plan and ability to build commitment to plan-making. Specific elements of organizational capability include the number of core staff responsible for writing the district plan, and procedural actions taken to improve consultation and review during plan preparation, and eventually plan implementation and monitoring its progress in achieving the desired environmental outcomes.

Staff Size

Those surveyed were asked to report on the number of full-time equivalent staff members (FTEs) in the core group responsible for writing the plan. Figure 7.3 shows considerable variation in the number of staff employed by district councils, but staffing appeared to be inadequate for most. Staff size in the core group was quite small for more than half of the councils, as they employed only two or less FTEs. Further, a substantial proportion (26 per cent) of councils committed less than one FTE. Another 30 per cent of councils employed more than two but less than six staff. The remaining 20 per cent of councils were relatively well staffed, as they employed more than six staff in the core group. Many staff interviewees acknowledged the stress that limited staffing had on plan-making. Councillors tended not to see this as a problem, even those sympathetic to plan-making. For example, one long-serving councillor with a keen interest in planning was self-critical upon reflection that, when staff requested delays or more resources, she thought they were dragging out the process to protect their jobs.

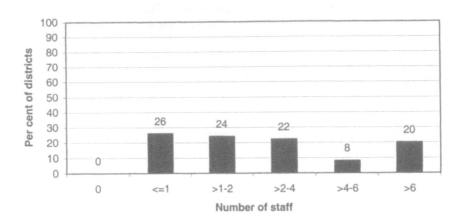

Figure 7.3 Number of staff in core group responsible for writing district plans

These findings are strikingly low for at least half of the district councils, given the substantial plan-making responsibilities assigned to them under the RMA. However, extensive use of consultants filled skill gaps in council planning teams. In some smaller councils, consultants prepared the whole plan, while in larger councils they contributed expertise on particular sections. Interviewees strongly endorsed the quality of input from external advisors. Despite this external support, however, where staff numbers are small the quality of plans is likely to be low, as issues cannot be fully identified and evaluated, let alone resolved satisfactorily.

Peer and Legal Reviews

As noted in the previous chapter, organizational capability also involves actions taken by councils to promote the critical review of their plans, particularly by peer groups and lawyers, which can be an effective means for raising the quality of district plans.

We found that only 39 per cent of district councils used peer reviews to improve the content and quality of their plans, while 83 per cent retained a lawyer to review their plans. The sheer uncertainty about the meaning of sustainable management and of key provisions in the Act induced councils to seek legal assistance, in order to avoid legal challenge to the plan. As well, planning under the new effects-based regime posed a major challenge by focusing more on environmental outcomes and on greater use of non-regulatory methods to guide market and public land-use decisions.

Many planners interviewed said that they attempted to achieve an effects-based approach. One planner summed it up when he said that they 'tried hard to keep faith with the Act, the idea of sustainable management, integrated management and section 32 and to minimize use of rules'. Another commented: 'No one has succeeded in producing simple plans within the legislative framework. I do not see much likelihood of this in the future because we have such a legalistic process.' District councils thus wanted some assurance that their plans were legally sound and complied with the RMA, an aspect of plan-making emphasized by the MfE when reviewing plans.

Enhancing Consultation

As we saw in Chapter 6, full public consultation in preparing policies and plans is a major goal of the RMA. We argued that efforts by councils to improve consultation can have a strong positive impact on plan quality, because the stakeholders can facilitate the identification of issues and formulation of policies in the plans, as well as the associated implementation actions.

Presented in Figure 7.4 is our evaluation of the effectiveness of the six common types of action for enhancing consultation during plan preparation. The most effective groups of action were: 1) involvement of councillors early on in the process of plan preparation; 2) establishment of focus groups; and 3) production of discussion papers. Nearly three-quarters of respondents rated these actions as effective (rating 4 or 5 on a 5-point scale, with 1 = not effective and 5 = very effective). Involvement of councillors early on in the planning process was considered an effective way to heighten their awareness and knowledge of key issues and the potential solutions available. Councillors were also more likely to be more committed to the plan and support its implementation. Use of focus groups was also favoured because they provided a formal forum for stakeholders to identify issues they considered important, and helped to formulate policy solutions that in their view were most suitable for addressing the issues. Production of discussion papers was also considered an effective means for raising awareness among interest groups about key issues, and facilitating incorporation of a better fact base into public debate about planning.

The other effective actions were: 1) establishment of early and ongoing consultation with interest groups; 2) establishment of iwi liaison committees; and 3) holding public workshops. Just over half of all respondents rated early ongoing consultation with interest groups as effective. Establishment of iwi liaison committees and holding public education workshops were considered effective by the lowest number of responding districts, less than half. This result is not dissimilar to that for the regional councils (Figure 6.4).

Figure 7.4 also shows the frequency of use of each type of consultation. The pattern of frequency of use is consistent with the pattern of effectiveness ratings — that is, those actions rated the most effective were used most frequently. Similar to regional councils, one exception to this pattern is focus groups. While this received the second-highest effectiveness rating (70 per cent), it was used by the second-lowest number of district councils (n = 30).

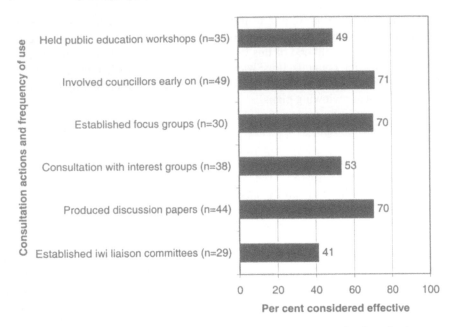

Figure 7.4   Actions considered effective in enhancing consultation during district plan preparation

Interviews confirmed the range of actions by councils to consult with their communities, adding insightful commentary on the nature and style of consultation. Some councils undertook demanding programmes of consultation with their communities, others did not. One councillor noted that 'officers began to realize how much energy and patience was involved, how slow the process was'. Several other councillors noted the merits of consultation and that 'it led to changing the values of councillors, reduced the number of contentious issues [and produced] marked changes in thinking'. One respondent noted the initial 'great enthusiasm' from community groups for the new planning process. However, when the plan took a long time to prepare, this enthusiasm 'later waned'. While early starts could seem problematic, some planners and councillors suggested that consultation had been a major strength of their plan preparation process. Some also acknowledged burn-out problems for the staff and councillors involved. In reflecting on their experiences, some councils suggested that consultation needed to be earlier, more targeted and 'made more accessible to the community'.

Clarity of Provisions in RMA Mandate

In order to prepare quality plans that comply with the RMA, its key provisions need to be clear to both planners and councillors alike. If not, a lot of resources are wasted in trying to figure out the intentions of the RMA rather than being used in actual plan preparation.

Ratings of the clarity of six key provisions of the RMA (sections 5, 6, 7, 8, 30 and 31) are indicated in Figure 7.5. Each section is described in Appendix 1 at the end of the book. Overall, the clarity ratings were mixed. As with the responses from regional councils, section 6 (matters of national importance) was the only provision to be considered clear by a substantial majority (65 per cent) of responding district councils (ratings 4 and 5 on a 5-point scale, with 1 = not clear and 5 = very clear). It appears that staff in district councils understood what was (and was not) a matter of national importance, as established practice and legal precedents in dealing with such matters were relevant and well understood.

Sections 7 (other matters) and 31 (functions of district councils) were considered clear by just over half of responding councils. There were several reasons for these lower ratings. First, for section 7, councils had to make their own judgements in deciding how responsible they were for addressing these matters. This raised some uncertainty over what was and was not an 'other matter'. The ratings for section 31 indicated uncertainty among many councils about their roles in carrying out the mandate.

The remaining three sections (5, 8 and 30) were viewed as clear by only a minority of councils. As for regional councils, section 5 (purpose and principles of the RMA) received the lowest rating of all sections, and for much the same reasons. Given the subtle but important distinctions between sustainable development and sustainable management, together with the enormous amount of rhetoric focused on these concepts, it is understandable that district councils struggled with translating the Act's definition of sustainable management into practice.

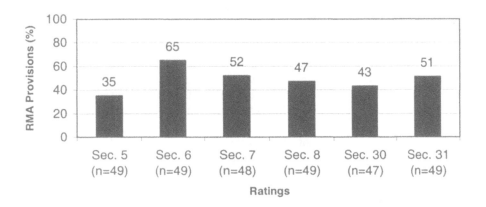

Figure 7.5   Sections 5, 6, 7, 8, 30 and 31 of the RMA rated as clear by district council staff

The rating of 47 per cent for section 8 (*Treaty of Waitangi*) is due mainly to the fact that, even where district councils did acknowledge the interests of Māori early in the plan, the general principles did not translate into substance in the various parts. Finally, the rating of 43 per cent for section 30 (functions of regional councils) reinforces the concern that district council planners had only limited interaction with regional council staff.

This was noted in interviews. One district planner stated that his council produced its plan early 'to avoid the regional council telling us what to do'. On the other hand, one regional manager commented that it was 'regional council policy to keep out of land use planning generally'. A regional council chairperson noted tensions in the relationship between the regional council and its districts. 'Council decided that the RPS would not be overly prescriptive in the spirit of 'buy-in'. The result is a positive relationship.

Usefulness of Central Government Organizations

The relationship between district councils and central government agencies is important in carrying out the RMA's intentions. Council planning initiatives can be greatly enhanced by receiving assistance from central government agencies.

The usefulness of information provided by central government agencies to district councils is quite low, as Figure 7.6 shows. The highest rated organization, the New Zealand Historic Places Trust, was considered useful by just under half of the councils (rating 4 or 5 on a 5-point scale, with 1 = not useful and 5 = very useful). The Trust has a comparatively positive relationship with some councils in providing information from their register of historic places. It targets district councils as critical users of historical information, as discussed in its strategic plan. The Trust also plays a useful role in advancing implementation of the RMA's bicultural mandate provisions. Māori support a strong relationship between the Trust and district councils over sacred sites and archaeological sites. Nevertheless, while the Trust received comparatively stronger usefulness ratings, most councils still do not consider the information it provides as useful for plan-making. This finding suggests that the Trust should seek more effectively to assist councils by making them more aware of its information base and more capable of using it in creating technically sound plans.

Four central government organizations were considered useful by a moderately low percentage of councils (29 to 42 per cent), including the Department of Survey and Land Information, Transit New Zealand, Department of Conservation (DoC), and Ministry for the Environment (MfE). These moderately low ratings can be traced to problems involving low awareness of the various information bases maintained by these organizations, the information being provided in a format that is not useful for local planning, and the inadequate technical ability of some councils to use the information. Specific consequences in some areas are still not resolved.

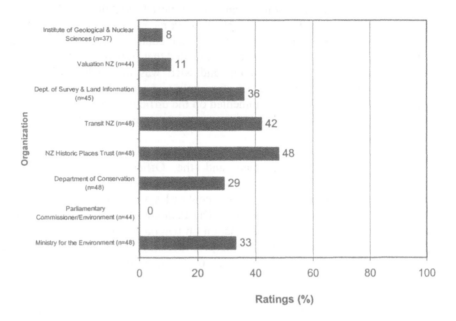

Figure 7.6 Usefulness ratings of central government organizations in providing relevant information for district plan preparation

A particular problem emerged in relation to the quality of information provided by DoC and the way it was then used by councils in preparing their plans. This drew critical comment in interviews. One respondent remarked that they 'were disappointed with DoC who withdrew from the ecological assessment [of their district] because of funding cuts'. Others, however, observed that 'individual officers were very helpful'. One councillor noted that 'relationship building with DoC occurred as a consequence of working through inaccuracies of DoC's information'.

Because of the lead role of the MfE in promoting planning under the RMA, additional attention to the reasons for the generally low ratings for this organization is necessary. First, while the MfE provided useful assistance (e.g. production of the 'orange booklet' on cascade matters, and in plan reviews), such assistance was often considered to occur too late in the process, or was too piecemeal and applied to only a limited number of councils. Second, the MfE wanted innovations to emerge from district councils, and so deliberately avoided supplying 'blueprint' guidelines. Finally, the MfE pushed the 'effects based' line of reasoning, while many district councils were comfortable with the old and well-established 'activities based' mode of planning.

These changes likely caused some disruption in councils, as well as ill feelings towards and distrust of the MfE. Not surprisingly, mixed views about the MfE's role were expressed by interviewees. Several planners and councillors noted that MfE was 'under resourced and under staffed', and thus, staff were 'very stretched'. Those who developed a close working relationship with MfE staff were much more positive about the quality of their contribution than those not able to engage their attention. One planner commented that 'it was hard to build a plan while the legal process was evolving and MfE was not really up to it'. A councillor observed that 'they didn't often help and were then critical when the plan emerged'. Some planners commented on the difficulties experienced by some councillors in understanding the language used by more junior MfE staff when visiting councils.

Finally, three organizations — the Institute of Geological and Nuclear Sciences, Valuation New Zealand and the Office of the Parliamentary Commissioner for the Environment — were considered useful by only a small percentage of responding councils (11 per cent or less). There are several possible explanations for these very low ratings. The Institute of Geological and Nuclear Sciences produces, on contract, information on natural hazards, especially seismic and volcanic hazard maps, that is likely to be very useful to many councils. However, the maps were characterized as too technical for citizens to understand and in an unsuitable format for planning purposes. The low rating for the Office of the Parliamentary Commissioner for the Environment is expected, since this organization's mission is to act as a watchdog, and not to provide technical information. Rather, it reports on the performance of councils in the environmental management system. Valuation New Zealand provides data on property values and little economic analysis was undertaken in plan preparation.

Māori Interests and District Plan Quality

Given the references in the RMA to the *Treaty of Waitangi* and the need for district councils to take Māori interests into account when developing plans, it was also important for us to assess the extent to which councils complied. We applied four of the eight plan-quality criteria to examine how well district plans addressed Māori interests: clarity of interpretation of provisions of the *Treaty of Waitangi*; issue identification; the fact base of the plan; and the plan's internal consistency.

The mean scores by the four plan-quality criteria and the total of these means are shown in Figure 7.7. (Derivation of an index score for each of the four criteria is explained in Annex 6.1.) A major finding is that plans generally received low scores, as indicated by the total mean score of 18.8 out of a possible 40, or 47 per cent, and the range was from moderate to very low. This result is not surprising, given earlier observations of a generally low understanding of, and sensitivity to, Māori by non-Māori (see, for example, Parliamentary Commissioner for the Environment, 1992; Ritchie, 1992).

The 'clarity of interpretation' of the provisions of the *Treaty of Waitangi* in the RMA received the highest score of 6.91 (out of a possible 10.0), followed by 6.0 for internal consistency. 'Clarity of issues' scored only 4.4, which indicates that in most plans issues related to local Māori concerns were either vaguely explained or not explained at all. Finally, the 'fact base' for Māori elements in plans was weak, as this criterion scored just 1.5.

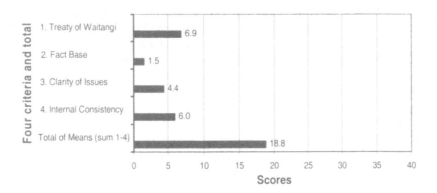

Figure 7.7   Mean score for each four criteria and for the total of means for district plans regarding Māori interests (maximum score for each criterion (1-4) is 10, and for total of means is 40)

Source: Berke et al., 2002: 126

The relatively high score for provisions of the *Treaty of Waitangi* was attributable partly to the involvement of many tribes in claims to the Waitangi Tribunal. Since Māori participation in planning has historically been denied or not sought, the focus of attention was on process and empowerment of Māori, not on substantive renewable resource issues. Iwi response to documents from councils on resource management issues (e.g. water discharges, land disturbances, natural and cultural heritage) usually focused on Treaty rights and consultation. Placing issues of consultation before matters of substance was understandable for Māori who at last saw the chance to advance their cause. The higher score can be attributed also to an awareness of the *Treaty of Waitangi* among planners. Planners had considerable experience of addressing issues relating to Māori under the *Town and Country Planning Act* (1977). Knowledge of Māori perspectives has also been an important part of the curriculum in tertiary education programmes on planning over the last two decades.

The low score for the 'clarity of issues' is likely due to the dominance of process, as planners experienced considerable difficulties in getting information from Māori on specific resource matters for the plan. There are several possible explanations for this. First, where no previous formalized relationship existed between councils and Māori, attention had to focus on process and relationship-building before substantive plan-making matters could be addressed. Second, for Māori, along with everyone else, the RMA's requirements were new and they were not able to produce the type of information the planners sought. Finally, councils did not resource planners to deal adequately with this matter.

The poor 'fact base', in particular, reveals the difficulties Māori and councils had in identifying culturally significant sites and developing management policies based on local Māori values and knowledge, as well as in obtaining scientifically relevant information on locations, conditions and threats to these sites. Māori occupation of many sites was disrupted during colonial occupation. Some groups were reluctant to identify sites publicly and prefer to use 'silent files' to ensure their protection. Many groups also had difficulty in devising standard criteria to reflect a unified view of what constitutes a 'significant site' (Parliamentary Commissioner for the Environment, 1998).

District plans, therefore, were moderately successful in explaining the national mandate for Māori interests and providing internally consistent management strategies for councils. However, as indicated by the low scores for issue identification and the fact base, plans were generally weak in translating the general intentions of the mandate into concrete actions. Strategies may be internally consistent, but they are not grounded on a strong fact base, and not targeted at resolving clearly identified issues. These findings make it clear that it is crucial to improve the understanding of factors that influence the extent to which plans support the specific requirements for Māori in the RMA. It is evident that the district councils that were most responsive to Māori interests tended to do better than other councils also in their citizen participation efforts, organizational capability, interpretation of the RMA mandate and relationships with regional planning, and had supportive contextual elements. These influencing factors are explained in detail in Berke et al. (2002).

The ranking of total scores of district plans for Māori interests is shown in Figure 7.8. Plans exhibited considerable diversity in scores, ranging from 5.0 to 35.7 (out of a possible 40). Nevertheless, district plans achieved better scores for iwi matters than they did overall.

Differences in how plans account for Māori interests are evident. The lower-scoring plans (e.g. Stratford, Lower Hutt and South Waikato) obviously did very little. Higher-scoring plans did more but were quite diverse in how they promoted Māori interests. Innovative features could be found in plans, such as for Waitakere City Council, which included an in-depth description of multiple group views that could be derived only from extensive networking and consultation with local Māori. Manukau City Council's district plan required applicants to pay a special permit (or consent) fee to financially support the review of resource consent applications that affect sites identified by Māori as culturally significant. Wellington City Council's plan included a formal 'declaration of understanding' and 'code of conduct' to guide council decisions in implementing the *Treaty of Waitangi*. The Christchurch City plan required adoption of Māori zoning provisions that specify management criteria for sites of cultural significance. It also audits its performance in meeting its statutory obligations to consult with Māori, monitors the conditions of natural resources claimed by Māori, and conducts annual surveys of Māori to track their compliance with resource conditions. These plans were not, however, representative of the mainstream plans produced under the RMA, which generally scored low in supporting the interests of Māori.

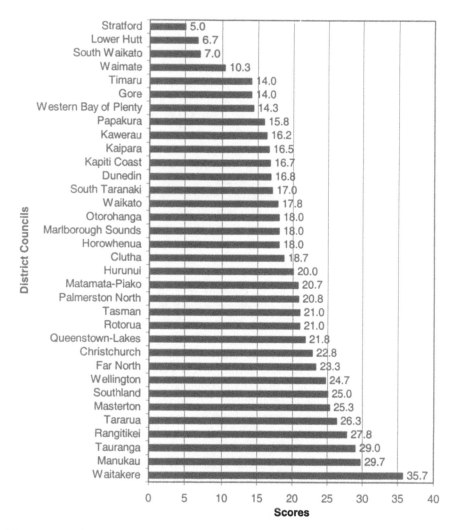

Figure 7.8 **Scores of district plans for Māori interests** (maximum possible score is 40)

Source: Berke et al., 2002: 125

Conclusion

The major conclusion to be drawn from our analysis of district plans is their mediocre quality. Results from the analysis of the eight plan-quality criteria indicated that district plans scored well for two (clarity of purpose and internal consistency). However, scores were fair to poor for five other criteria, and very poor for the fact base criterion. Overall, two-thirds of plans were well below half of the maximum possible score for plan quality.

Our research also showed that councils made slow progress in the recognition of Māori interests. Variable willingness and capability of councils to take initiatives and develop partnerships with Māori hampered progress. Some councils were not hindered by uncertainties over Treaty obligations, however, and simply got on with the job of developing relationships with Māori in their district. Other councils struggled to develop constructive working relationships, and consequently in this regard plans were weak.

Despite the generally negative results, some councils (Tauranga, Christchurch, Queenstown Lakes and Tasman) were able to produce good plans. These more commendable efforts suggest that it is possible for a district council to have the capacity and support to produce a reasonably sound plan under the RMA planning regime. Other councils (Waitakere, Manukau and Wellington) did well in promoting Māori interests.

However, our survey of district councils revealed some evidence as to why the overwhelming majority of district councils were unable to produce good plans. First, staff size of the core planning group was quite small for many councils. About a quarter of district councils employed less than one FTE, and another quarter between one and two FTEs only. Given the intensive demands of producing effects-based plans, these staffing levels were wholly inadequate.

Second, only 39 per cent of district councils used peer group reviews, yet more than 80 per cent sought legal review. It appears that, while legal reviews helped to make plans more resistant to lawsuits, they did not bolster plan quality. To strengthen plan quality, perhaps more effort should be directed towards peer reviews, especially with districts that produced good plans. Third, in addition to peer and legal reviews, district councils were quite active in using various other consultation techniques (e.g. involvement of councillors, establishment of focus groups, preparation of discussion papers). These techniques, however, did not appear to make a difference in improving the quality of district plans, possibly because consultation faded over time. This is unfortunate, since consultation is most useful in testing methods and rules prior to public notification of the district plan (see case studies in Part IV).

Fourth, councils had mixed results in understanding the intent of key provisions of the mandate. Three provisions were considered to be clear by a majority of responding councils, while the other three were not. Many councils have thus floundered in attempting to understand what they should do to produce good plans and to fulfil the intentions of the mandate.

Fifth, when asked to rate the usefulness of information provided by eight different central government organizations, the majority of district councils did not consider any of these organizations to be useful in their plan-making efforts. Possible reasons for such low ratings included low awareness by councils of the existence of various sources of information, lack of technical skills among council planning staff in using the information, and information that was not in a usable format for local planning purposes.

In sum, the findings on the planning efforts by district councils (and in Chapter 6 regional councils) begin to reveal some reasons that might explain why plans have not been well developed under the RMA. Small sample evaluations based on the natural-hazard components of five district plans (Berke, 1995) and the bicultural components of ten district plans (Nuttall and Ritchie, 1995) show initial corroboration for our findings. These earlier studies, however, focus only on particular dimensions (i.e. hazards and iwi) of plan quality, and provide evidence from only small samples in explaining why local governments produce good and poor plans (Becker and Johnston, 2000).

What is lacking is more rigorous statistical modelling using data from larger samples of councils. The intent would be to test how organizational capacity, mandate clarity and inter-organizational relations independently influence plan quality. Such analysis would offer more insight into identifying which factors provide the most powerful explanations of variation in plan quality. The results would then provide policy-makers with an understanding of which factors need attention and have the most influence on local councils in producing better plans.

Chapter 8

Influencing Factors: Linking Mandates, Councils, Capability and Quality

We now make a comparative analysis of the quality of policies and plans produced by regional and district councils. The main aim is to identify the differences in plan quality between regional policy statements and district plans, and the most influential factors on the preparation of plans and their resulting quality, in order to improve the quality of the next generation of policies and plans.

Comparative Analysis of the Quality of Regional Policy Statements and District Plans

Several key findings are derived from the comparative analysis of plan quality. The first is that the overall quality of regional policy statements is not significantly different from that of district plans (t-test = 0.68, $p > 0.1$), as the sum of means of all criteria is 37.22 for regional policy statements and 35.37 for district plans. However, there is some variation between individual criteria. Comparisons of mean scores for plan quality between regional policy statements and district plans are shown in Figure 8.1. Mean scores for three criteria (interpretation of the mandate, fact base and identification of issues) were significantly higher ($p < 0.1$) for regional policy statements, while means for two criteria (organization/ presentation and monitoring) were significantly higher ($p < 0.1$) for district plans. (For details see Annex 8.1.) No significant differences between the two groups were detected for the three remaining criteria (clarity of purpose, integration with other plans/policy statements and internal consistency).

The differences in scores for criteria can be explained by differences in the functions of regional and district councils, and in the two types of planning instrument (regional policy statements and district plans). Regional councils have a mandate to manage biophysical resources. They are amalgamations of agencies that were primarily staffed by scientists and engineers with a history of managing water and soil resources over several decades. Thus, an effects-oriented fact base can be readily generated for preparing regional policy statements. Further, in the early stages of plan preparation, staff from the various regional councils met several times to discuss templates for preparing regional policy statements. Consequently, regions were in a stronger position than districts to identify issues, and to provide clarity in interpreting substantive provisions of the RMA mandate for their purposes.

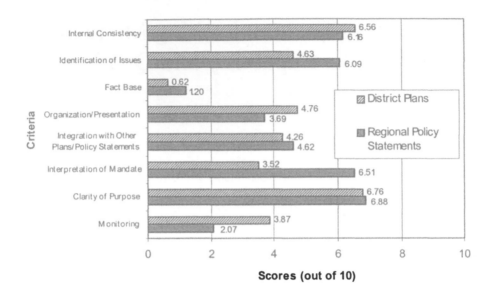

Figure 8.1   Comparison of scores for regional policy statements and district plans by plan quality criteria (maximum possible score is 10)

Source: Berke et al., 1999: 653

District councils were built on a land use planning culture of several decades. The wider range of issues to be dealt with in district plans (e.g. amenity, urban development and subdivision) did not always lend themselves as readily to an effects-based approach, and district councils did not have the research orientation of regional councils. Thus, it was not as easy for district councils to assemble a fact base. Further, district plans deal with the effects of activities at a more specific level of implementation. Inevitably, they are more complex than regional policy statements. There were also more district councils with considerable variation in local circumstances. As a result, it was not as easy to develop consistent views on identification of issues and interpretation of key RMA provisions, given the required shift in thinking from prescriptive to effects-based approaches. Finally, district council staff, mostly planners, had more experience at plan writing than regional council staff. Regional councils had little history in plan preparation. Consequently, district council staff had more skills in organizing and presenting information in formats that were more readable and comprehensible for lay and professional people. District staff may also have had a stronger understanding of how to arrange monitoring of plan performance in terms of linking indicators to plan objectives and policies, and in assigning organizational responsibility to monitoring specific indicators.

As discussed in Chapters 6 (regional policy statements) and 7 (district plans), another major finding was that both groups of documents received somewhat low scores. Recall that we calculated the scores for each criterion as having a possible range from 0, indicating no mention of a criterion in a plan, to 10, indicating that a plan incorporates all the dimensions of a criterion we scored. The 'clarity of purpose' criterion received the highest score for both groups of plans, but was only 6.88 for regional policy statements and 6.76 for district plans out of a possible score of 10. Further, the overall total of the means of the eight criteria for both regional policy statements and district plans was less than half of the possible total score.

In sum, while there was some variation in quality of regional policy statements and district plans for individual plan quality criteria, there was no major difference in the composite mean scores between both groups. However, as noted in Chapters 6 and 7 the overall scores for both groups were lacklustre. These findings indicate a need to improve the understanding of factors that influence plan quality, especially those over which policy-makers and planners have control (e.g. mandate clarity, number of council staff devoted to planning and central government technical assistance strategies).

Findings on Factors that Affect the Quality of Regional Policy Statements and District Plans

The results for factors that influence plan quality were derived from a series of regression analyses. Regression analysis is a statistical modelling technique that allows for determining the strength of the influence of one factor on plan quality while the influence of other factors is held constant. For example, the independent impact of district council staff capability to plan on district plan quality can be determined while holding constant all other factors predicted to influence plan quality, such as clarity of the RMA and median home value (to indicate resources available). Details about measurement construction and sources of data for the factors that influence plan quality are indicated in Annex 8.2.

Figure 1.3 in Chapter 1 illustrates the theoretical framework linking mandate, organizational capability, policies and plans, and socio-economic characteristics of councils. Figure 8.2 shows the results from testing the strengths of these hypothesized links derived from regression modelling. (Additional information about regression results is provided in Annex 8.2 and in note 1 in Annex 8.3.) The figure shows the strength of paths linking the RMA planning mandate and implementing actions of central government organizations with regional council capability to plan and regional policy statements, and with district council capability and district plans. In addition, the figure shows how the paths from socio-economic factors of influencing councils (regional and district population size and median home value) are linked to regional council capability and regional policy statements, and to district council capability and district plans.

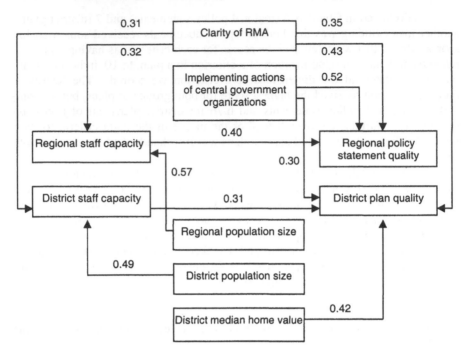

Figure 8.2   Factors affecting regional policy statements and district plans

Source: Berke et al., 1999: 655

The larger the coefficient assigned to each path in Figure 8.2, the stronger the influence of a given factor on a subsequent factor (Annex 8.3, note 2). Each coefficient indicates the number of unit changes in plan quality associated with a one unit change in the factor associated with the coefficient holding all other coefficients constant (Annex 8.3, notes 3 and 4). For example, a one unit increase in the clarity of the RMA will result in a 0.43 unit change in the quality of the regional policy statements. Coefficients range from 0 to 1.0. Only paths with coefficients that are comparatively strong are displayed. Thus, coefficients of less than 0.25 were considered weak and are not displayed in the figure. Other planning studies have used similar values for rating the strength of standardized regression coefficients (Burby and May et al., 1997; Rohe and Bates, 1984) (Annex 8.3, note 5).

To determine the total impact of each factor, the direct and indirect impacts were summed. To derive the indirect impacts, we computed the influence of coefficients through different paths portrayed in Figure 8.2 (e.g. clarity of the RMA to the quality of regional policy statements through regional staff capability). This was done by multiplying the separate coefficients (i.e. using the above example, 0.32 × 0.40 = 0.13). The total impact for each factor was obtained by adding the direct and indirect impacts (0.43 + 0.13 = 0.56). The following discussion presents two sets of findings: one examines factors that influence the quality of regional policy statements; the second, factors that influence the quality of district plans.

Assessing Influences on the Quality of Regional Policy Statements

When mandate provisions were clear they had, as indicated in Figure 8.2, a positive indirect influence on the quality of regional policy statements through regional staff capability. If key provisions in the national mandate were clearly understood, then staff capability was likely to expand. Councils that were able to clearly interpret the mandate were more likely to agree with its intentions, and therefore more likely to assign more staff to formulate policies and plans that comply with it. Increased staff capability, in turn, allowed councils to give more time, attention and expertise to policy and plan preparation.

Furthermore, clearly understood mandate provisions directly contributed to the likelihood of producing high-quality regional policy statements. This finding was consistent with our expectation that, when councils were able to clearly interpret the key mandate provisions, they were better able to infuse the intentions of the mandate into regional policy statements. The total impact of the clarity of mandate factor on the quality of regional policy statements (indirect plus direct impacts) was greater than any other factor ($[0.32 \times 0.40] + 0.43 = 0.56$).

The linkage in Figure 8.2 also shows that the usefulness of information provided by central government agencies as seen by council staff had a direct and strong positive influence on the quality of their regional policy statements. As expected, central government agencies were considered to play a crucial role in helping regional councils produce good regional policy statements. Primary information-transfer actions taken by central government agencies included: participating in meetings to provide advice to council personnel; peer review of draft documents; providing legal advice when asked; and making submissions on notified regional policy statements. Thus, when good information was provided, it was infused into regional policy statements and improved their quality.

Because regional staff capability was positively related to the quality of regional policy statements, the provision of information provided through the implementing actions of central government organizations in building regional staff capability was important. We theorized that, as information improved, certainty in identifying issues and in setting the best policy alternatives would improve and thus councils would be more supportive of planning and willing to expand planning staff. However, our findings were unexpected, as the impact of information provision by central government on regional policy statements was weak and inconsequential.

We considered population size as another factor to serve as a proxy for resource availability of regional councils to build staff capability to prepare high quality plans. We thus expected that population size would have a direct and positive influence on regional staff capability. Figure 8.2 indicates that the overall resource context, as indicated by the proxy population size, had a positive indirect influence on the quality of regional policy statements through regional council staff capability $(0.57 \times 0.40 = 0.23)$. This factor had a stronger influence on staff expansion than any other factor.

Looking at the case for iwi, scores for recognition of iwi interests were better than overall scores for regional policy statements. There was a positive relationship with the number of planners on staff, clarity of the mandate provisions for the Treaty of Waitangi, and population size.

In sum, these findings provide insights into how to improve the general lacklustre quality of regional policy statements. First, the clarity of key mandate provisions has an important indirect and direct influence on the quality of regional policy statements. This suggests that emphasis should be placed on helping planners to better understand the intent of the national mandate and how the key provisions should be integrated into regional policy statements.

Second, information provision by central government organizations had a direct effect on the quality of regional policy statements. Emphasis should thus be placed on enhancing the ability of central government agencies to provide useful information for plan development, such as the matters of importance identified in sections 6-8 of the RMA.

Third, the resource context factor (i.e. regional population size) has an indirect positive impact on the quality of regional policy statements through regional staff capability. As a result, smaller regions are more likely to lack staff capability, and thus lack the ability to produce high-quality regional policy statements. Hence, special attention should be given to smaller regions in building their staff capability to plan.

Assessing Influences on the Quality of District Council Plans

As expected, Figure 8.2 shows that the clarity of key provisions in the mandate had a direct positive impact on the quality of district plans. When key provisions of the mandate were clear, there was more certainty about how to infuse the intentions of the mandate in plans. Mandate clarity also had a positive indirect impact on the quality of district plans through district staff capability. When district councils understood the intentions of the mandate, they were more likely to be committed to assigning staff for plan preparation. Similarly to the findings on mandate clarity and regional policy statements, this factor had the most powerful influence on the quality of district plans $([0.31 \times 0.31] + 0.35 = 0.45)$.

The results only partially supported the hypothesized importance of information provision for the quality of district plans. On the one hand, Figure 8.2 indicates that this factor had a direct positive influence on the quality of district plans. Most councils did not have adequate information bases for preparing high-quality district plans to support issue identification and policy and rule selection. Thus central government organizations can make a difference when they provide information useful for district planning purposes. On the other hand, the expected positive influence of information provided by central government on district council staff capability was not supported. The coefficient for this relationship was small and had an inconsequential impact (see Annex 8.2). This factor was expected to positively influence district plan quality through district staff capability because our model hypothesized that central government builds local capability.

Unexpectedly, in terms of the hypothesis of co-operation and partnership between district and regional councils, we found that the quality of regional policy statements and regional staff capability had no association with district staff capability and district plan quality (see Annex 8.2). While we expected that regional council staff would have a strong influence on district staff capability and district plan quality (e.g. through provision of technical information, reviews of draft plan for consistency with regional objectives and policies, guidance in interpreting mandate provisions, and so on), in practice their activities had not extended that far.

Interviews with district council staff planners revealed several reasons for this disconnection. One was the tension in political control and territoriality between the two types of council, primarily involving issues to do with the hierarchy of responsibility. In some cases, land use policy directions at the district level had not been well supported, even opposed, by regional councils. Other reasons included perceived and actual differences in mandated functions, and lack of resources (financial, technical and personnel) for co-operative activities.

Moreover, policy direction in the RMA did not foster close ties between regional policy statements and district council plans. Instead of requiring consistency between both documents, the Act simply requires that district plans be not inconsistent with regional policy statements. This did not foster close inter-council collaboration, as district councils were not giving much attention to regional policy statements in district plan preparation. Apart from the district plans that were prepared ahead of or at the same time as the regional policy statement, many policy statements did not provide sufficient direction.

District median home value (as a proxy for wealth and a source of funds to support planning through the property rating base) is shown in Figure 8.2 to have, an expected, direct and positive influence on district plan quality. That is, greater wealth makes available greater resources for developing the plan. This variable may also reflect the political support for plans. Wealthy districts that had high median home values may have been more likely to support high-quality plans that carefully manage and control the effects of land development. Compared with wealthy districts, poor districts tended to be less politically organized, had less access to formal authority structures, and thus had less ability to create plans that prevent adverse effects generated by powerful development interests (Logan and Motoloch, 1987).

We expected the overall resource context, as indicated by population size, to be positively related to the quality of district plans. Our findings partially supported this expectation. As shown in Figure 8.2, population size had a positive indirect influence on the quality of district plans through district council staff capability ($0.49 \times 0.31 = 0.15$). This factor also had the strongest direct influence on district staff expansion compared with other factors.

Conclusions and Implications for Improvement

Findings from our analysis of the quality of regional policy statements and district plans, and of the factors that affect their quality, begin to explain the generally unimpressive quality scores, as discussed previously. We theorized that the planning mandate and implementing actions of central government organizations could directly and indirectly improve district council capability and district plans through improvements in regional council capability and regional policy statements. However, we found a major gap in the intergovernmental hierarchy — regional council staff capability and the quality of regional policy statements had no relationship with district capability and district plan quality. Regional and district planning were thus operating independently, with weak inter-organizational co-ordination and variable policy directions. This suggests that, in spite of the partnership roles of regional and district councils, there was little, if any, integration and co-ordination between them. Lack of resources, turf protection and conflict generated by uncertainty in the roles at each level of government were likely reasons for this disconnection.

Policy-makers should thus work towards closing this gap. A priority would be to revise the RMA to include a vertical consistency requirement whereby district plans must be consistent with regional policy statements. The current 'not inconsistent' requirement is too vague and lacks clear policy direction. A follow-up requirement would be for the lead central government agency, the Ministry for the Environment (MfE), to review and certify regional policy statements and district plans to ensure that they meet a basic quality threshold and are vertically consistent. Central government should also explore how common expectations between regions and districts could be improved, whereby districts are allowed to create plans that fit their own needs while at the same time meeting regional needs.

The lacklustre quality of regional policy statements and district plans could also be improved through various actions by central government. First, the clarity of key provisions in the mandate was the most important factor in positively influencing the quality of these documents. This factor had both a direct influence, as seen with iwi matters, and an indirect influence through regional and district staff capability. However, results from the postal questionnaires of regional and district councils indicated that only about a quarter to half of regional councils and a third to half of district councils rated five out of six key provisions in the RMA to be clear. Given the importance of mandate clarity, how could central government and its agencies enhance the clarity and therefore understanding of the RMA? Previous efforts designed to explain how provisions of the Act could be translated into practice have focused on meetings and the distribution of technical guidelines by the MfE, but were considered by practitioners to be 'too little too late'. That so many practitioners had so much difficulty in interpreting key provisions of the RMA suggests that redrafting the legislation, and improving initiatives for education and dissemination of information, should be priorities.

Second, while information provided by central government agencies had no influence on regional and district staff capability, it did have a direct positive impact on the quality of district plans and regional policy statements. However, results from the postal questionnaire indicated that central government agencies were generally not considered to be useful by staff in regional and district councils. For example, only about 20 per cent of regional councils considered the two leading central government agencies, the MfE and the Department of Conservation (DoC), to be useful (see Chapter 6), while just under 50 per cent of district councils viewed the MfE as useful and just over 10 per cent viewed DoC as useful (see Chapter 7). How can central government agencies provide more useful information to improve the quality of regional policy statements and district plans? One step is to enhance the role of these agencies in this regard. Information generated by central government agencies may not be provided in a form useful for preparing plans. The agencies need to work closely with regional and district staff to improve their understanding of the sub-national policy and data requirements needed to produce better plans. Also, in some cases, regions and districts may have received useful information, but were not able to effectively use it. Thus, attention should be given to building capability to use technical information more effectively during plan preparation.

Third, staff capability had an important positive direct influence on the quality of regional policy statements and district plans, but many councils did not give sufficient resources to support staff. Postal survey results showed that many district councils placed a bare minimum of staff on the core planning group, with about 50 per cent of them having one or less FTE staff member (see Chapter 7). Only 6 per cent of regional councils employed one or less FTE, but another 44 per cent employed just two or less (see Chapter 6). How can staff capability be expanded? The obvious answer is that the government should change the RMA from an unfunded planning mandate to one that provides resources for expansion of regional and district staff. In addition, central government could enable sub-national governments to raise revenue by other means. Staff capability has not been commensurate with the demands for plan preparation and implementation under the RMA.

Two other factors were important for staff capability and plan quality. Population size (i.e. proxy for available council resources) was a major determinant of regional and district council staff size, and median home value (i.e. proxy for wealth) was an important predictor of district plan quality. Consequently, smaller regional and district councils were less able to support a sufficient number of staff, and poorer districts were less able to produce high-quality plans. Thus, more support is needed for smaller regional and district councils to build up their staffing, and for less wealthy councils to improve district plan quality. How can staffing of less advantaged councils be improved? Possibilities for lifting support include: central government might invest in the professional development of their staff and councillors; other councils could assist smaller and less well-off councils with technical assistance and advice; small councils could be amalgamated; and councils could pool resources so that similar activities (e.g. monitoring and research) were not being inefficiently replicated across adjoining councils.

This analysis of the intergovernmental planning system of the RMA has revealed the key factors that must be improved if the obstacles that constrained plan quality are to be overcome. Results from other studies of planning mandates suggest that our findings should not be too surprising. An examination of British planning by Cullingworth (1994) found that implementation of local plans suffered primarily from lack of clear goals and policy guidelines, which had seriously undermined national planning legislation. A study of American state planning mandates by DeGrove (1992) found major weaknesses that constrained effective implementation of inter-governmental planning programmes. The weaknesses included lack of staff and funding resources for planning, poor vertical co-ordination among state, regional and district governments, and vague and incomplete mandate goals and policies to guide local planning efforts.

Finally, this assessment of the RMA as a planning mandate should be extended to include a longitudinal research design through case studies. In-depth evaluation of the links between mandates, organizational capability, the quality of plans, implementation of plans and their environmental outcomes would reveal the type and effect of key actions taken by different levels of government during the process of plan preparation and implementation. This research should examine how various activities (e.g. education campaigns, consensus building, information dissemination, public consultation and enforcement activities) influence the quality of plans and their implementation.[1]

---

[1] Research on the quality of plan implementation under the RMA was undertaken by the authors between 1998-2002 (Day et al., 2003).

Annex 8.1

Details on statistical findings associated with comparison of means between regional policy statements and district plans

Criteria	Regional Policy Statements mean (std. dev.)	District Plans mean (std. dev.)	t-tests	prob.*
clarity of purpose	6.88 (3.10)	6.76 (2.43)	0.13	ns
interpretation of the mandate	6.51 (1.69)	3.62 (1.21)	6.36	p<.01
integration with other plans	4.62 (1.80)	4.26 (2.02)	0.63	ns
organization/presentation	3.69 (1.14)	4.76 (2.02)	2.96	p<.01
fact base	1.20 (1.02)	0.62 (0.88)	2.05	p<.1
identification of issues	6.09 (3.29)	4.63 (2.69)	1.55	p<.1
internal consistency	6.16 (1.44)	6.56 (1.87)	0.83	ns
monitoring	2.07 (2.43)	3.87 (3.49)	2.11	p<.05
overall total of means	**37.22 (7.87)**	**35.37 (9.32)**	**0.68**	**ns**

* ns = not significant

Source: adapted from Berke et al., 1999: 653

Annex 8.2

Factors affecting organizational capacity

Regional Policy Statements

	Standardized regression coefficients[a]	
	regional staff capability	regional policy statement quality
Mandate: clarity of RMA	0.32	0.43
Central government agencies: usefulness	-0.09	0.52
Staff capacity: in regional council	-	0.40
Context[b]		
regional population size	0.57	0.01
regional median home value	0.16	0.01
n	15	15
Adjusted R^2	0.56	0.42

[a] Significance levels are not included for regressions on regional staff and quality of regional policy statements because the sample is the population of all regional councils.
[b] The natural log of the standardized scores for regional population size and regional median home values were used to reduce skewness.

District Plans

	Standardized regression coefficients[a]	
	district staff capability	district plan quality
Mandate: clarity of RMA	0.31*	0.35*
Central government agencies: usefulness	-0.02	0.30*
Staff capacity:		
regional staff capacity	0.17	-0.01
district staff capacity	-	0.31*
Regional policy statement: quality	0.18	0.03
Context[b]		
district population size	0.49**	0.17
district median home value	0.12	0.42**
n	29	29
Adjusted R^2	0.19	0.44
F ratio	2.10	4.13
Significance	0.09	0.01

Note t-test significance levels: * $p < 0.1$; ** $p < 0.05$.
[a] The natural log of the standardized scores for district population size and median home values were used to reduce skewness

Source: adapted from Berke et al., 1999: 656

Annex 8.3

Notes on statistical analyses

Note 1:

The overall quality scores for regional policy statements and district plans were used, rather than individual indices that represent each plan quality criteria (see Chapters 6 and 7 for scores of regional policy statements and district plans respectively). Statistically, the eight criteria by which the quality of plans was measured were determined to be internally consistent — i.e. we found that plans that score high (or low) on one criteria also tend to score high (or low) on others.

We used the Cronbach's alpha statistic to test for internal consistency. This analysis revealed an alpha score of 0.74 for regional policy statements and 0.76 for districts plans. Alpha scores of 0.60 are generally considered to achieve a reasonably high level of internal consistency. Thus, we were confident that a summed overall score of all principles was a reliable measure of the quality of regional policy statements and district plans.

Note 2:

We did not include regression results of the consultation factor (an index that represents the number of actions taken by councils to enhance consultation, as discussed in Chapter 6). This factor scored well below the 0.25 cut-off for standardized coefficients for regional council data, and was insignificant at the 0.1 level for the districts, and was thus not considered an important factor in explaining the quality of regional policy statements and district plans.

Note 3:

Since standardized coefficients are used, two paths leading to the same factor can be compared to indicate which is stronger. For example, the coefficient of 0.44 associated with the arrow from district median home value to district plan quality is nearly one and a half times stronger in effecting plan quality than the 0.30 coefficient associated with the arrow from clarity of the RMA to the quality of district plans (see also Annex 8.2).

Note 4:

Since we use standardized regression coefficients, the unit of analysis referred to here is the standard deviation.

Note 5:

While interpreting coefficients from the population of regions only requires examination of the value of the coefficient, we used both the value and significance of coefficients for the sample of district councils, since significance tests are most appropriate for data derived from a sample of a population. For the sample of districts, paths with coefficients whose values were less than 0.25 also had t-values that were statistically insignificant at the 0.1 level.

Source: adapted from Berke et al., 1999: 663-64

PART 4
LOCAL CASE STUDIES

Chapter 9

Far North District: Resisting Innovation

The Far North District is the most northern in New Zealand. Though predominantly rural, it is one of the country's fastest-growing districts. Within its area lies a rich natural and cultural heritage, being the cradle of both Māori and European cultures in New Zealand and containing one of its most diverse and unique biosystems. This makes the district an important destination for tourists and lifestylers. Some areas along its eastern coastline display considerable economic wealth, but overall the district contains a disproportionate share of the country's poor and unemployed, most of whom are Māori. It also has the highest number of threatened indigenous species of any district in New Zealand.

This mix of factors poses a serious challenge for plan development. Although our assessment scored the *Proposed Far North District Plan* (October 1996) highly in relation to other district plans, intense dissatisfaction among some landowners over attempts to protect significant natural and cultural areas pressured a fractious council to withdraw it and start again. In this chapter, we examine why this occurred, with particular reference to significant natural areas (SNAs).

Geographical Background

The Far North District has an area of 7505 square kilometres (Figure 9.1), the tenth largest in New Zealand. In 1996, nearly half the population lived in eighteen urban areas, only seven of which contained more than 1000 people; the largest were Kaitaia (5280), Kerikeri (4161), and Kaikohe (4107). The total resident population in 1996 was almost 53,000, an increase of 11.5 per cent in five years, whereas for New Zealand the increase was 7.2 per cent. About 45.5 per cent of the total population was Māori, whereas for New Zealand the figure was 15.1 per cent (Far North District Council, 1997). The Far North stands out as one of several areas of deprivation in New Zealand. Almost all but the south-eastern part of the district from Kerikeri to Kawakawa falls within the top three deciles of deprivation (Crampton et al., 2000).

The district has some of the most distinctive ecological areas in New Zealand, with exceptional biological diversity offering habitat to a wide range of indigenous fauna. Some species are endemic to the district and are at risk from continued land development and invasion by exotic weeds and pests (Department of Conservation, 1996a). Maps comparing pre-European vegetation with current indigenous vegetation for the south-east corner of the Far North District, where development pressures are greatest, are shown in Figure 9.2.

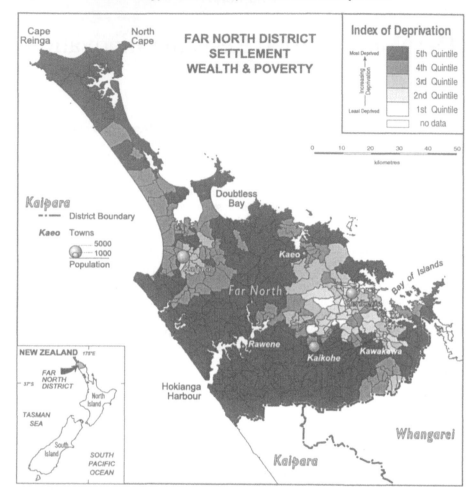

Figure 9.1   Far North District: settlements, wealth and poverty

Source: adapted from Crampton et al., 2000

Since the start of European occupation in the early nineteenth century, the district has lost to agriculture and forestry 70 per cent of its indigenous forests, 95 per cent of freshwater wetlands, 80 per cent of dune-lands and 90 per cent of podzol kauri gumfields. Never very extensive in recent centuries, the swamp forests and swamp shrublands are now very small scattered patches, with only a few flecks of the former remaining along the southern margins of the district (Figure 9.2). The continual loss of habitat threatens many species, including kiwi. Some species have become extinct.

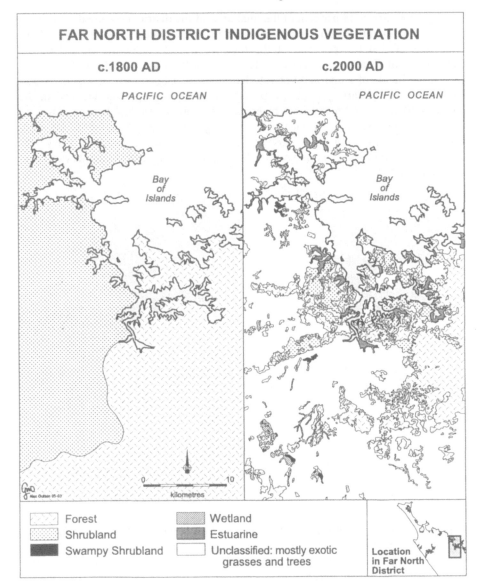

FAR NORTH DISTRICT INDIGENOUS VEGETATION

| c.1800 AD | c.2000 AD |

Figure 9.2   Indigenous vegetation changes in the Bay of Islands around 1800 and 2000

Source: 1800 extract from Descriptive Atlas of New Zealand, Map 14, 1959; 2000 extract from Department of Conservation, Northern Conservancy, SNA Map, 1997, with permission

By 1996, about 38 per cent of the total area of the district contained significant natural vegetation that also harboured important and threatened fauna (Figure 9.3). Over half of that (21 per cent of the total area) was under the control of the Department of Conservation (DoC). The more scattered remainder (17 per cent) was in the hands of some 2200 private landowners, a lot of which was on Māori land. Returns for farm products, especially beef and sheep, had declined in the previous 30 years, and many farmers, especially mortgagees, had either subdivided where demand allowed, or cleared remaining indigenous vegetation for farm expansion or exotic forests. This was easier to do when exotic forestry was undergoing rapid expansion in Northland in the 1970s and 1980s than its quiescent state in the 1990s.

Figure 9.3   A significant natural area (SNA) on Karikari Peninsula, Far North (This private property at Paeroa-Knuckle Point was recently purchased by the Crown through the Nature Heritage Fund and made a scenic reserve)

Source: Department of Conservation, Northern Conservancy

These sorts of pressures on the natural environment, and SNAs in particular, in areas under its jurisdiction posed a serious challenge to the council, which was required (within limits) by section 6 of the RMA to protect it from the adverse environmental effects of development (Waitakere Ranges Protection Soc. Inc v Waitakere CC & Eggink A89/2000).

Institutional and Organizational Arrangements

The Far North District is one of three included within the Northland Regional Council, which has its offices in Whangarei. The Northern Conservancy office of DoC is also in Whangarei, while the office of the Northern Region of the Ministry for the Environment (MfE) is in Auckland further south. The Far North District Council (FNDC or council) deals with seven iwi (which manage about 16 per cent of the total land in the district), mainly via the Te Kotahitanga o Te Tai Tokerau Resource Management Committee forum.

A review of organizational arrangements indicated that FNDC's structure and resources for preparing a good plan were satisfactory — given its limited rating (property tax) base. Early on, amalgamation issues and ongoing arguments over making the council a unitary authority, together with organizational and managerial changes, meant that councillors and staff did little on the district plan before 1993. Until then, the plan had one staff member and three councillors developing it in a working group that reported infrequently to the council. Limited in-house resources led the council to employ consultants to help develop information for the plan.

Restructuring in 1993 gave plan-making a measure of importance on the council agenda, facilitated by the creation of a new Projects Department aimed at intensifying efforts to develop the new plan. Thus, the council organization had been relatively stable for three years before public notification of its plan in October 1996, and the plan-making effort overseen by a seven-person Policy and Planning Committee working with two staff and consultants.

Annual funding for plan preparation doubled to $322,000 in 1993/94, when the Projects Department was established, and climbed every year to reach almost $1 million in 1997/98. The total cost of the original district plan up to the hoped-for completion date in 2002 (when it would become legally operational) was $4.45 million, of which $1.8 million had been spent by 1996, when it was notified. This level of funding meant that the council could purchase capable personnel for carrying out the research, policy analysis and consultation necessary for a good plan.

In 1993, the new Projects Manager saw the need to consult the public and interest groups more effectively than previously, and later saw the need for filling gaps in information by employing more consultants. At the start, two staff were given the primary role of preparing the district plan. By the final year, three consultants were employed to help with aspects of the plan, such as iwi and landscape values, and three staff were involved in plan-writing, together with a planning consultant and a solicitor.

Chronology of Plan-Making Process

The FNDC followed the rational-adaptive approach to preparing its plan, aiming at an effects-based plan based on research and consultation. This is evident from the listing of main activities in Tables 9.1 and 9.2. There was, however, a tendency to invert the needs so that consultation was stronger at the start than towards notification, and research was stronger towards notification than at the start. The problems this caused will become evident later in the chapter.

Table 9.1 Main plan preparation actions to first notification, 1991-96

1991	Review and synthesis of existing information.
1991–92	Preparation of four discussion documents (A, B, C, D) on resource management issues of the district, including those of concern to Māori, and call for submissions on documents (section 75(a)).
1992–93	Consultation with general public and meetings with community and special interest groups over the resource management issues. Preliminary consultation with tangata whenua (clause 3, 1st Schedule).
1993–94	Formal consultation with seven iwi through Te Kotahitanga o Te Tai Tokerau (clause 3, 1st Schedule).
1994	Preparation of discussion document on draft objectives and policies and call for submissions (section 75(b) & (c)) (document published October 1994).
1994–95	Consultation with general public and meetings with community and special interest groups on draft objectives and policies. Consultation on documents, through meetings and correspondence, with nine statutory organizations (clause 3, 1st Schedule). Identification of needs for further technical information (sections 6(b), (c) & (e); 31(d) and 7(e)) or planning studies (10 reports published between June 1995 and April 1996).
1995–96	Identification and assessment of alternative methods for implementing policies (section 32). Preparation of draft objectives, policies, outcomes and methods and call for submissions (section 75(b), (c), (d), (e) & (g)) (published Feb 1996).
1996	Consultation with tangata whenua in 20 workshops (January–April) and four formal hui-a-iwi (May). Consultation with general public and meetings with community and special interest groups (to March). Development of plan provisions (January to August). Evaluation of current zoning and determination of new zoning. Data entry to GIS and creation of plan maps. Decision not to produce draft plan for public comment (June 1996).
1997	Completion of the *Proposed District Plan* (notified 31 October 1996).

Research and Policy Analysis

Council employed experts in various fields to carry out research on issue identification and specific topics. They also helped develop policy and methods for the cascade of elements within the plan, as indicated in the series of discussion documents, papers and reports listed in Table 9.2.

The research and policy analysis can be seen in four phases. Early in the process of plan preparation, consultants were employed to extract from existing documents what was known about various issues in the district. This information was organized around coherent themes, such as primary industries, economic and social conditions and services, the natural environment, and Māori interests. They emerged as discussion documents between November 1991 and August 1992.

Following discussion of the issues with a range of interested groups and communities, council staff fashioned sets of objectives and policies aimed at dealing with the issues. This resulted in a discussion paper for public consideration in October 1994. From this process, it became clear to the council that further information was needed on a range of key issues, and several consultants were employed to carry out the research. Thus, over the next 18 months to June 1996, ten research papers were published for public review, dealing with both the natural environment and social concerns (Table 9.2). During this third phase of research and policy analysis, methods were being developed to implement the emerging policies for achieving desired environmental outcomes. These methods and outcomes were included in a discussion paper entitled *Objectives, Policies, Outcomes and Methods*. It was published in January 1996 amidst a bevy of research documents and nine months ahead of publicly notifying the plan.

Consultation and Consensus-building Actions

Staff in the council were well aware of the need to consult with central, regional and local government agencies that might be affected by the district plan, as well as tangata whenua and the public at large (Figure 9.4). Consultation centred on the series of discussion documents, discussion papers, and commissioned reports released for public comment (Table 9.2). It involved four rounds of consultation in the five years to plan notification, during which time the council received over 200 informal written submissions from various stakeholder groups and individuals.

By the time the plan was publicly notified, council records showed that staff had had 127 meetings with nearly 100 different groups, including three meetings with the most vociferous critics of the notified plan — Federated Farmers.

The first two rounds focused on objectives and policies outlined in discussion documents, while the last two rounds focused on papers and commissioned reports, and involved key interest groups (Table 9.2). In the third round (1994–95), over 50 meetings were held on objectives and policies. These led to 80 written submissions, which helped the council to refine methods for achieving them. The fourth round (1996) focused on the second discussion paper (covering objectives, policies, outcomes and methods) and included some examples of rules, as required by section 75 of the RMA (Table 9.2).

However, because councillors wanted the plan notified by October 1996, there could be only limited consultation on the methods to be included in it. Nevertheless, eighteen meetings were held with various interest groups. Since the council had a statutory duty under the RMA to consult iwi, rather more meetings

(24) were held with Māori. In 1992, the council employed Māori consultants to develop a detailed iwi consultative process and they also helped the council follow it through. The Māori consultants also helped develop the tangata whenua chapter for the plan.

Table 9.2 Key documents prepared by council for the district plan of 1996

Issues Discussion Documents 1991–1992
Document A (21 November 1991)
Issue 1: rural development and agriculture
Issue 2: exotic forestry
Issue 3: tourism and tourist facilities
Issue 4: transportation networks
Document B (January 1992)
Issue 1: economy and employment
Issue 2: settlement and subdivision
Issue 3: water and waste services
Issue 4: refuse
Issue 5: communication and energy networks
Issue 6: noise
Issue 7: hazardous activities
Document C (10 August 1992)
Issue 1: natural environment (i.e. landscapes and habitats of flora and fauna, water and coastal resources, mineral extraction, and air and sunlight)
Issue 2: cultural heritage
Issue 3: recreation and reserves
Document D (10 August 1992)
Māori issues

Policy and Methods Discussion Papers 1994–1996
Objectives and Policies (October 1994)
Objectives and Policies, Outcomes and Methods (January 1996)

Special Topics Research Reports 1995–1996
1. *Evaluation of notable trees* (1995)
2. *Landscape study* (1995)
3. *Urban areas study* (1995)
4. *Reserve study* (July 1995)
5. *Noise assessment* (July 1995)
6. *Heritage study* (September 1995)
7. *Significant natural areas surveys* (December 1995)
8. *Productive land study* (January 1996)
9. *Financial contributions and analysis* (April 1996)
10. *Kerikeri study* (June 1996)

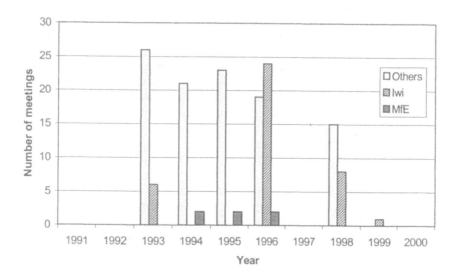

Figure 9.4   Meetings held between FNDC and various groups, including iwi, general public, interest groups and MfE (Note: from April 1998 onwards, a multi-stakeholder committee re-wrote the plan in regular meetings open to the public. See Annex 9.1)

Source: compiled from Far North District Council records

Throughout the five-year period to plan notification, the council held consultative meetings and corresponded with nine statutory agencies, including MfE, DoC, New Zealand Historic Places Trust, Ministry of Forestry, Ministry of Agriculture and Fisheries, Transit New Zealand, Northland Regional Council, Whangarei District Council and Kaipara District Council. Judging by our review of documents, these consultations were constructive and generally useful.

Key Stakeholder Input

Having to consult stakeholders ensures that a range of views are taken into account when identifying and dealing with important issues in a district plan. Often contrary views — like those of environmentalists and developers — have to be reconciled by the plan-makers. Similarities and differences between key stakeholder input into the Far North plan up to notification are indicated in the following summaries.

Environmentalists   Environmental groups made submissions on the various public discussion documents put out by the council between 1992 and 1996. The main groups were the Northern Branch of the New Zealand Royal Forest and Bird Society, Bay of Islands Coastal Watchdog, North Watch, and Eastern Bay of

Islands Preservation Society Inc. Their responses were consistent in wanting the council to use a variety of incentives and methods, including rules where necessary, to deal effectively with protecting indigenous flora and fauna, to ensure that development would not adversely affect the aesthetic quality of natural or heritage landscapes, including the coast, and to minimize pollution of air and water by human activities. For example, in its 1996 submission the Eastern Bay of Islands Preservation Society 'endorses as positive initiatives for the protection of natural flora and fauna [and] strongly supports the provision of incentives to achieve these, such as subsidies and rates relief'. It also encouraged better recognition of 'cross-boundary issues' and 'joint actions with other councils', as required by the RMA.

Federated Farmers   From the outset, land developers, like farmers and foresters, showed interest in the district plan through the submission process and in meetings with council staff. We focus here on Federated Farmers because they feature strongly later in the charge against the notified proposed plan. Their interests were represented by the highly organized Bay of Islands District of Federated Farmers (and its ten local branches), one of three districts within the Northern Province of Federated Farmers. It is important to highlight the consistency with which the farmers put forward to the council their main concerns about the plan. In its various submissions, the Bay of Islands District of Federated Farmers expressed the view that 'if significant features are on private freehold land, then there must not be any protection imposed without compensation to the landowner concerned' as it would be unfair for them to 'meet undue costs for activities for which the wider community benefits e.g. those which have benefit nationally' (Bay of Islands District of Federated Farmers, 1992: 3-4; 1994: 1-3; 1996: 4). As early as 1992, they stated that 'a blend and hierarchy of mechanisms must be found to promote sustainable resource use [such as] performance standards developed by the community; education and information involving research and monitoring, incentives where appropriate, and rules and enforcement where persistent breaches occur'. While they recognized 'resource management issues of extreme importance and concern to iwi', the farmers felt that 'no one group should have access to greater consultation than any other [and] planning must treat all land owners equally' (Bay of Islands District of Federated Farmers, 1992: 9-10). Meetings between Federated Farmers and council staff followed each submission (council records, 1992–96).

Iwi agencies   Generally, where development conflicts with conservation, Māori tend to see the former as more important (e.g. interviews, Te Runanga o Te Rarawa, 1999). Nevertheless, all iwi wanted sites of cultural significance protected on all land. Of prime importance, however, was the need to settle land claims and develop their own lands in order to improve their economic and social well-being. Thus, responses to public discussion documents showed that iwi were first and foremost concerned to ensure that their views on the Treaty and their lands were enshrined in the plan. They were less concerned about using scarce resources to comment on specific environmental topics. Strongly expressed by tangata whenua was the need for 'power-sharing' in resource management decisions and for a better deal than in the past, such as by replacing what they saw as an 'iniquitous rating system' (Far North District Council, 1992; Te Runanga o Te Rarawa, 1992).

Ministry for the Environment   The Ministry for the Environment (MfE) made submissions on all public discussion documents and papers.   Early on, it emphasized the need for effects-based planning and alternative methods to regulatory controls (Ministry for the Environment, 1992a, 1992b).   Later, it emphasized the need for clear definitions and linkages in the cascade of issues, objectives, policies and activities, and for cross-boundary matters, especially with the regional council, to be adequately addressed (Ministry for the Environment, 1994).   The council's document on methods for achieving desired environmental outcomes drew a ten-page response from the Ministry (Ministry for the Environment, 1996).   It clearly supported the philosophical approach proposed for the plan, and its comments were aimed mostly at 'increasing the clarity and certainty over values the plan intended to protect'.   Generally, it supported the methods proposed.   For indigenous flora and fauna, it stated that provisions 'show an excellent appreciation of the methods available to council outside of the district plan process'.   For landscape and natural features and heritage, it showed council how its proposed incentives to landowners for protecting particular sites and features could be extended (Ministry for the Environment, 1996).

Department of Conservation   DoC commented on all discussion documents, and provided information on SNAs (Department of Conservation, 1996a).   Its submission on objectives, policies, outcomes and methods was positive.   For indigenous flora and fauna, it stated that 'The council is congratulated on the range of methods chosen (rules, compensation, information, incentives, and planting programmes) for they are likely to result in extremely positive conservation outcomes'.   Importantly, it notes that while the document 'provides a very positive direction for the forthcoming proposed district plan [it] will not be until the plan includes rules and assessment criteria at notification ... that the plan can be properly assessed' (Department of Conservation, 1996b).

Northland Regional Council   As expected, the Northland Regional Council (NRC) commented helpfully on all relevant FNDC documents, although staffing changes made their involvement in the ongoing process disjointed.   From the outset, it emphasized an integrated approach, pointing out to the FNDC the need for making the public aware of the responsibilities and functions of each local authority and the 'interrelated nature of district and regional planning' (Northland Regional Council, 1992).   Later, it advised the council that desired 'environmental outcomes be realistic' and commented helpfully on proposed policies and methods for achieving them, including those for SNAs (Northland Regional Council, 1996).   Provision of technical information and expertise to the district council was always well intentioned, but rather slow to arrive.

Comparative Views

The views of farmers and environmentalists are often portrayed as being in conflict. Yet their submissions on documents before the plan's notification show a surprising degree of unanimity over key concerns, such as using not only rules, but other methods for achieving compliance, like education, incentives, rates relief and compensation to landowners. These views were generally supported by central government agencies too.

Given this useful and reasonably coherent feedback from a range of key stakeholders on its developing district plan, how sensibly did the council incorporate it into the final draft of the notified plan?

Pre-notification Review Process

Reviewing the plan before its public notification ensures that all councillors know what it contains and how readily it will fit both their constituents' interests and the mandate. It also allows statutory consultees to review the final product in light of previous comments.

Within the council, monitoring the progress of the plan's development was left in the care of its Policy and Planning Committee. This, in turn, relied heavily on council staff who controlled the plan preparation process and work allocations. Drafts of chapters were parcelled out to staff and consultants to develop, and the overall team met periodically to review them. A month or two later, the chapters would be rewritten. The first chapters were ready for review in March 1996, but most were not ready until July, three months ahead of notification. Members of the Policy and Planning Committee did not review each chapter together, but in a more diffuse manner.

While the full council had opportunities to review parts of the plan at any stage, it did not do so until a full plan was available shortly before notification. By keeping to its October 1996 deadline, the council did not have the time to send the plan to statutory consultees, like MfE and DoC. Did truncation of the peer review process materially affect the quality and acceptability of the resulting notified plan?

Evaluation of the Plan Produced from the Process

When assessed according to our eight plan-quality criteria, the *Proposed Far North District Plan* scored 44.9 out of a possible 80 (56 per cent). Although only a fair plan in general, it ranked seventh best of the 34 notified district plans we evaluated, and can therefore be considered relatively very good (see Figure 7.2). The *Proposed Northland Regional Policy Statement* of Northland Regional Council (1993), to which it relates, scored only 29 per cent — the second worst in the country (see Figure 6.2). Because the *Proposed Far North District Plan* became so controversial, it is important to note that we coded it in early May 1997, before intense opposition was aroused, but after plan Variation No.1 of February 1997 (Far North District Council, 1996; 1997a; 1997b).

Planning Approach

The relatively good score for the plan shows that its makers really thought a lot about the mandate and how to make it work using an effects-based approach. Consequently, they did not produce an 'off-the-shelf' plan, but one tailored to the requirements of the RMA in general, and to the district's resources in particular. The approach taken allows any resource use activity to occur in the district subject to performance standards intended to minimize the adverse environmental effects of activities. Restrictions therefore applied only to environmental effects.

Provisions in the plan were structured in two main tiers. The first dealt with individual resources found throughout the district (amenity and open space, landscape and natural features, indigenous flora and fauna, soils and minerals, natural hazards, heritage, air, water, hazardous substances, and energy). The second tier dealt with the environmental effects of activities within four separate 'environments' (urban, rural, recreation/conservation, and coastal), each of which contained sub-zones. As well, there were separate provisions for subdivision, financial contributions, transportation, designations, and utilities.

The effects-based approach to provisions for resources, environments and subdivisions resulted in performance standards of four types: design, development, community, and critical. If a proposed activity complied with all four standards it was *permitted*. If it exceeded: a design standard, it was *controlled*; a development standard, it was *restricted discretionary*; a community standard, it was *discretionary*; a critical standard, it was *non-complying*. There were also some prohibited activities for which resource consents could not be requested. This threshold approach therefore created a sieve or net through which the adverse effects caused by activities got strained when a land use or subdivision consent was being sought. For land use consents, the applicant first had to locate the property on the relevant environmental zone map and apply the four standards, then do the same again with the resource maps. How this was to be done was clearly and thoroughly explained step by step in a users' guide (pp. 21–26), one of the few we encountered in assessing 34 notified district plans. There was also considerable detail on information requirements for resource consent applications (pp. 35–41).

While the plan-makers aimed to create a plan that would enable development of the district's resources for the social and economic well-being of its constituents, they emphasized the need to protect nationally important and locally significant resources as directed by the RMA. They used the unusual technique (in relation to other plans evaluated) of stating the desired environmental outcome immediately after each objective and ahead of the policies and methods for achieving it. The range of methods in the plan included alternatives to regulation, but prescriptive rules dominated for achieving nationally and locally important resource objectives, such as for SNAs.

Testing the approach through to application of rules was outside the scope of our formal plan evaluation, but the plan coder suspected that applying the tiered performance-based plan might pose difficulties. Subsequent testing by professional planners showed that to be the case. For example, while applauding the threshold approach in the proposed plan, it was noted by one professional planner that the 'mesh (the bureaucratic control measures) is so fine that very little material is able to pass through the sieves and qualify as a permitted (or even as a controlled) activity'. His solution was not necessarily to reduce the number of sieves, but to make the mesh coarser. Other planners produced substantial guides to help track resource consent applications through the myriad of possibilities (Bay of Islands Planning Ltd., 1997a; 1997b).

Applying the Evaluation Criteria

The plan coding protocol used for evaluating district plans is included in Appendix 3 to the book. It reflects the criteria set out in Chapter 2, Table 2.2. Four of the eight criteria for evaluation relate to the overall plan, while the remaining four were used to evaluate selected topics: natural hazards, iwi interests, and for the Far North, significant natural areas (SNAs). We deal with each set of criteria in turn, below. We emphasize SNAs because of the unique northern environment and the controversy they generated after the plan was released.

Overall Evaluation

For the four criteria used to evaluate plans overall (clarity of purpose, application and interpretation of the RMA, integration with other plans, and organization and presentation), the Far North plan scored reasonably well. It scored highly on clarity of purpose. The plan sought a coherent overview of environmental outcomes, with discussion on how social and economic matters affect those outcomes. Issues for the district were clearly detailed and demonstrated not only good knowledge of the district's resources, but consultation over what is important to protect (if not how).

The plan scored less well in applying and interpreting the RMA. On the one hand, matters of national importance and other matters (sections 6 and 7) and the duty to gather information and monitor (section 35) were clearly explained in terms of the resources of the district, and therefore scored highly. On the other hand, explanations of provisions of the Treaty of Waitangi (sections 6, 7, and 8) and duties to consider the benefits and costs of alternatives (section 32) were limited. This was particularly surprising given strong chapters in the plan on tangata whenua and monitoring, and that a separate section 32 analysis was prepared.

Although the need to consider the relationship to other plans was noted early in the proposed plan, integration with other plans with suitable explanations for their inclusion was rather poorly done. Of eleven plans listed, and referred to throughout the proposed plan, only one had a clear explanation (regional air quality plan). Four other plans had limited explanation (own strategic plan, own annual plan, other district plans, and regional water and soil plan). The remaining six had no explanation at all (New Zealand Coastal Policy Statement, Northland Regional Policy Statement, regional plans for reserves, coasts, and pests, and iwi management plans). This is surprising, because submissions on pre-plan documents by several stakeholders emphasized this need.

The plan scored fairly well for its organization and presentation. It was well-structured and easy to read, if not to use. It was well integrated in the way sections 6 and 7 of the RMA were dealt with throughout the plan, being clearly sequenced in Chapter 4 and elsewhere in the plan as needed. Standards and discretions were clearly laid out in the rules sections to each chapter. We have already commented on other features required of a high-quality plan: a clear explanation of approach coupled with a users' guide. Other essential features present included: a glossary of terms; cross-referencing of issues, objectives, policies and rules; and the ability to readily identify individual properties on maps.

The haste in pulling together the plan is reflected in the lack of other details that would help users, such as a detailed table of contents (not just a list of chapters), a detailed index to locate specific rules and policies (not just page references), and a clear indication of supporting documents.

Topic evaluation: natural hazards and iwi interests   The four criteria used in evaluating selected topics in the plan were: identification of issues, quality of the fact base, internal consistency (the cascade of links — explained in Chapter 2, Figure 2.3 — from issues, objectives, policies, methods, anticipated results and indicators), and monitoring. The fact base in the plan for iwi interests and natural hazards was weak, sharing between them only half of the six methods we believe to be important to any plan. Where facts were used, explanations relevant to issues, objectives, and policies were, however, sound. For issue identification, the situation was no better. Coincidentally, six issues were identified for each topic. For iwi interests, four of those issues had no explanation, and for the two that did the explanation was rather poor. For natural hazards, the six issues identified had no explanation at all of the management of effects. Where they applied, however, both topics had good internal consistency: the relationship between each of the items in the cascade was clear and strong. The proposed plan had a separate chapter (13) for 'monitoring' the achievement of objectives and policies for each topic, but the topic chapters themselves did not refer the reader to it. A further weakness was that monitoring contributions sat in a separate Volume II, whereas the ideal would be to have it in each chapter. The iwi topic not only identified specific indicators to be monitored and the relevant databases, but also the key agencies responsible for monitoring the indicators of environmental results, and so scored maximum points. The natural hazards topic was vague on relevant databases and did not identify responsible agencies, and so scored less well.

Topic evaluation: significant natural areas   The SNAs section of the plan scored well, although some elements did less well than others. Only one item could be scored for the fact base behind planning for SNAs: significantly, it related to the Northland-wide survey of such areas, and included brief descriptions of some results.   From this useful base, five SNA issues were identified. We have (randomly) selected two of them, in order to trace the links through the policy cascade: Issue 2, the role of vegetation clearance and incompatible land use activities causing loss or degradation of indigenous habitats; and Issue 3, habitats under threat from many sources.

Both issues were clearly described, but all of the threats listed under Issue 3 were from human activities. Important threats to indigenous habitats from pests and weeds were omitted, perhaps due to the plan-writer's use of abbreviated definitions of section 6(c) of the RMA. Nevertheless, for both issues, the policy cascades demonstrated strong relationships among all their various parts, so that each linkage in both cascades scored maximum points. Particular strengths, compared with many other plans, were: 1) the philosophical rigour of the effects-based approach set out in the mandate; 2) the highly systematic and logical connections within the cascade; and 3) the completeness of the cascade — e.g. five indicators were listed whereas many plans did not specify any indicators at all. However, the plan scored less well on monitoring. The SNA rules were well targeted to the issues and objectives, but the plan did not develop the identification of issues far enough, and mapping errors were significant.

Evaluating Community Reaction to the Plan

Our evaluation of the *Far North District Plan* showed it to be better than 80 per cent of the 34 notified district plans we examined. It achieved good 'environmental fit' by dealing effectively with all matters of national importance within the council's jurisdiction. Yet, within months of its notification, the *Far North District Plan* had achieved local notoriety and national attention. The council was being asked to withdraw a plan that had taken six years and $1.8 million to produce. Clearly, it had poor 'community fit'. What went wrong, and why?

The Disaffected

The public had five months, to 31 March 1997, to make submissions on the notified plan. This date was extended twice, to 30 May 1997 and to 25 July 1997. The first extension was to allow Variation No.1, aimed at tidying up the numerous errors and omissions in the notified plan. The second extension — against the advice of the council's lawyer and four members of the Policy and Planning Committee — was to accommodate mounting public pressure over some provisions within the plan, especially for SNAs. The concern was over not so much those rules which affect all property owners evenly, but those which impact unevenly on some owners and not others, the former thereby unfairly bearing the cost of protecting environments on behalf of others.

By May 1997 there were around 200 submissions on the plan, but within two months there were nearly 4000 covering 35,000 items. A month earlier, hundreds had marched on council headquarters in Kaikohe, led by Federated Farmers, *Campaign 98* (twelve ratepayers' associations wanting less rates) and Grey Power (retirees). By the time submissions closed at the end of July, calls were being made to 'can the plan', 'sack the council' and have the Minister for the Environment 'intervene'. The case had drawn prominent media attention (see box), including featuring on national television news and in a documentary. Two months later, another protest march coincided with a 3,700-signature petition to Parliament.

Newspaper Stories
From May to September ... and on to November, 1997

The mayor must go: landowner claims undue influence of DoC on significant natural areas in the plan and that their protection will adversely affect his property values (*Northland Age*, 22 May)

SNA 'horror story' pushes plan back: FNDC extends submission date for second time owing to councillors' grave concern at 75 per cent error rate for marked-up significant natural areas (*Northern News*, 3 June)

Farmers battle district plan: Federated Farmers prepared to take legal action to get plan back to draft form (*New Zealand Herald*, 9 June)

Carter seeks scrutiny of council: Northland MP John Carter asks Local Government Minister to investigate performance of FNDC (*Northern News*, 17 June)

Feds promise a fight over Far North plan: Federated Farmers says that council treats people like peasants and the community has had enough (*Northern Advocate*, 20 June)

Meeting wants plan scrapped: A Federated Farmers organized meeting of 100 calls for FNDC to scrap plan. Guest Owen McShane reported as saying the plan is 'asset theft on a massive scale' and would damage the Far North's economy (*Northland Age*, 24 June)

Angry residents march: Hundreds of residents to march on FNDC headquarters, Kaikohe, to 'can the plan' organized by Grey Power, ratepayer groups, and Federated Farmers (*Northern Advocate*, 27 June)

Balloon goes up as Feds try to can plan: Federated Farmers supported by ratepayers took balloons to Kaikohe FNDC to begin Can the Plan Campaign (*Northland Age*, 1 July)

Ruling 'breach of faith': The council decision to hold in abeyance withdrawing three controversial sections of the plan was reported as 'a deliberate breach of faith. ... It simply shows the council can't be trusted.' ... The 'facilitator process the council initiated yesterday was not going to achieve anything' according to Northland Federated Farmers leader, Ian Walker. (*Northern Advocate*, 26 September)

Farmers aim to throw out plan and council: Despite council's efforts to resolve concerns over the plan, Northland Federated Farmers' president, Ian Walker, is reported saying 'If we have to throw the council out to throw the plan out, so be it.' (*Northern News*, 30 September)

Local Government under attack: FNDC has developed a reputation for arrogance by failing to attend to needs of ratepayers [and] does not work in partnership with the community. The FNDC proposed district plan is one of the worst in New Zealand, wrote Muriel Newman, ACT MP, in a three-column assault on local government and the RMA.

Far North petition will be presented: A 3,700 strong petition calling for government to force withdrawal of the plan was being presented in Wellington. It was reported that Federated Farmers Northland president Ian Walker said ACT MPs Newman and Owens would receive the petition for presentation to the Minister for the Environment, Mr Upton. (*Northern Advocate*, 18 November)

Upton heads north next week: Minister wants answers: Upton to attend a special workshop of FNDC to get answers to his concerns that council voted to continue with the planning process (*Northern News*, 27 November)

The antagonists, led by strong-minded individuals, especially within the Northland Province of Federated Farmers, wanted the plan to be withdrawn and rewritten as a draft for public comment. Farmers, especially on poorer land to the west and north (including Māori), felt aggrieved because rules protecting significant natural and cultural areas for the 'public good' were seen as adversely affecting their livelihoods and property rights. The protagonists, including environmental groups, iwi and the council, wanted the statutory planning process to continue, in order to keep faith with the 4000 people who had made submissions on the notified plan.

In the face of further challenges to its decision to continue with the hearings process, the council resolved in September 1997 to review the future of the plan once public submissions and cross-submissions were summarized in December.

Key Issues

Much disaffection with the plan focused on Chapter 4 (Natural and Physical Resources of the District), dealing with activities that adversely affected amenity and open space, landscape and natural features, indigenous flora and fauna, and water. The most intense objections were to rules for the use and development of SNAs (that is significant areas of indigenous vegetation and significant habitats of indigenous fauna) on private land, especially since the draft policies of 1994 included encouraging their voluntary protection and conservation. There was also great dissatisfaction with the council over the consultation process during 1996 and with some of the research on which the rules were based, as well as imprecision of the relevant planning maps.

Quality of data   Not all data in reports proved to be ideal, especially where the plan was dealing with new issues, such as SNAs and outstanding landscape values. For example, the council paid DoC to identify SNAs in its district, wishing to use the information in developing maps and methods for protecting them in the plan. Limited resources and a tight time-frame caused DoC to use rapid field-survey techniques and some outdated information (Department of Conservation, 1996a: 1-2). An alternative would have been to fly new aerial photographic surveys, estimated at that time to cost around $50,000. Later checks of DoC's data revealed an error rate for boundaries of 5-10 per cent.

Raising expectations   At the outset, DoC had sent a letter to land-owners in 1994 informing them of the reasons for the survey of SNAs, such as helping landowners preserve important habitats on their properties and providing information for the council's district plan. It said that permission would be sought from landowners if more detailed information was needed (Department of Conservation, 1994a). An accompanying news release from DoC expanded on the reasons, adding that identifying important natural habitats would 'assist landowners to *voluntarily* manage and protect key sites' (Department of Conservation, 1994b, emphasis added). The resulting report stressed a course of action for the council to follow before notifying its proposed district plan.

The Department believes it will be critically important for Council to follow up on this earlier advice (letter) and prior to completion of the District Plan, consult with and explain to landowners that SNAs identified on their lands will be listed within the proposed plan. The reasons for the proposed schedule of Significant Natural Areas together with any proposed incentives and regulatory provision should be explained. Landowners should be given the opportunity to discuss the proposed provisions and the values and boundaries of SNA identified on their property. [All this could be explained in] another letter [to landowners] as a follow-up to the original and signed by Council, [including] a map of the actual SNA.... Follow-up meetings may be necessary to address landowner concerns (Department of Conservation, 1996a).

While this was sound advice, it blithely ignored the overwhelming logistical and financial costs for the council in obtaining accurate data and mapping relevant to the 2200 affected property owners — an exercise that DoC had itself not been willing to commit dollars and time to doing well. On top of that, the council would have to consult with a large number of landowners. Later council efforts in 1998 to ground-truth the SNA data and meet with the relevant landowners showed the cost for just 80 properties to be $150,000, a cost it believed central government should bear to protect a resource of national significance (meeting between FNDC and Government ministers, September 1998).

Problems with mapping   Overall, the mapping of information proved very problematic for the council. First, there were difficulties getting the council's computer storage and manipulation system for geographic information (GIS) running effectively. Second, as with other councils, there were difficulties matching data on maps from multiple sources. Property boundaries (and hence zoning) were based on the national Digital Cadastral Database (DCDB). Resource features, such as significant landscape areas, valued landscape units and natural hazard-prone areas, were digitized from national topographic data (1:50,000 NZM Series). The cadastral data did not match the topographic data, but this was not realized by FNDC until its planning maps were published (Figure 9.5). Indeed, if co-ordinates are field-plotted on GPS (the satellite geographical positioning system), the margin of error with the cadastral base is between two and five metres — enough to mean that vegetation was shown to be on the wrong property. The council did not do ground-truthing ahead of notifying the plan.

A significant problem was that the DoC data had taken so long (about eight months from receipt) to integrate into the council's GIS that no maps of the SNAs could be produced until shortly before public notification of the plan. This must have affected the development of SNA rules for the plan. By then, however, councillors were committed to a notification date that did not allow time for a re-appraisal, let alone ground-checking of even a handful of SNAs. In addition, the planners were not aware that the mapping done by DoC was so unreliable. What is more, a council decision in 1993 not to have aerial photography of the district flown for budgetary reasons denied planners an ideal tool for checking the information. These problems exacerbated issues that came under debate after the plan was notified.

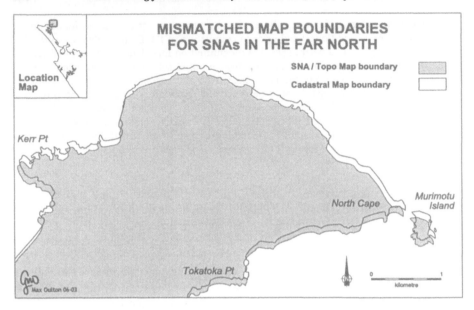

Figure 9.5   Mismatch between cadastral (DCDB) coastline boundary and topographic boundary for significant natural areas (SNAs)

Source: Far North District Council Proposed District Plan, Map A4, 1996 and 1997a

Choice of methods The choice of methods (maps and rules) in the 1996 plan was prescriptive. It therefore relied on certainty when mapping the SNAs. As we have seen, the plan-makers did not know that the SNA data was unreliable, and had no time to check it anyway, but had already set the rules. Where data is uncertain and/or unreliable, the best option is a general clearance rule. For example, landowners cannot cut down more than x hectares of trees over y metres in height without a resource consent. Only a change in the status quo gets attention under this approach. Instead, as we have seen, the council tried mapping SNAs on specific properties in relation to prescriptive rules, but got into difficulty because of faulty data and mapping procedures.

Consultation on methods   These problems were exacerbated by the council's not consulting with affected landowners ahead of notification — as recommended, somewhat naively, by DoC (1996) in its report on SNAs. What is more, DoC decided against helping to fund consultation with landowners anyway, and certainly did not offer to help pay for aerial photography which would, after all, have helped to identify nationally important flora and habitats for fauna. Whether or not the councillors were aware of all this is beside the point. They pushed their staff to notify the plan within just eight months of having received SNA data from DoC.

The Bay of Islands District of Federated Farmers in its competent, clinical and concise presentation drew on all these materials, and more, when writing to three Ministers in mid-1997 outlining problems associated with the proposed district plan and recommendations for action (Bay of Islands District of Federated Farmers, 1997). Given their consistent viewpoint on development of the plan since their first submission in 1992, it is easy to see why farmers became so angered. Not only did the plan contain prescriptive rules affecting their properties when there was some indication from both council and DoC that voluntary methods might be preferred, but the council had ignored their wish that they be consulted at least as well as iwi, and that a draft of the proposed plan be made available for public review. The council did none of that and suffered the consequences.

Other issues There were, however, other legitimate issues that caused the adverse reaction. The lack of readability and accessibility of the innovative threshold-based plan, with its new terms and jargon, created problems for those who read it. Another issue was that some people on contract to the council and DoC to help prepare data and/or policy for the plan came from iwi and environmental groups, like the local branch of the New Zealand Forest and Bird Society, that were active in urging enforcement. Land developers perceived this as causing a bias in the methods and rules used in the plan — a problem that might have been averted had a farmer, for example, also been hired as a consultant to the plan preparation team, or if more consultation had occurred over the rules and methods before plan notification.

Misinformation

Beyond these issues, there emerged in the rhetoric of heated public debate a host of half-truths, deceits and outright lies about the plan which many with a predisposition to oppose did little or nothing to verify before climbing on the band-wagon of discontent. A few strong-minded individuals with political aspirations took advantage of the situation. Not an inconsiderable amount of public comment was injudicious in the extreme, and it is surprising that some individuals were not sued for libel. One can but wonder how many of those who made public utterances about, and written submissions on, the plan had actually read it.

As misrepresentations of the plan proliferated both by word of mouth and in print, the council employed someone to identify each misrepresentation with a view to publicly repudiating it. Councillors and staff became so paralyzed by the controversy that they did not use the information and the problem escalated.

Why Did the Council Err so Badly?

There were several interrelated problems, both technical and political, that made the council culpable for such an adverse public reaction to its innovative plan. They relate to both the pre-notification and post-notification periods of plan preparation.

Pre-notification

Two broad and interrelated factors account for the council not getting the 'community fit' right for its notified plan: unrealistic expectations of planning staff in the final phase of plan-writing; and a poor plan review process adopted by the council.

Unrealistic expectations   Over most of the five year plan preparation period, councillors had unrealistic expectations of what staff could accomplish and placed unrealistic deadlines on them for producing documents for public review. When staff could not meet the deadlines, disappointment grew within and beyond the council. Two months before notification, staff verbally recommended to the Policy and Planning Committee a delay so that they could test some new rules, especially through consultation with landowners over matters of national importance, gain better definition of heritage boundaries on the resource maps, and carry out peer review of the proposed plan. Staff were instructed to produce the plan or 'heads will roll'. (Ironically, staff complied, the public reacted, a new council was elected and heads rolled anyway.)

Staff acceded to councillors' demands that the plan be produced by October 1996 when research reports for it were still being prepared as late as June that year. While staff did foresee the consequences of not consulting more widely over rules and alternative methods, they were not able to convince councillors to provide more time for doing so.

More specifically, trying to meet the deadline adversely affected the plan-makers' efforts to integrate the resource maps with the rules to which they pertained. Indeed, the consultant on SNAs, for example, did not have any maps to appraise when developing the rules. As already noted, where data quality was insufficient to warrant specific rules, as for SNAs, the policy should have been adjusted so that a more generic approach was taken. Instead, staff gave the appearance of having a mind-set (perhaps influenced by environmentalists and DoC) that protecting significant environmental values could be achieved only by prescriptive, site-specific rules.

Perhaps the plan-makers (being pushed by MfE and its Minister for an effects-based plan) were expecting too much of the council as a whole and the public at large that they would readily accept a proposed plan that was such a radical departure from earlier versions. It was not only effects-based but employed an innovative threshold approach, which introduced concepts and language that was difficult for the uninitiated to access. It therefore required time for the council to put in place an educative process aimed at helping to ensure community understanding and acceptance of the plan. There was, however, no budget for public relations and education over the plan. And, staff members were unable to prevail upon the council to delay notification and/or to produce a draft version first, or to do more consultation on methods and rules before notification.

Poor plan review process   Policy and Planning Committee members did not spend much time discussing the plan, and certainly were not taken through it chapter by chapter with the consultant and staff before it went to the council as a whole. They did not, therefore, scrutinize the cascade of requirements in the plan at all. Indeed, some members report that 'generating discussion of the plan in committee was difficult to achieve'.

Because the council as a whole wanted to stick to its earlier decision to have the plan notified by October 1996 — in spite of warnings from staff and the planning consultant not to do so — they had very little time to read, let alone critically review, the plan when it reached them at the eleventh hour. That being so (together with other problems and recommendations already identified), it was irresponsible for councillors to have publicly notified it, especially without its review by MfE, DoC and other statutory consultees as required by the RMA.

Draft plan for informal review   In light of the above, councillors may well have erred in not producing a draft plan for informal public comment ahead of its statutory notification — as recommended by many involved in the consultation process — especially as there was neither case law nor existing models to help guide the council in its development of an innovative effects-based plan. Acting on behalf of constituents in late 1995, one member of the committee raised the possibility of producing a draft for public review and comment. Staff considered the advantages and disadvantages of this verbally, but advised against it (there was no written report).

Our interviews with planning staff revealed that they were influenced by a consulting firm's report recommending councils not to produce a draft. A major reason was that publicizing specific rules for the plan ahead of their formal notification would provide the opportunity for individual resource users and developers for pre-emption, because the draft would have no status in law and its intentions could be subverted. That is a valid concern. A less valid reason for not having a draft plan for public comment was that it would extend the time and effort, and therefore the cost, of producing the notified plan.

The reasons for not producing a draft plan may seem unconvincing in light of the later withdrawal of the proposed plan and then rewriting a new one, including its release as a draft for public comment (see Annex 9.1). Indeed, as staff grappled with writing the original plan around mid-1996, they had changed their view and recommended to the committee that the plan be published as a draft. Members resolved not to do so, and stuck to their October deadline, because they felt the plan preparation had gone on long enough. An alternative suggested by the consultants was external peer review before notification, but staff did not take this option up with councillors.

Post-notification

Councillors reflect the schisms and tensions that exist among the electorate, and this seems especially so in the Far North, where land developers and environmentalists hold strong and seemingly opposing views. With such polarized views, councils often operate in a fractious manner. When public discontent is aroused over an issue like prescriptive environmental rules, many councillors too readily subjugate their corporate responsibility to the council and all its constituents to their personal predilections for the vested interests of particular groups. This certainly happened in the Far North, where too many councillors failed to take ownership of the plan that they had just pressured staff into publicly notifying. They seemingly failed to understand, or to have confidence in, the legitimate process of public submissions and further submissions on the plan. This behaviour helped to fan the discontent among some disaffected constituents, whereupon some councillors blamed their own staff and each other for the fiasco. All this suggests that too many councillors were ill-trained for, or ill-suited to, the job. They seemingly lacked knowledge of planning and the RMA, and the ethics of good governance.

By extending the deadline for submissions not once, but twice, the council enabled the opponents of both the plan and its councillors to fan public opinion in support of a campaign to 'can the plan' and 'sack the council'. In essence, extending the deadlines was a fatal concession to those seeking power for their various causes. For example, *Campaign 98*, which aimed to replace councillors with those it blessed, 'had no interest in the plan prior to its notification', but then used the adverse public reaction for advantage by joining with Federated Farmers in the clamour for sacking the council. This is a case of pure political opportunism. They eventually achieved their goal.

What Support from Central Government?

Given that some groups had petitioned the Government in 1997 to get rid of the plan and councillors, how did the Government react to the growing crisis? This question is all the more important because the council had met the demands of the Government's mandate — in particular embracing innovation and effects-based planning.

As friction in the Far North caused heat down south in Parliament, relevant government ministers were forced to respond. Under an early 1997 headline, 'Councillors buckling under RMA — Upton', the Minister for the Environment was reported as saying that too many councillors up and down the country were taking the easy path by re-creating activities-based plans of old, rather than the effects-based plans required by the RMA (*Rural News*, 24 March 1997). One would therefore expect him to speak out strongly in support of the FNDC's proposed effects-based plan, one that his Ministry had encouraged the council to prepare (Ministry for the Environment, 1992a; 1992b; 1994; 1996).

In late May 1997, the minister made a detailed submission (prepared by his staff) on the council's proposed plan as part of the public review process. The decisions he sought did not amount to a damning of its plan. The fact that he made a submission acknowledged that the normal planning process should proceed (Ministry for the Environment, 29 May 1997).

As the campaign against the plan hotted up in July, the minister was reported as criticizing the council's 'heavy handed and non-consultative way' in placing restrictive and costly rules for SNAs on landowners (Simon Upton, 'Media

Release', 24 July 1997). This surely aided the opponents of the plan and helped intensify the debate, making it that much harder for the council to get on with the hearing.

Meanwhile, the Minister of Conservation, Nick Smith, had already acknowledged that the council had a particularly tough job because of the high count of indigenous flora and fauna in the district. He explained that 'The amount of data required is huge, and we need to take that extra amount of time to ensure accuracy of information' (*Northern News*, 3 June 1997: 3). There was, however, no offer of financial help to the council, or his department, to do this, even though the issue involved matters of national importance. His comments were therefore like 'locking the stable door after the horse had well and truly bolted'.

Six months after bemoaning the lack of effects-based planning in district councils, the Minister for the Environment had over 3000 letters from Far North residents seeking his intervention in a planning process that had produced an innovative effects-based plan. The minister informed the district mayor that he wanted her council to 'reconsider its recent vote to continue the plan process' through the hearing of submissions. This was because he believed that the depth of opposition to the plan indicated there was no realistic chance of addressing key concerns in the short term. He also believed a report to the council was deficient in downplaying advantages in withdrawing the plan, and he was critical that it had not evaluated the advantages of a variation to withdraw the more contentious parts of the plan (Planning Consultants Ltd., 1997). These views were aired in national and local media ahead of his planned meeting with the council to discuss the crisis (*Northern News*, 27 November 1997; *EnviroNet*, 7 November 1997). Clearly, the minister's publicized views could have served only to undercut what the council had legitimately planned to do, and to further stoke the fires of discontent.

Meanwhile, the Government's difficulties in dealing with disaffected constituents in the Far North were a godsend to the new right ACT Party, which hoped to garner their support.

In short, the Government was unhelpful. It was good in making formal submissions on the proposed plan, thereby accepting its veracity, but it failed to support the council when that was most needed — to proceed with the statutory planning process. This seems all the more reprehensible because the council had done its best to produce the kind of plan for which the Minister for the Environment had pushed since introducing the RMA to Parliament in 1991. The Government also failed the council in not providing extra material and/or financial support for helping to improve the database after the SNAs' survey had been completed, or helping to defray costs to landowners who were being required to protect resources of national importance. Clearly, the devolved co-operative mandate was not working well in the Far North (Claridge and Kerr, 1998; Davis and Cocklin, 1997; Salmon, 1998).

Lessons and Epilogue

The unique cultural heritage of the Far North District is matched by a unique and complex natural environment. Much is of national importance and at threat from future development in a fast-growing district. It is therefore an area where Part II provisions of the RMA are particularly relevant. It is also an area with polarized interests: Māori and Pākehā; development and conservation; great wealth and poverty. More than for many councils, the character of the Far North District raises serious issues for plan-makers in seeking to satisfy competing interests when meeting mandated demands. There are some lessons for the major players in this case.

Lessons

The key lesson for the Government is that effective policy development requires walking the talk. Jawboning compliance with its mandate is meaningless if it then responds to the first major challenge with a U-turn, fails to ensure that central agencies have the resources to provide adequate and effective help to councils, and expects poor councils to subsidize on its behalf landowners that are adversely and unfairly affected by its mandate.

For local politicians (councillors), the case clearly shows that it is sheer folly not to understand the mandates under which they operate, to use election cycles to push staff unreasonably to produce complex plans that meet unrealistic deadlines, to fail to review their plan adequately before public notification, and not to have the moral fibre to own their plan when negative reactions emerge among some elements of the electorate.

The staff associated with planning in this case were crucified by their employers, the media, and some segments of the electorate for having produced what was a good-quality, effects-based plan that met the demands of the mandate. For planners, the lesson is that producing a good plan counts for little unless they have the buy-in of landowners, unflinching support of councillors, and the backing of central government agencies and ministers if the heat comes on.

Epilogue

Dealing with conflict over the notified plan led the FNDC to a series of decisions resulting in the preparation of a new plan (Far North District Council, 1998; 2000). It took from February 1998 to April 2000 to prepare, but the 1996 plan was not formally withdrawn until 19 October 1998, soon after an accord was signed with farmers and foresters giving SNAs interim protection. By this stage the administrators found that the 1996 plan had lost its credibility and needed to go.

The new plan was prepared under the guidance of an independent facilitator and a Plan Review Committee that included not only councillors, but also key stakeholder groups (farmer, environmentalist, professional and commercial). All meetings were open to the public, and the committee reported directly to the council as a whole, and a council newsletter kept people informed.

The steps in preparing the new Plan 2000 are detailed in Annex 9.1. It cost $1.2 million to notification, whereas the previous plan had cost $1.8 million. In the meantime, the council was further restructured and had significant changes in personnel, but continued to be somewhat fractious and indecisive.

In short, developing the new plan included key stakeholder groups in plan preparation, provided for open consultation over methods, and allowed the necessary time and resources for tasks to be completed (including producing a draft for public comment).

Did it result in a better plan? Plan 2000 is less effects-based than Plan 1996. Having lost the sieve or threshold approach, it is far less innovative than its predecessor. It is much weaker on monitoring, and does not have indicators for any topic — a weakness identified by an expert retained by the council to review the new proposed plan ahead of its notification. Plan 2000 is probably more honest than Plan 1996 as it does not pretend that there will be extensive monitoring when the council cannot afford it. It identifies priority issues and the section 32 report concentrates on those, as will the monitoring, eventually. Plan 2000 is more readable than before. Finally, it is more acceptable to the community. Overall, however, we scored Plan 2000 lower than Plan 1996.

Ironically, the council entered into a partnership with Northland Regional Council in 2000 to do an aerial photographic survey of the district, costing $200,000 — four times more than when its consultant recommended it be done in 1993. Had DoC done an aerial survey in 1993/94 instead of relying on outdated data and curbstone judgments, the research and analysis on SNAs may well have resulted in far better outcomes, and the rabid opposition to Plan 1996 may not have eventuated.

Annex 9.1

Steps in preparing the second Far North District Plan, 1996-2000

1996	First Proposed District Plan publicly notified (October).
1997	Plan Variation No.1 to correct obvious errors, such as on maps (February).
	Plan submission closure date of 31 March 1997 extended to 30 May to allow for public review of Plan Variation No.1.
	Plan submission closure date extended to 25 July 1997 in response to public march on Kaikohe town hall in June and calls for the plan to be withdrawn.
	Council resolved in September to review plan after submissions and cross-submissions summarized in December 1997.
	Council appointed independent facilitator on 25 September to identify way forward.
	Facilitator gave the council in early November the costs and benefits of five options for dealing with the plan.
1997	Council resolved to rewrite the plan using information from submissions, workshops and public consultation, and issue the new plan as a draft for public comment and notification in June 1998. Did not withdraw 1996 plan yet, but intended to eventually.
1998	Council resolved in mid-January that the plan be rewritten by an independent Draft Plan Review Committee under its direction.
	Council ran workshops from February to April for people interested in the new plan.
	Council in April appointed members to an independent District Plan Committee for reviewing and re-writing the plan (environmentalist, farming, professional, and urban rate payers, plus two councillors, one consultant planner and the facilitator).
	Council signed Accord with farmers and foresters over SNAs to provide interim protection, September.
	Council withdrew Proposed District Plan on 19 October 1998, ahead of local body elections in late October.
	October local body elections forced new councillors to reflect on make-up of District Plan Committee.
	New Draft Proposed District Plan released for public comment on 6 November.
1999	Comments on draft plan closed 15 January.
	District Plan Committee revised draft plan in light of public comments (April-August).
	Revised draft of plan given to councillors for their evaluation in August;
	Council took next 8 months to review the issues (e.g. SNAs) with the new plan. (i.e. councillors had to take ownership of the plan).
	Council appointed an expert to review the economic impact of the new plan, and had its various committees review the plan's regulatory and infrastructural implications.
2000	Second Proposed District Plan notified, 28 April.

Chapter 10

Queenstown Lakes District:
Development Meets Environment

Nowhere in the country are the conflicts between resource development and environmental protection so sharply defined and the consequences so far-reaching as in Queenstown Lakes District. The area is renowned internationally for its outstanding natural environments of lakes, rivers, mountains and basins. The district was one of the fastest-growing areas in the country in the 1990s, and the district council came under intense pressure to develop these resources to service international and domestic tourism. There is much at stake for the district and its people, and it is not surprising that the *Proposed Queenstown Lakes District Plan* (Queenstown Lakes District Council, 1995a) ignited strong and passionate views about how growth should be managed in the midst of competing economic and environmental interests. This case provides an opportunity to examine the development of a plan for a biophysical environment seemingly tailor-made for the RMA. In a community that returned councils during the 1990s that swung politically from left to right of centre, there is also an opportunity to assess the effects of ideology on plan-making.

Geography of the Area

Queenstown Lakes District is located in the heart of Central Otago (Figure 10.1). Lying on the eastern side of the Southern Alps, most of the district comprises steeply sloping mountains, with only about 5 per cent of land low-lying. Peaks rise to 3000 metres and provide the headwaters of the largest river system in the South Island, the Clutha River. It is notable for its three major lakes (Wakatipu, Wanaka and Hawea), fast-flowing mountain streams and steep gorges.

The district has a diverse range of habitats, including forests, shrublands, alpine herb fields, tussock grasslands and bodies of water, which support an array of indigenous plants and animals. The uplands to the west still contain much natural vegetation. Near the main divide, beech forest predominates, with snow tussock on the upper slopes above the bush line and alpine plants on the tops.

There are many areas of outstanding landscape value which contribute not only to the district's distinctiveness, but also to the region and the nation. They include mountain slopes, terraces, flats, fans and deltas. As well, DoC has identified over 100 sites of significance in the district where plant and animal communities and habitats are representative, rare or unique (Figures 10.2-10.3). Given the distinctive nature of its biophysical environment, it is not surprising that about 61 per cent of land in the district is owned and managed by the Crown for scenic, historical and recreational purposes.

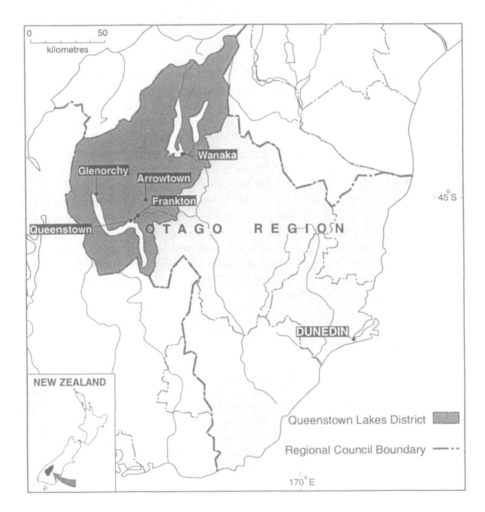

Figure 10.1   Location of Queenstown Lakes District within the Otago Region

**Figure 10.2   Outstanding landscapes: Frankton Arm (Lake Wakatipu) and
the Remarkables** (not far from Queenstown)

Source: Department of Conservation, photographer John McMecking

The district economy is dominated by tourism, which has become increasingly internationalized. In the mid-1990s, 620,000 visitors came to the district each year, staying an average of 2.5 nights. The daily average over the whole year was 4250 visitors, and 700,000 domestic and international visitors were projected annually by the year 2010 (Queenstown Lakes District Council, 1995a).

At the 1996 census the resident population was 14,283, an increase of 43.1 per cent over the 1991 census, compared with the national rate of 7.2 per cent. This was the largest growth rate of any district in the country. Ninety per cent of people were of European ethnicity while 5.9 per cent were Māori, much lower than the national figure of 15.1 per cent. At 8467 square kilometres, the district is 55th in size out of the 74 districts and contained 11,520 rateable properties, which returned $14.8 million in income to the Queenstown Lakes District Council in mid-1998, placing it mid-way in relation to other councils.

Figure 10.3   Outstanding landscapes: Queenstown, Lake Wakatipu, and mountains (The Wakatipu Basin — an area of high visual and landscape value)

Source: Photograph, Claire Gibson

The main urban areas (shown in Figure 10.1) in 1996, soon after the *Queenstown Lakes District Plan* was notified, had small populations, including: Queenstown (7491), Wanaka (2133) and Arrowtown (1428), plus a half dozen or so smaller towns. Population growth varied within the district with greatest pressure being placed on the Wakatipu Basin, containing about 13,000 people. Queenstown was characterized by one planner as a 'gold-rush town with an entrepreneurial dynamic environment'. This in itself signals that any plan aimed at protecting outstanding and significant landscapes through regulatory controls would cause controversy.

Beyond the towns, pastoral farming dominated the uplands, while diversification was occurring in the Wakatipu and Wanaka–Hawea basins. In most of the Wakatipu Basin (between Queenstown/Frankton and Arrowtown) land was subdivided into small farming or rural lifestyle properties, and grape production was increasing. Some small plantation forests for timber production were evident, especially in the Wanaka–Hawea Basin.

Key Planning Issues

The major planning issue confronting the Queenstown Lakes District Council (QLDC, or the council) when it started preparing its plan in 1992 was how to accommodate urban and rural growth without compromising the environmental qualities for which the district was renowned. Specific issues included the urgent need to address rapid growth in tourist accommodation in Queenstown, provide for residential expansion, improve and upgrade existing infrastructure (such as water supply and sewerage reticulation in some parts of the district), and meet the demand for small lot rural subdivision adjacent to towns and settlements. For the latter, a reduction in commodity prices meant that rural landowners were increasingly looking for alternative means of generating income — particularly in the Wakatipu Basin.

Concern about this basin was great enough to produce a call in some quarters for a moratorium on growth of visitor accommodation so that a more strategic view of growth management could be taken. Specific issues included localized effects of runoff and contamination of water bodies close to settlements, potential for groundwater contamination by increasing numbers of septic tanks in the Wakatipu Basin, reduction in the productivity of farming land, problems of waste disposal in environmentally sensitive areas, over use of waterways, and effects of developments on the visual and landscape resource. Some plant and animal species were perceived to be at risk. These concerns were less about the current environmental degradation of resources, and more about what thresholds should be retained in relation to the continuing pressure for development of those resources (such as potential modification of rural landscapes and rivers) so important to the local economy. In other words, could Queenstown continue developing as a successful tourist resort?

According to the planning lawyers and planners we talked with and local papers reviewed, community opinion on various issues in the district changes quickly. Even so, in a district where the local economy depends on the promotion of its natural resources, resource management issues captured much public interest.

As one planner said: 'Council was the biggest interest in town.' In contrast, there was 'no common ground in the community' about how to deal with the district's future. A simple grouping identified on the one hand conservative permanent residents and on the other a more transitory group who came to make money and move on. In fact, there was a multiplicity of views on key issues that the council had to grapple with in preparing its plan. More than one planner observed that these views resulted in 'a very tough environment for planners to work in'. Clearly, developing a plan for the district was going to provoke controversy in the community and would serve to polarize views on developmental and environmental issues.

Institutional and Organizational Arrangements

The Queenstown Lakes District was created in 1989 from three former councils (Vincent, Arrowtown and Queenstown Lakes) and was one of five within the area of the Otago Regional Council (ORC), headquartered in Dunedin. In late 1994, an ORC branch office was opened in Queenstown. In 1998, the number of elected representatives to the four wards and one community board that made up the new QLDC was reduced from fifteen plus the mayor to twelve, including the mayor.

The statutory consultees of most importance to preparation of the Queenstown Lakes plan (in 1993–95) were MfE, which early on had a regional office in Dunedin, but later relied on its Christchurch office, and DoC, through its regional conservancies headquartered in Dunedin (Otago) and Christchurch (Canterbury).

The area (rohe) of the Kai Tahu iwi (tribe) included most of the South Island, and therefore the Queenstown Lakes District. Thus, plan-makers had only one iwi to deal with. Within the Otago region, most of the small Kai Tahu population lived in coastal areas. Kai Tahu had, therefore, been largely invisible as a significant interest group in the Queenstown Lakes District until quite recently.

Council Structure

The QLDC's organizational structure changed significantly during the 1990s. Plan preparation was started in 1992 under a left-of-centre 'greenish' council and ended under a 'new right' minimalist council that supported growth and privatization of its assets. At the beginning of plan preparation, there were five units within the council: corporate services, operations, works business, regulatory services, and the district planner. By November 1997, these had been reduced to three: operations and infrastructural assets, corporate and community services, and resource management and regulatory services, with a total staff of 84. Further restructuring in 1998 saw a major loss of staff with contracting-out of services to the private sector. All regulatory services (building, planning, and so on) were contracted out to Civic Corp, a private contractor comprising former council staff.

The reduced staff complement of just over 30 included only one planner, as policy manager. This position was cut in 2000, as total staff fell to 20.

Council Capacity

No money was set aside for developing the new plan, so the council took out a $1 million loan in 1992 over ten years (later extended to fifteen) to pay for it. The first set of consultants was contracted by the council to write the plan to notification for a fixed sum of $250,000 in association with its District Plan Review Committee. The seven committee members were described as being prepared to 'throw money' at further studies which were necessary during the plan's preparation. However, the cost of additional studies was relatively small, and the total cost of plan preparation to public notification was $400,000. Towards the completion of public hearings in mid-1998, about $2.3 million had been spent on the plan (Queenstown Lakes District Council, pers. comm., 1999), and it was estimated that the final cost of having it operational in about two years hence was around $4 million, considerably higher than the mean for district plans ($1.495 million, cited in Chapter 4).

Table 10.1   Planners involved in plan preparation, 1993-99

Activities	Consultants	Council Staff
Plan writing, up until notification (1993-95)	1-2	1
Post-notification (1995-96)	4	-
Hearings on submissions (1996-98)	3	3
Post-council decisions* (1998-99)	4-5	0.2

* i.e. prior to Environment Court hearings

No planner was employed for the area until 1990. Council had so much difficulty retaining planners and attracting senior staff to guide review processes and plan development that it decided to commission consultants to prepare the plan (Davie Lovell-Smith and Partners, based in Christchurch). They worked directly with the council's District Plan Review Committee and staff between 1993 and 1995, during which period the local economy and applications for resource consents boomed. This put heavy pressure on the very small number of planning staff and created conditions for a high staff turnover. The council therefore increased the number of planning staff to include five for consents and three for policy by 1995, and felt confident it could handle the post-notification period internally. However, the sheer number of submissions and cross-submissions on the notified plan made that impossible, and the council had to buy in additional capacity to cope with RMA processes. Thus, since its start in 1992, a succession of four groups of consultant and staff planners was involved in preparing the plan, as shown in Table 10.1 (above). At any time in the six years to 1999, at least four full-time equivalent (FTE) planners were involved in preparing the plan as either consultants or staff.

Table 10.2   Chronology of key actions, 1992–95

1992	Consultants appointed to prepare plan.
1993, April-June	The following reports published: - *Settlement Strategy* (Constantine Planners Ltd., 1993a) - *Heritage* (Constantine Planners Ltd., 1993b) - *Waterways* - *Public Open Space and Recreation Areas* (Davie Lovell-Smith and Partners, 1993) - *Wanaka-Hawea-Makarora: Planning for Landscape Change - Visual Impact Assessment* (Bennett and Davie Lovell-Smith and Partners, 1993) - *Geology for Resource Management Planning* (Riddolls Consultants Ltd., 1993) - *Floodplain Management Report* (Otago Regional Council, 1993b) - *Queenstown Lakes District: Issues and Options* (Queenstown Lakes District Council, 1993).
September	Drafting of plan began, consultation undertaken.
1994, September	Council proposed Plan Change 99 re rural subdivision.
1995, early	Plan delayed pending Christchurch City Council. Appeal to the Environment Court.
April	Report on *Development Strategy – Frankton* (Queenstown Lakes District Council, Davie Lovell Smith and Partners, and Thomas Consultants, 1995).
June	Report prepared on *Ecological Issues* (Lucas Associates, 1995).
August	Hearings on Plan Change 99, decisions inform draft plan.
September	*Queenstown Lakes District Strategic Plan* published.
October	District Plan notified just before council elections.

Chronology of Plan-Making to Notification

Preparation of the plan got under-way at a time when the district was perceived to be at a crossroads in dealing with intense development pressures and the need to protect its recognized environmental qualities. Plan development fell into two distinct phases, preparation to 1995 and hearings to 1998, and while we emphasize the former in our evaluation, salient features of the latter are also reviewed.

Council Actions

The main actions taken by the council in preparing its *Proposed Queenstown Lakes District Plan* (Queenstown Lakes District Council, 1995a) are listed in Table 10.2 (above).

The council's District Plan Review Committee and its newly appointed consultants met first monthly, then weekly, to prepare the plan. A councillor who was a solicitor with a specialist interest in resource management law chaired the committee. A small group of councillors on the committee, with quite different backgrounds, was influential in driving the process, although this did not mean that their views were always congruent or co-ordinated on particular issues. They had a strong intellectual commitment to developing the plan in the spirit of the RMA. They were demanding of the professionals' time, and frequently requested additional meetings to discuss particular aspects of drafts. As new issues for consideration arose, the council commissioned reports on them (Table 10.2).

Planning Approach

Being a 'greenish' council, committee members saw a need for strong environmental controls, particularly to 'slow down development' in the rural areas and to control visitor accommodation in Queenstown itself, according to a councillor involved with plan-making. Methods adopted were largely drawn from past practice, such as minimum sizes for rural subdivisions, and extended for new topics, such as the use of contour lines to demarcate outstanding and/or significant landscapes to be protected.

At first, the council's consultants succeeded in having adopted the cascade approach from the *Christchurch City Plan* (Christchurch City Council, 1995). As for the *Far North Plan*, it used a sieving mechanism to assess whether the effects of an activity or various components of it met specified performance standards. During reviews of the plan by councillors in 1995, this cascade approach was substantially modified.

Council decided to prepare the plan *in camera*, because of concern that release of proposals in a draft plan would encourage speculative investment in a boom period, and that open meetings would be inaccurately reported in the media. As well, where resource consent applicants feared plan changes, they might have wanted to seek certainty that their activities complied with the plan by requesting a certificate of compliance, which is deemed a resource consent under section 139 of the RMA. The council saw a potential for its staff to be overwhelmed with thousands of requests for certificates of compliance.

Thus, the process of drafting the plan was tightly controlled by a small group of councillors with peer review by MfE and DoC, the planning consultants, and a lawyer to ensure it was legally sound. This *in camera* approach later fuelled views that there had been a lack of consultation on some critical aspects of the plan.

The committee was determined to see the plan notified before the October 1995 local body elections, to avoid further delays and costs should a new set of councillors disagree with it. They made it with four days to spare.

Research and Policy Analysis

Little data was available within the council to assist consultants with the various issues, apart from a few earlier reports by other consultants on the relationship of the town centre to the lake, and landscape issues. Thus, early in 1993, the council predetermined the issues to be addressed in background reports before any consultation with the community. It then requested its consultants to prepare reports on urban settlement, iwi, geology, water resources, soils, landscape, visual assets of the district, use of waterways, and recreation. As plan preparation unfolded in 1994–95, the council realized that further information was needed on other important issues, such as indigenous ecosystems and the development of Frankton (Table 10.2).

A section 32 report, which documents the analysis of policy development done during plan preparation, was not prepared. There was some uncertainty over how section 32 requirements should be implemented by councils, and the consultants advised that it was not necessary to produce a separate report. Nevertheless, there was careful attention to detail in the internal consistency of issues, objectives, methods and rules in the plan. However, not all data drawn on for policy development proved to be reliable. As explained later, this was important, particularly in matters of national importance as defined in the RMA, such as areas of outstanding landscape value and significant indigenous vegetation and habitat of fauna, as well as natural hazards.

Consultation and Consensus-building Actions

The first point of formal consultation with the community was publication of the issues and options report in June 1993 (Table 10.2). By plan notification time, council staff, councillors and consultants had participated in 200–300 meetings with residents and community groups, including informal sessions with sector groups.

However, consultation with iwi started earlier with a joint hui (meeting) at Otakau marae in Otago in September 1992. It included all Kai Tahu rūnaka (decision-makers) from the southern regions having an interest in the district. A core rūnaka committee emerged and a year later met with the council, following which the committee prepared the iwi section of the plan in conjunction with the plan-writers. In 1994, the council funded the rūnaka committee's presentation of the iwi section, and this then became the section *Part 4 District Wide Issues 4.3 Takata Whenua*.

For the wider public, key issues that emerged as publicly contentious in plan preparation (and after notification) centred on proposals for new controls in the rural areas, dealing with expansion of visitor accommodation in Queenstown, and how far the council should direct urban expansion. Also contentious were the council's efforts to protect outstanding landscape values and significant natural areas, as required in section 6 of the RMA. Some of the issues that arose had not been addressed in previous plans — for example, the siting and design controls for buildings in rural areas (including the proposal for a colour palette) and restrictions on vegetation clearance. The council's planning committee deliberately brought some issues, like rural subdivision, to a head by floating various options for dealing with them in the public consultation process.

A major issue for some stakeholders, and one that the council had been grappling with before plan preparation, was how the council would deal with the pressure to create lifestyle blocks and the expansion of settlements. Perceived legal reasons for delaying notification of the plan in early 1995 over this issue irritated councillors, as they were hoping that public debate on the new plan would assist in resolving this and some other outstanding matters. This led them to alternative action — public notification of a change to the transitional plan (Queenstown Lakes District Council, 1995c).[1] Known as Plan Change 99, it would significantly restrict further rural subdivisions by proposing a minimum of 150 hectares for new rural subdivisions, as well as proposing several new planned settlements. This change aimed to stimulate public debate over provisions for rural subdivision — the figure of 150 hectares was selected by one of the councillors. Up until then, the rural provision required applicants to show that new lots created by subdivision were economic units.

Reactions of Stakeholders to Plan Preparation

With strong views for and against development and the environment, it is obvious that the somewhat 'green' stance of the council in preparing its plan would provoke strong reactions to the major issues. One developer commented that 'the

---

[1] A plan change enables a council to test new policy by providing opportunities for public involvement in the process of submissions and hearings. Plan Change 99 floated a very restrictive approach to subdivision to test public reaction.

community regarded resource management as a sport'. A plan which proposed new provisions for controlling rural and urban land use as a means of averting adverse environmental impacts was always likely to generate conflict in an area where economic development depended heavily on the use of highly valued natural resources, and where there had not previously been stringent controls. A planner summed up the dilemma as 'People here are philosophically averse to having the landscape dotted with development, but when they know they can make huge gains by developing their own property, they want to be allowed to do it' (*North and South*, 1994: 97). Reactions of the main stakeholders are summarized below.

Rural Landowners

The proposed change in subdivision rules created considerable angst in the rural community, which saw it as imposing unnecessary controls on landowners. The landowners affected spearheaded 85 per cent of the 975 submissions the council received on this issue. In consequence, decisions on submissions to the transitional plan change made in August 1995 were put into the proposed new plan, providing for a minimum of 20 hectares for rural subdivision, and the proposal to provide for new settlements was dropped.

Nevertheless, the notification of Plan Change 99 (while the district plan was still being drafted) did add more statutory opportunities for community participation — over and above those provided in the preparation of the plan. This reaction informed the subsequent development of provisions for rural subdivision in the plan.

In spite of this plan change, rural landowners were critical of the extent of the consultation process for the district plan. One commented: 'Council should have come out to farms and talked to people and asked ... what's important to them ... would have solved lots of issues.' One planner said that this view was put forward by people 'who did not get what they wanted', and that there had been considerable consultation with rural people on issues affecting them. One rural landowner ventured to suggest that consultation was in the interests of the consultants as it meant that the process was spun out, resulting in 'more money for consultants'. As the consultants had been hired on a fixed-price contract, this was not in fact the case.

Despite the criticisms by rural landowners that there was a lack of consultation, their representative body, Federated Farmers, was sent a confidential copy of the draft rules for their comment. However, landowners were not always consulted individually, particularly on newly identified sites of significant indigenous vegetation, and were unaware that some activities would in future require resource consent.

Pastoral Runholders

For the pastoral runholders (also represented by Federated Farmers), the identification of significant vegetation for protection was confounded by a similar process being undertaken through the Pastoral Lands Tenure Review, whereby pastoral leasehold land was assessed by government agencies for its freeholding potential. As a consequence, sites of particular conservation value could be excluded from land being freeholded. Land held under pastoral lease was already subject to strict conditions, and the district plan review process was perceived by the runholders to be replication, adding yet another layer of restrictions.

Residential Developers

While the proposed rural provisions in the plan drew strong responses in meetings, other aspects drew more favourable responses. A major residential developer commented favourably on their involvement in the process and the council's response. A lawyer acting for a number of urban clients was highly supportive of the consultation undertaken by the council about small townships such as Glenorchy. On the other hand, the same lawyer was highly critical of the consultation undertaken with local moteliers, and referred to one meeting between council staff and key sector interests as almost 'clandestine'. He perceived reluctance by the council to talk with representatives of the industry, which he regarded as surprising, given the proposed moratorium on visitor accommodation pending the development of a growth management strategy.

There was considerable debate over the extent to which the council should direct, rather than manage, the future expansion of towns such as Queenstown. This debate lay at the heart of the RMA and involved questions of a philosophical nature, such as whether the management of growth was a strategic issue which lay outside the concerns of the RMA. The notion of limiting growth through the plan was dropped before it was notified.

Environmentalists

Groups like the Wakatipu Environmental Society, Royal Forest and Bird Protection Society, and the Upper Clutha Environmental Society represented environmental concerns. There were also several residents' groups from the small towns with interest in protecting amenity and heritage values. Kai Tahu also shared the environmental concerns of other groups. Much of this concern focused on the pressures created by increasing intensification in rural areas due to subdivision and the consequences for degrading water quality, landscapes and visual amenity, natural conservation and wildlife values. There was a strong call from these groups for the community to determine its own rate of growth, rather than having it 'dictated by development', as one environmentalist described it.

As noted earlier, there was an extensive process of consultation undertaken by the council around the district and through community workshops. Environmental groups had a number of opportunities, both in respect of developing the new plan and in proposed changes to the transitional plan, to put their views forward and did so. It is fair to say too that the issues and options report published in 1993 had foreshadowed many of the issues raised by these groups. Thus, their views endorsed the direction of council policy as it was later developed by a council committee that was strongly environmentally-focused. As the notified plan revealed, concerns about the impact of rapid growth on the resources of the district had been taken on board by the council and significantly influenced the shape of the plan.

Statutory Agencies

MfE, DoC and ORC staff contributed to the plan by providing data, attending meetings and commenting on relevant parts of drafts. Council staff and consultants found these agencies to be helpful, but some staff to be more so than others. There was a shared view that, if the RMA could not work in this district, it would not work anywhere.

Because it was one of the first plans in the region to get under-way, MfE staff took a keen interest in its development, and attended over a dozen meetings on it in Dunedin, Queenstown and Christchurch. Strong input also came from DoC staff through its Otago and Canterbury conservancies. DoC was commissioned by the council's consultants to prepare information on significant natural areas, and its staff attended several meetings on proposed provisions in the plan with councillors and representatives of Federated Farmers. However, the data on significant natural areas was challenged once the plan became publicly notified. Since inadequate (or in most cases no) consultation had been carried out with farmers and the information shown was not accurate, all sites of significant natural areas on private land were withdrawn from the plan (Queenstown Lakes District Council, 1998e).

The establishment of an ORC office with an experienced staff member in Queenstown in October 1994 assisted immeasurably in bridging the 'Dunedin/Queenstown gap'. While political relations between ORC and the QLDC were fraught at times, particularly when the latter sought unitary authority status in the early 1990s, relationships between professional staff of the two agencies were seemingly unaffected.

While there was significant co-operation between the plan-writers and staff in these government agencies, the plan was drafted in a considerable policy vacuum at regional and national levels — as the evaluation of selected topics (such as areas of landscape importance) shows below.

Quality of Plan Produced from the Process

The *Proposed Queenstown Lakes District Plan* rates as 'good' when evaluated by our plan quality criteria. It scored 50.1 out of 80 (i.e. 62.6 per cent), the third-highest among the 34 district plans we assessed (see Figure 7.2, Chapter 7). Since the plan was coded before the council's decisions on submissions, that suggests considerable expertise on the part of its authors. But how well did the plan meet the environmental mandate and community needs?

Planning Perspective

The plan viewed the protection of the environment, particularly its unspoiled state and amenity values, as the most critical objective, because 'Decisions in the district have too often been the result of reacting to the effects of development rather than addressing the issues and establishing the objectives, policies and environmental outcomes sought'. The plan attempted to accommodate urban development without compromising environmental values, and to protect significant features of the rural areas from the adverse effects of development.

The reasoning behind provisions in the plan provided a strong basis for assessing resource consents, which should lead to well-informed decision-making. In this sense, the authors put together a coherent plan which contained robust reasoning from the explanation of issues through to reasons for rules. It worked effectively at two levels: the district as a complete entity; and with attention to the townships and their particular needs.

The plan was effects-based, and drew strongly on the principles of the RMA to establish its objectives and policies. However, a prescriptive approach was adopted for implementation. Thus, the 'front part' of the plan met the spirit of the mandate, while the 'back end' was more traditional in its prescriptive or rules orientation. The plan therefore had a strong resource management orientation, although rules, as opposed to other methods, were the key means of implementation. This approach was justified, given the development circumstances of the district and its nationally important environments.

Applying Evaluative Criteria

After evaluating the overall plan, we focused on areas of landscape importance in our topic evaluation because of their great importance for the district and nation, and the controversy they caused when the plan was notified. This was in addition to the two topics selected for all four case studies; natural hazards and iwi interests.

Overall evaluation   The Queenstown Lakes District scored quite well on the four criteria used for evaluating its plan overall (clarity of purpose, explanation of key provisions of the RMA, integration with other plans/policy statements, and organization and presentation). The plan displayed a clear understanding of what sustainable management meant for the district and how it might give effect to this, and demonstrated extensive knowledge of the physical and natural environment and a commitment to protecting the district's outstanding features. The plan also discussed how these matters related to the council's responsibilities under the RMA. Thus, for clarity of purpose and identification of issues the plan scored highly.

Assessment of the plan's explanation of key RMA provisions was more mixed. The interpretation of section 6 (matters of national importance) and section 7 (other matters) was a particular strength in the way these were related to features of the natural environment, such as landscape. The plan was clear about the difference between section 35 (monitoring) and section 75 (contents of district plans), but other provisions of the RMA were not explained as well. The plan was vague in discussing the Treaty of Waitangi, which is surprising given the close collaboration between its authors and Kai Tahu representatives. The plan did not, as already noted, address section 32 requirements, and the functions of the district (section 31(a)) were not explained at all.

The plan did, however, provide a full explanation of the relationship between it and some other council instruments, such as the annual plan, but was less well 'integrated' with respect to the regional policy statement and conservation management strategies, probably because those documents were prepared at the same time as the proposed plan. Their later publication did not, however, provide an adequate explanation for why 'cross-boundary issues' were not well addressed in the plan. The publication of the council's strategic plan in September 1995 suggested that some important strategic thinking was prompted by the district plan process rather than informing its development, particularly in issues relating to the future management of visitor accommodation and expansion of urban settlements. Although one of the plan's strengths was the excellent in-depth discussion of issues and of explanations and reasons for adopting objectives and policies, a weakness was that it became repetitious, both within the parts and across the plan, where the same themes emerged.

The notified plan was over 900 pages long and organized into 20 chapters on key topic areas, such as sustainable management, district-wide issues, rural areas, urban growth, residential areas, and so on. A second volume contained 14 appendices on topics, such as areas of outstanding landscape values, significant indigenous vegetation and ecological structure. The maps were in a separate A3 volume (Queenstown Lakes District Council, 1995a).

Of the ten items we identified as useful in evaluating 'organization and presentation' of material in a plan, this plan had five (table of contents, glossary, users' guide, clear illustrations, and readily identifiable properties). Some reorganization and editing of material would have made the plan more user-friendly, but, according to a key plan-maker, the rush to notify before the council elections precluded further editing.

Topic evaluations: iwi interests and natural hazards We now apply the other four criteria used to evaluate plans (identification of issues, quality of the fact base, internal consistency, and monitoring) to the selected topics of iwi interests, natural hazards, and areas of landscape importance.

For iwi interests and natural hazards, the quality of the fact base was weak, although the council did undertake further studies while the plan was being drafted. Thus, the issues identified for planning were not underpinned with supporting information. For example, the plan acknowledged that information on natural hazard risks to sites that might be adversely affected by events would be collated later through implementation of policies and methods — a weakness later noted in challenges to the plan through the submission and hearings process.

On the other hand, internal consistency in the treatment of iwi interests and natural hazards was strong, with robust reasoning for the identification of issues through to the explanation of rules. Each issue may have had its own set of objectives, policies and methods, but where multiple issues were grouped together the objectives were framed to include all the issues to which they related. The plan was reasonably strong on provisions for monitoring and the indicators for topics were extensive, although lacking detail on how to measure them. The plan demonstrated a good understanding of what comprises an anticipated environmental result, being among the strongest in all plans we studied.

Topic evaluation: areas of landscape importance   One of the most controversial topics in the plan was the identification of areas of landscape importance. These areas were shown on the planning maps as additional layers over zones and signified that methods additional to regulatory rules applied to landscape protection. The explanation in Part 4: Landscape and Visual Amenity 4.2 made it clear that the council saw the social and economic well-being of the district as depending largely on the quality of the landscape image. Provisions were included to avoid development that would detract from the general landscape image, special landscape features, open character of the landscape, and amenity of lake shorelines and adjoining hillsides.

Three areas of the district were identified as having a high visual and landscape value: Wanaka–Hawea–Makarora area, the Wakatipu Basin, and the Upper Wakatipu area. (See Figures 10.1–10.3). Special landscape features within those areas were also identified. In the Wakatipu Basin, the 400-metre contour was chosen as the lower boundary for the areas of landscape importance. The plan contained one objective from which several policies followed:

4.2.5: Objective and Policies
Activities and development being undertaken in the District in a manner which avoids potential adverse effects on landscape values, particularly areas of landscape importance, being the skyline and upper slopes of the hills, the semi enclosed rural valleys, prominent topographical features and areas, the open character of the rural area and the shorelines of the lakes and rivers.

The policies aimed to avoid the adverse effects of development on distinctive landforms and landscape features, preserve visual coherence, restrict major new residential development outside areas identified in the plan, manage mining and forestry operations, clearing and burning, amenity planting, and so on. Methods for policy implementation to achieve the objective included:

- identification of areas of landscape importance
- control of the height and location of buildings in the rural areas
- provision for the design and appearance of all new buildings in the rural zones to be a controlled activity in respect of colour and location
- provision for the design and appearance of buildings to be a discretionary activity in areas of landscape importance
- inclusion of a colour palette for all new buildings in the rural areas and in many of the urban areas
- new roads and tracks to be a controlled activity in rural areas, and discretionary in areas of landscape importance
- management of the end appearance of structures around the lake shore
- provision of guidelines to encourage development compatible with the landscape.

In Areas of Landscape Importance, both subdivision and buildings were viewed as activities that did 'not comply' with the plan (with certain exceptions), and earthworks and tree planting could happen at the 'discretion' of the council when application for a resource consent was made to it.

Clearly, these provisions placed additional constraints on landowners with property that had been identified as falling within areas of landscape importance. Some properties also carried other notations where specific controls were proposed, such as flood hazard areas and areas of significant nature conservation value.

Our evaluation of the plan quality on areas of landscape importance gave results similar to those for the iwi interests and natural hazards. There is strong identification of issues, followed by strong internal consistency of issues, objectives, policies and methods. However, the fact base was stronger than for the other topics, probably due to the specific studies done by the council.

Did the Plan Meet Expectations?

Having assessed the plan as relatively good, we now examine how far it fitted not only the environmental requirements of the RMA, but also the community's expectations. In so doing, we also review intergovernmental efforts for ensuring production of a plan that dealt effectively with matters of national importance.

What was the Influence of New Councils?

With six councillors retiring and a fair level of disaffection with the plan from some quarters, the October 1995 local body elections saw in a largely new, right-leaning council. Only four councillors remained from the previous term, but none was involved in plan development. There were strong suggestions during the election campaign that some candidates wanted to scrap the plan if elected.

Some people were concerned about having a new group of councillors overseeing the process of submissions and making decisions without detailed knowledge of the various rationales underpinning the plan. Others saw this as an advantage, in that the councils hearings committee would be objective in dealing with submissions.

Most people did, however, perceive the new council to be 'pro-development' and thought this would lead to considerable changes to the notified plan, possibly even withdrawal. The plan was retained, but major changes were made to its objectives and policies as a result of submissions. Interestingly, a professional with experience of both councils later observed that the 1992–95 planning committee had 'the zeal to reform', while the 1995–98 hearings committee was driven by 'the zeal to make it work'. In dealing with a range of issues below, we help to answer whether this was so. Key actions taken by the council following plan notification are summarized in Table 10.3.

Table 10.3   Chronology of key post-plan notification events, 1996-2001

31 Jan. 1996	Submissions on plan closed with 23,000 points (4000 rural).
Sept. 1996	Hearings on submissions began.
Oct. 1996	Hearings 'fast-tracked' by not having planners' reports.
Nov. 1996	Fast-track decision reversed; consultant employed to manage hearing process and pull in extra planners to prepare reports.
1 July 1998	Contracting out of regulatory activities to Civic Corp began. Hearings committee decisions released to public in stages (7th, 14th, and 21st).
Sept. 1998	Closure for references to the Environment Court.
Oct. 1998	Local body elections and new council.
20 July 1999	Appeals to the Environment Court began with the Wakatipu Environmental Society case.
Oct. 2001	Local body elections and new council.

What Were the Facts of the Matter?

The original version of the *Proposed Queenstown Lakes District Plan* was developed during a boom period in the local economy by a committee determined to protect the natural resource base of the district, which they saw as under threat. The plan embraced the RMA vision for sustainable management and was internally consistent, but had significant weaknesses. Critical to its passage through the hearing process was the weak fact base, in part due to the limited resources and time for research and analysis made available by the council. The weak fact base in turn constrained the ability of planners to devise robust methods for protecting matters of national importance. That in turn caused uncertainty when the plan was tested through the submission and hearing process. Ultimately, it provided the new council with reasons to retreat from hard issues when challenged by developers wanting unfettered options for growth.

How Good was the Consultation?

Better consultation before notification of the plan may have averted some of the opposition to it, and may have alerted the council to the need for more robust methods founded on better data. Instead, the council failed to deal adequately with conflict and to consult with key stakeholders at a level required for the satisfactory resolution of issues not addressed in the earlier district scheme plans. This became evident in conflicts which emerged once the plan was notified, and demonstrated its poor 'community fit'. Of several important issues to which this applied, we have selected outstanding natural features and landscapes to illustrate.

Outstanding Natural Features and Landscapes

Maps identifying areas of outstanding natural features and landscapes and rules for protecting them drew one of the largest groups of submissions on the plan. When preparing the plan, the council did have access to reports on the three main areas of landscape importance (Boffa Miskell Partners, 1991; 1992; Bennett and Davie Lovell-Smith and Partners, 1993) and to three Environment Court cases supporting its decisions to decline applications from developers on the basis of the adverse effects their developments would have on the landscape (Thomson and Anor v QLDC 2NZRMA 1992 189; Design 4 Limited v QLDC 2NZRMA 1992 161; Campbell v Southland District Council W114/94).

    While the council drew in part on those reports, our interviews with those involved in drawing up the plan revealed that 'there was little professional input' from landscape architects in identifying areas of importance on the planning maps. In fact, it was the councillors and staff who drew the lines on the maps. One ex-councillor on the District Plan Review Committee acknowledged that 'the landscape analysis was not good enough ... more work was needed'. Little wonder that this part of the plan drew such strong opposition from those affected.

The professionals writing the planning report for the hearing on this landscape issue drew on the reports and court decisions mentioned above (Queenstown Lakes District Council, 1998c). After reviewing the submissions and various methods for dealing with the issue, the planners recommended to the council that the best method for protecting outstanding natural features and landscapes was to identify relevant areas on maps and provide rules in the plan. Some modifications to the existing maps would, however, be needed after further investigation. For example, some areas were thought not to require the level of protection afforded to other areas.

The hearings drew further arguments from the opponents. As culled from submissions, the key points were:

1. the areas of landscape importance were at the cost of the productive use of land and therefore at the expense of the property owner;
2. farming operations should be included as part of the landscape;
3. the areas of landscape importance were an encroachment on property rights;
4. areas of landscape importance were interventionist and hindered farming;
5. the council should remove and replace interventionist methods with effects-based criteria;
6. there was insufficient consultation in the identification of the areas.

In particular, the use of the three landscape studies of the early 1990s came under severe criticism, because they were not done for the purpose for which they were being used — that is, they had not been prepared for identifying 'outstanding' landscapes of national importance, as required by section 6(b) of the RMA. The studies had used different methodologies, and therefore there was a lack of consistency in how various parts of the district had been handled. In fact, there had been considerable changes since the studies had been done. Furthermore, the studies had not developed guidelines for enabling what development could go ahead without adversely affecting the landscape. Opponents also complained that the studies had not included sufficient consultation, although there had been some public input into the early studies. Many submitters argued that the whole of Queenstown Lakes District was an outstanding landscape and it was arbitrary to put lines on a map. They also argued that the Environment Court cases showed that the council and the court were able to decline applications on the basis of the objectives and policies in the RMA, without the need for areas to be specifically identified as requiring protection.

In the face of intense public pressure, and severe criticism of the basis on which the plan's section on areas of landscape importance had been drawn up, the hearings committee deleted it (Queenstown Lakes District Council, 1998d). They accepted the arguments that the studies should not have been used for that purpose, that the council could still decline applications under the RMA, and that the whole district was considered to be important for landscape protection. The committee retained section 4.2 of the plan (quoted earlier) as a policy framework for landscape and amenity values, and stated that additional methods should be used, such as education programmes, advocacy to other agencies, and incentives offered to individuals for protection or enhancement of outstanding natural features and landscapes. There were, however, no criteria for assessing what constituted an outstanding landscape, no schedule of outstanding landscapes, nor any indication that these might be included in the plan at some future time.

This topic of outstanding landscapes is one of several showing that the attempts by the 1992–95 council planning committee to achieve a stronger 'environmental fit', in keeping with the philosophy of the RMA, were substantially changed to accommodate community demands. As with other topics, such as significant natural areas and natural hazards, much argument centred on the adequacy of the fact base and proposed controls on land use, which submitters claimed impinged on their property rights. The 'new right' shift in the council after 1995 resulted in a pro-development interpretation of the RMA, which allowed individuals to do anything they wanted on their land as long as it did not harm the environment. For important landscape areas this included all productive land uses.

How Well did the Government Help?

While the lack of resources and time for fact-finding and analyses by the council goes some way towards explaining the lack of robust methods for protecting matters of national importance in the notified plan, it is not in itself sufficient. Capability-building agencies, both central and regional, had a responsibility under the co-operative RMA mandate for ensuring that the council was provided with appropriate technical support and advice on policy direction and on methodology and data.

Central Government

As already demonstrated in earlier chapters, there was a failure by the central government to provide the necessary policy frameworks and technical advice, so that, like other councils, the QLDC was left to address national issues in a policy vacuum. This included iwi interests, natural hazards, significant natural areas, and outstanding landscape areas.

The conflict at hearings over these matters underscored the complete lack of policy direction from central government. It also showed that, as happened in the Far North, data on significant natural areas provided by the under-resourced DoC and methods for identifying them were challenged once the plan became publicly notified. Here, however, we focus on the all-important areas of outstanding landscape value.

There is a strong statutory obligation for councils to have regard to landscape matters through sections 5 and 6(b) and clause 2 of Part II of the Second Schedule of the RMA. Apart from the limited wording in the RMA, central government agencies, such as MfE and DoC, provided little guidance to councils on how these sections should be interpreted — as, for example, through a national policy statement. Nor did they provide robust methods for identifying matters of national importance, like areas of outstanding landscape value, for local councils like QLDC to use.

Despite the lack of national policy, DoC staff did draw on policies it was already drafting to advise the council. These policies were later published in the *Otago Conservation Management Strategy* (Department of Conservation, 1995) in October 1995 — the month the district plan was notified. Even so, staff in the two DoC conservancies serving the QLDC gave varying views on the extent of development which should be allowed in areas identified on the plan's maps as being important.

Lack of clear policy and commitment from central government on matters of national importance left local councils reliant on partisan community views. As already shown, the council weakened considerably the plan's provisions in the face of opposition from particular groups in the community. In the short term this may have damped down community opposition, but in the longer term a major resource management issue has been inadequately addressed.

Regional Government

There was poor vertical co-ordination and integration between policy development at regional and local levels. Both the district plan and regional policy statement were being prepared at the same time, which did not provide much opportunity for the plan to take relevant regional policies into account.

In its land chapter, the *Otago Regional Policy Statement* (Otago Regional Council, 1993a) did provide an objective, 'to protect Otago's outstanding natural features and landscapes from inappropriate subdivision, use and development' (5.4.3), and a range of policies to provide for their protection. Importantly, the explanation stated that the recognition and identification of such features should be based on 'objective criteria and undertaken in consultation with the community', which as we have seen was not done by the QLDC. The method for achieving this objective was the preparation of an inventory of regionally significant outstanding natural features and landscapes, in consultation with various parties. Other than this, the regional policy statement provided little guidance for district councils in its region.

The withdrawal of the areas of landscape importance from the Queenstown Lakes plan raises important questions for plan preparation. Given its outstanding natural environment, if the QLDC cannot prioritize landscapes in its district in some way, what would it take to put the statutory obligations of the RMA into effect? Should there be central government assistance in the form of guidelines and technical support, given that the landscape is a matter of national importance? How realistic was it to rely on the council's objectives and policies being determined on a discretionary basis where only a few decisions are contested before the Environment Court? Would the council follow through as promised with education programmes, advocate to other agencies the need for protection, and offer incentives for individual cases?

Is Minimalist Neo-liberalism Effective?

Following the elections in October 1995, the new council was led by the former Minister of Local Government, Warren Cooper, well known for supporting minimal government and for privatizing public goods and services where possible (see Chapter 4). The councils he led assiduously pursued these neo-liberal principles.

For example, although faced with 23,000 points of reference from 2500 submitters on the notified plan, the council sought in October 1996 to reduce the time and costs of the hearings process by fast-tracking it. This was to be achieved by doing away with the convention of having reports prepared by planning professionals, which provide substantive advice and recommendations on how submissions should be dealt with by councillors. There was an outcry locally and nationally, underscoring fears that the quality of decisions would be badly impaired. This caused the council to rescind its decision, which it followed up by employing a new consultant to manage the hearing process and oversee the preparation of planning reports. This required increased resources and the allocation of several more planners to prepare 54 reports.

Nevertheless, as already noted, by 1998 the council's operating units had been almost halved, staff numbers quartered, and all regulatory services contracted out to private consultants. Some other councils, such as Papakura District Council and Thames Coromandel District Council, adopted this model of contracting out services, or variants of it. Thus, there was national interest in how these new contracting arrangements by local government were working in practice.

In 1999, the Audit Office reviewed and reported on the process used by QLDC for contracting out its services as a means of developing generic steps for good practice in the local government sector (Controller and Auditor-General, 1999). Its report noted that frustration with the costs of developing the new plan had been a factor in the council's examination of alternatives to using in-house services. However, the report made some critical observations about the tendering and contracting processes used by QLDC. In short, the council had performed poorly. The Audit Office did commend the council for appointing a policy manager in recognition of the potential for high-level conflict of interest issues where the contractor was involved in both preparing the plan and its administration. However, as already noted, this position was later axed.

We have also seen that, as the hearings unfolded, regulations in the plan for protecting matters of national importance were trimmed and non-regulatory methods proposed, such as the provision of information and education aimed at improving environmental behaviour. In the two years after that decision, there was no expenditure on implementing the alternative methods. Rather, attention focused on informing the public about the district plan through a communications programme. Apart from small expenditure on a few pamphlets, non-regulatory methods have not, however, been pursued by the council.

The constant changing of staff, and the failure to address potential conflict of interest issues and to follow up the plan with an education programme, reflected the minimalist and cost-cutting approach of the Cooper-led council. This behaviour highlighted the way in which the neo-liberal agenda with its constant drive for economic efficiency, undermined the effectiveness of the plan-making process, and inevitably the plan's implementation.

It is clear in retrospect that the council at the outset had little appreciation of the likely costs of producing a plan and, like many other councils, 'muddled along', driven by looming election deadlines, community resistance to change, and minimal help from central government, especially over matters of national importance. This was exacerbated by the preparation and amendment of the plan over four election cycles of councils (to 2001), and by four different sets of consultants and staff.

Lessons and Epilogue

With so much at stake, it was not surprising that community views were characterized by wild swings from the green end of the political spectrum to pro-development, and back again. The stunning natural environments of the Queenstown Lakes District created huge tensions in plan-making. These tensions were played out over Part II provisions of the RMA in particular. Some key lessons for the major players can be drawn from this experience.

Lessons

It is essential for the central Government to accompany new policy with adequate capacity building for implementation. This is particularly important in areas where councils do not have a strong planning capability. Given the national importance of its landscape, the QLDC should not have been expected to deal with the landscape issue on its own. There is no point in decreeing a new national mandate without following it through with adequate guidance, such as national policy statements and methods for identifying outstanding landscapes as a matter of national importance. To do so sets a poor example for local government.

The case demonstrates a key lesson for local politicians in that election cycles should not be used to push a plan through when it is not ready. In the end the submission process on the plan becomes a huge cost, not just for the council but for all the submitters, and adds years to the time it takes to get the plan operative. Councillors need to understand the mandates they are required to implement, the role of professionals, and the impact of major U-turns in policy on the plan-making process.

The planners involved in developing the plan changed several times in response to crises. That in itself demonstrates that Queenstown is a tough political environment for planners to operate in. An important lesson for planners is to get it right at the outset and carry out sufficient research and consultation when embarking on a new approach to plan-making. It is essential to have council and community buy-in and support from central government when developing new policies in an environment which is so contested. Without that backing in an environment like Queenstown's, the role of a planner can become untenable and the plan of little value.

Epilogue

The plan as amended by council decisions on submissions in 1998 was reduced to about half its original length (Queenstown Lakes District Council, 1998a; 1998b). A considerable amount of explanation was culled, such as for issues and rules. There was not always the same sense of completeness in the treatment of issues, and the plan seems like a shadow of its former self. It was, however, more manageable in its layout, and thus a more user-friendly document. Our plan coding of the revised plan showed that its score was unchanged, but there was some variation across the plan quality criteria. For example, overall integration within the plan improved while some explanation of key statutory provisions and identification of issues had been reduced.

The decisions of the hearings committee were released in August 1998, and 200 references (appeals against the decisions) were subsequently lodged with the Environment Court. This was a relatively high number of references, particularly in a rural district. It was, however, consistent with the extensive interest taken by locals and others in planning matters.

Since 1996, the population of the Queenstown Lakes District continued to grow rapidly, by 19.3 per cent (although that is only half the growth in the previous census period). Concerns of local residents about the effects of continued subdivision in rural areas, such as the Wakatipu Basin, also grew, culminating in a testy exchange between Sam Neill (an internationally known actor whose home is in the basin) and then Mayor Warren Cooper. This nationally reported exchange highlighted for other local residents the implications of the 1998 revised plan, and may well have helped a major change of representation on the new council elected in October 2000. While Warren Cooper chose not to stand again as mayor, Barry Lawrence, a member of the 1992–95 council and leading proponent of the 1995 plan, was re-elected.

The Environment Court began hearing references lodged in respect of the plan in 1999, and was still proceeding in 2001. It had made several key decisions having considerable implications for development in the rural area, particularly with the assessment of landscape matters (C180/1999; C74/2000; C186/2000;

C75/2001; C129/2001; C162/2001; and C191/2001). These decisions required the council and various parties involved to collaborate extensively in creating a new approach to the assessment of development in the rural general zone. A fully discretionary approach replaced what was proposed in the 1998 plan. Consent is generally required for all buildings and subdivision in the rural general zone and applications are evaluated according to assessment criteria. There is no minimum lot size in this zone. The plan attempts to deal with issues of the form and density of development and cumulative effects through a raft of assessment matters, including a requirement to look at alternative locations within a 500-metre radius of proposed building platforms, regardless of ownership or whether development is contemplated.

These provisions have produced a much more sophisticated and rigorous assessment of landscape values than was proposed in the 1998 revised plan, and this is more in keeping with the spirit of the 1995 plan. Time will determine the effectiveness of this approach. There is still considerable potential for ad-hoc approaches, given the discretionary nature of provisions in the plan. Thus, the quality of its implementation will rest in part on the calibre of political decision-making, and council commitment to it.

The plan had cost the council around $6 million dollars to 2001, and should become fully operative in 2003. Ironically, a decade after work began on the first version of the district plan, it is being completed by a new council that is as environmentally focused as its 1992–95 counterpart. However, had all the research and analysis been carried out when work on the plan began in the early 1990s, and been accompanied by a more focused programme of consultation in the year prior to notification, a speedier and much less costly process of plan development would surely have resulted.

Chapter 11

Tauranga District:
Policy Coherence on the Coast

Tauranga District is a predominantly urban area which has experienced rapid growth over the last 20 years. Before European colonization, it was densely settled by Māori because it had a benign climate, good soils, access to seafood and thermal springs. These same advantages explain the area's current attraction for retirees, lifestylers, holiday-makers and the agricultural, horticultural and forestry industries that underpin its economy.

Rapid urbanization, particularly along the coast, has inherent risks for the environment. There are also risks for coastal developments that are exposed to erosion and inundation from storm surges and rises in sea-level. Urban expansion also threatens the remnants of ancestral land still owned by Māori. These and other threats require good management if services are to be provided in a timely fashion. The comparatively high-quality plan notified by Tauranga District Council (TDC) in early 1997 was not buffeted by disaffected stakeholders to the same extent as in our other cases, primarily because a draft plan was produced, although the policies and rules dealing with coastal hazards did attract some last-minute criticism. The plan was, however, written within an organization that was undergoing managerialism and frequent restructuring. While we examine the effects of these on the rational–adaptive model of plan-making, our main focus in this case is on integrated management as manifested through the hierarchy of policies and plans which the RMA provides for the coastal environment.

Geography of the District

Tauranga District is located in the western Bay of Plenty on the east coast of the North Island (Figure 11.1). It has a land area of 12,742 hectares and is situated at the eastern end of a fine natural harbour with a deep-water port capable of handling large container ships and many places suitable for launching pleasure-craft. A large sandspit occupied at the seaward end by a volcanic cone, Mt Maunganui, forms the eastern arm of the harbour. This spit has a lengthy and exposed ocean beach that offers many residential and recreational opportunities, but is at risk from coastal erosion and inundation (Figures 11.1 and 11.2). The inner harbour has extensive mudflats and is the southernmost extent of mangroves within New Zealand. Thermal and mineral springs are common. While the sandspit is flat, the rest of the district occupies low hills and valleys.

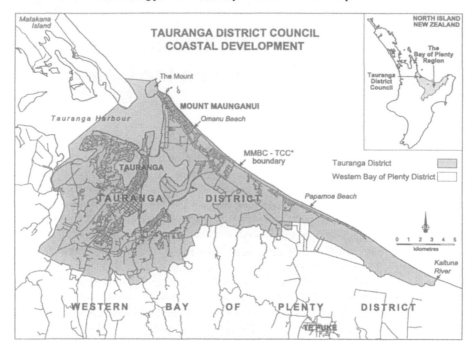

Figure 11.1 Tauranga District: built area and coastal development

Source: Adapted from map supplied by Tauranga District Council

The climate is temperate and consistently mild all year, with one of the highest totals of sunshine hours in New Zealand. As this is a long-settled area, the natural vegetation has been largely modified by human occupation. The native pohutukawa tree is still found in coastal locations and there are sedges and other sand-loving plants on the dunes. Remnants of native forest are scattered throughout the district. Mostly, there are lush, well-tended gardens and parks.

The Tauranga city centre is augmented by several suburban service centres serving extensive housing. There is a large industrial area near the port where businesses associated with forestry, agriculture, and horticulture are located. Many of these serve the hinterland, which is particularly noted for horticultural production such as kiwifruit and avocado. Rural land in the adjoining Western Bay of Plenty District is in high demand for residential or 'lifestyle' use because of its proximity to the city and Tauranga harbour.

Tauranga's population in 1996 was 77,778, an increase of 16.9 per cent over the previous five years, compared with 7.2 per cent for New Zealand as a whole. This growth rate was the second-highest in New Zealand, after Queenstown Lakes District.   The rise in the number of new dwellings in Tauranga is therefore high (19.3 per cent over five years) and compares favourably with more populous cities like Auckland and Christchurch. A high proportion of Tauranga's population is retired (17.2 per cent are over 65 years old, compared with 12.1 per cent for all New Zealand). There is also a substantial Māori .population (15.7 per cent), with some communities still living on small remnants of their ancestral lands, which are under pressure from urbanization.

Figure 11.2   Development at risk from coastal erosion, Papamoa (An aerial view looking eastwards across Papamoa Beach towards the Kaituna River mouth [see Figure 11.1 for location]. Note: Figure 11.4 shows the eastern end of this built-up area)

Source: Photograph supplied by Tauranga District Council

A pleasant climate, access to the coast, and excellent views afforded by sloping land, all contribute to the outdoor lifestyle enjoyed by Tauranga residents and help to explain its high growth rate. These factors also attract a large number of holiday-makers and international visitors, especially during the summer.

The main industries are property and business services (28 per cent of all businesses). A significant number of the 34,000 full-time equivalent employees are in the manufacturing and construction industries. Average household income was $38,034 in 1996 compared to $40,300 for New Zealand (Statistics New Zealand, 1999). The Port of Tauranga is New Zealand's largest export port (logs, wood chips, dairy products and kiwifruit) and third-largest importing port.

Tauranga District had 34,172 rateable properties in 1996, returning an annual income of almost NZ$23 million. The council's operating revenue in that year was NZ$58 million (Tauranga District Council, 1997a) and is now $86 million (2002/03 year). Residential sites comprised 94 per cent (33,928) of all rateable properties, while business premises comprised 6 per cent (2175).

Institutional and Organizational Arrangements

Institutional Setting

In 1989, local government reform led to boundary changes which ensured that Tauranga City and the areas destined for its future expansion were brought mainly under one local authority — Tauranga District Council (TDC). Thus, the whole of what were then Tauranga City and Mt Maunganui Borough was amalgamated with urban Papamoa, formerly part of Tauranga County, an adjoining rural council. The rural areas of Tauranga County were included in the new Western Bay of Plenty District, whereas the eastern half of the Bay of Plenty came under Whakatane District. Environment Bay of Plenty (EBOP — otherwise known as the Bay of Plenty Regional Council) was made responsible for providing a regional overview of resource management for these three new district councils.

Although Māori have a long association with the Bay of Plenty, very little ancestral land formerly owned by the three iwi within the boundaries of Tauranga District is still in their hands. This is due to the wholesale confiscation of property by the Crown in what it regarded as punishment for Māori participation in the New Zealand Land Wars (1864–65).

The main central government agency involved with plan preparation was MfE, through its regional office in Auckland. Given that the district is mainly urban, DoC and other departments responsible for the primary sector did not have a high profile, although pressure from DoC did help to initiate studies on the coast. Except for the coastal and harbour margins, EBOP did not have a high profile although it did contribute to urban area structure plans.

Organizational Changes

During its first seven years, the organizational structure of TDC changed several times.

In 1989, a mayor and fourteen councillors, representing five wards, were elected to the new council. They hardly had time to settle before organizational changes began. In 1991, new technical and administrative systems were put in place to meet the requirements of the *Building Act* (1991) and the RMA. At the same time, road-building, pipe-laying and parks' maintenance work, which had traditionally been done by the council's staff, was transferred into a new company, Aspen Contractors Ltd.

In 1994, there was a further re-organization, which separated the council's operational and business interests from its regulatory activities. The number of wards was reduced to four, but there was no reduction in the number of councillors. In this period, the team writing the district plan reported to the Planning and Environment Committee. In 1995, the council's Chief Executive Officer (CEO) resigned and further minor organizational changes occurred.

After a new CEO was appointed in 1996, yet another re-organization took place. Its main objectives were 'to improve the delivery of services to the community via a customer services centre and to provide an emphasis on the development of a long-term strategic plan' (Tauranga District Council, 1997a: 5). The district plan team was separated from the planners responsible for processing resource consents and became part of the Strategic Planning Department, which reported directly to the CEO. Other changes further consolidated the separation of service delivery and policy development from regulatory activities. About this time, the Corporate Strategy Committee (the whole council) assumed oversight of the district plan, reflecting the new institutional focus on strategic planning.

Table 11.1 Departmental responsibilities in Tauranga District Council, 1996

1. *Strategic Planning*: includes the Corporate Plan, Long-term Financial Strategy, District Plan, Significant Corporate Policy and the Strategic Plan.
2. *Environmental Services*: consents, Land Information Memoranda, licensing processing, monitoring, property consents and regulatory matters.
3. *City Services*: asset development, engineering, community assets, transportation, waste resources and water services.
4. *Corporate Services*: accounting, administration, customer service centre, rates.
5. *City Enterprises*: business units, Historic Village and Museum, libraries, Toll Plaza (the harbour bridge) and Baycourt (theatres, etc).

For service delivery, the council set up Tauranga Civic Holdings Ltd, a holding company for the various trading enterprises and property companies. The Corporate Support Unit, dealing with human resources and personnel matters, was also formed. Both reported directly to the CEO. The sequence of restructurings from 1991 to 1997 resulted in five departments with functions as in Table 11.1.

Capacity for Plan-Making

During the period of council restructurings, staff changes were frequent and disruptive, resulting in loss of internal advice. This affected the district plan-making effort, which the lead plan-makers said 'was a huge task'. Nevertheless, there seems to have been an attitude in the council that a plan was needed and had to be adequately funded.

Preparing the plan took five years, during which deadlines kept getting pushed back: 'Personnel were under-skilled in professional knowledge and experience and young planners needed a lot of supervision.' Some reports were done within the council, mostly by other departments, and consultants were also employed to do technical work for the plan. Most of the plan writing was done in-house, particularly the integration of technical reports with plan policy.

At its peak in 1995, the district plan team comprised three land use and environmental planners, two GIS experts, an engineer, a recreation planner and two part-time editors. Just prior to notification of the plan in 1997, between three and four staff were involved in the plan-making effort. By then, 'the majority of the plan-writing team had changed at least twice in five years, due to staff losses'. Team leadership did not change throughout.

It is difficult to work out the plan-preparation costs, but they may have been about $1.1 million. A planning councillor believed that the 'council should have put more resources into plan-making at the beginning to get the job done sooner' and 'paying more would have attracted more and better planners' to do the job.

The Plan-Making Process

TDC took over the administration of three existing district schemes (plans) covering its area of jurisdiction and had made them operative by 1991. At the time, it was known that the RMA was forthcoming and that a new plan would have to be written dealing with the whole district. With the RMA in force by April 1991, the council aimed at starting work on the new plan in 1992.

Approach to Plan Preparation

The process 'began in 1991 by setting out a programme to review the main pressure points and issues' in the district. There were 'some strategic overviews of business, residential, infrastructure, growth and heritage', which formed 'modules for research, consultation and discussions'. At this stage, there was 'no idea for the plan design'. The approach to plan preparation changed as the processes unfolded through to 1993, adjusting in a 'flexible, but unsystematic, manner to the realities' of the council's larger agenda. Nevertheless, there was 'good political buy-in and support' and by-and-large 'planning staff were allowed to get on with the job', even though disrupted by frequent structural changes in the organization. Councillors were kept informed of the process through workshops and committee reports. Plan preparation itself got under way in 1992, but the process changed when it was decided to prepare a draft plan. There was general agreement in the council that a 'practical and cautious approach' be taken to preparing the plan and that it should portray a 'mix of effects-based and activities-based planning reflecting an 80:20 split'. Writing the draft plan started in 1994.

Need to Control Growth

Before planning was under way, the council had been pressured in 1990 by 'large land owners wanting to cut up kiwi-fruit farms for residential lots'. There was, therefore, a sense of urgency in the council about plan preparation because of the high rate of peripheral urbanization and the need to manage this effectively. The urgency for addressing urban growth was also explained, in part, by the need to establish a justifiable financial contributions regime — that is, the need to recover some of the costs of installing the infrastructure required for new developments.

Thus, the council produced reports on urban growth and financial contributions. The recommended strategies were incorporated into *Proposed Plan Change No.1 to the Transitional District Plan*, which was publicly notified in 1992 (Tauranga District Council, 1993). While this may appear to be a positive experience for preparing the district plan, in retrospect a planning committee councillor thought that 'they ought not to have got diverted by it. Rather, council should have held off on the plan change until the new district plan was done'. Focusing on the problem of urban growth had affected progress on the district plan, but it certainly facilitated significant growth.

Plan Preparation Actions

The main tasks of the planning team were to manage the plan-making process, including consultation and technical input, write the policies, and prepare the zoning maps. The main actions taken to achieve this in the seven years to plan notification are listed in Table 11.2.

Research and Policy Analysis

The council produced around 55 preparatory documents of various kinds between 1991-96 (Table 11.2). All but five were either reports, surveys, or assessments of specific topics, like growth (Beca Carter Hollings and Ferner Ltd., 1991), water, heritage, coastal hazards, landscapes, noise, pollution, transport, housing and so on. Clearly, the reports flowed from issue identification. Some reports aimed at developing responses for dealing with issues. Thus, a lot of resources went into researching the main issues to be dealt with in the district plan.

Table 11.2 Main plan preparation actions to notification, in 1997

1990
Mt Maunganui section of the *Transitional District Plan* made operative
Tauranga Urban Growth Study
1991
Tauranga City and Tauranga County sections of *Transitional District Plan* operative
Confirmation of Tauranga Urban Growth Strategy
Discussion paper on Financial Contributions Strategy
Formation of Commercial Strategy and Design Workgroup resulted in draft commercial strategy
1992
Publicly notified Proposed Plan Change No.1 to the Tauranga section implementing the recommendations of the Tauranga Urban Growth Study, including development impact fees (financial contributions). Research, consultation and preparation of new district plan began in earnest
Nine reports, surveys, studies and discussion papers
1993
Special sub-committee for Māori land issues set up within council (not a district plan initiative)
Nine reports, assessments, studies, surveys, strategies, and issues papers
1994
Strategic planning process initiated with extensive public consultation
Council makes submissions on the *Regional Policy Statement*
Draft District Plan due in June but deadline extended
Eight reports, studies, assessments and discussion documents
1995
Ngaiterangi iwi completed their resource management plan
Draft District Plan due in June, but deadline extended
Proposed Regional Coastal Environment Plan notified
Fourteen reports, studies, assessments, surveys, reviews
1996
Draft District Plan issued on 12 February
'Comments' closed on 29 March; 141 parties contributed
Key themes noted in report to council (11 April) included: coastal hazard policy areas (18), future zoning in the Bethlehem growth area (15), performance standards in the Residential A zone (130) and roading matters (16)
Informal hearing of 'comments', 22-24 May
Eleven studies, assessments, and surveys
District Plan Urban Growth Structure Plans reviewed and redrafted
Natural Heritage Management Techniques
1997
Council wrote to affected parties giving background information on coastal hazards policy prior to releasing the *Proposed District Plan* (9 January)
Proposed District Plan publicly notified on 18 January
Review of flood levels for Tauranga Harbour completed
1998
Draft strategic plan completed, included long-term financial plan (February)

Consultation and Consensus Building

The council did not set out to consult widely on issues, objectives and policies early in the plan-preparation process. Rather, it did some focus group consultations on a few strategies and topics, like financial contributions and commercial strategies (1991), the growth strategy (1992), residential environment and industrial strategies and performance issues (1993), waterfront policies (1994), heritage strategies and iwi land (1995).

Public and interest groups The need for extensive public consultation for much of the plan-preparation process was averted by the council's early decision to release the district plan as a 'draft' for public comment. It was thought that public comments would provide feedback on the proposed approach and test the findings of the many technical studies. Nevertheless, before writing the draft plan, council did consult with interest groups and the public generally. 'Public meetings were useful', but a councillor thought 'small neighbourhood meetings (home-groups) were best', while a senior planner thought 'sector think-tanks' were as well.

The council aimed to have the 'draft plan' ready for public comment by June 1994 (and a 'proposed' plan by mid-1995), but the draft did not emerge until 12 February 1996. Some focus groups, such as those with architects and developers, were set up by the council to examine the draft, or aspects of it, and contributed effectively.

The 'long and costly task of producing the draft plan' did lead to useful reactions and was 'the correct approach', according to a planning committee councillor. Key players in the overall consultation process were sector groups and affected property-owners. Some vested-interest groups, like the port company, watched the whole planning process closely and applied pressure as they saw necessary. Input by environmentalists was mainly through the Forest and Bird Society.

Responses arising from the draft plan process led to further research on several important issues while its re-writing proceeded. For example, when the deadline for notification was looming, planning and surveying consultants were asked to review structure plans for parts of the urban periphery and present more detailed costing schedules for amendments to the financial contributions regime.

On the other hand, dissatisfaction with proposals for managing coastal hazards along the Mt Maunganui and Papamoa beaches caused the council to write to residents nine days ahead of notifying the plan, inviting them to make submissions on it (TDC letter, 9 January 1997). The letter summarized management options for four levels of risk, and identified the extensive research by EBOP (Bay of Plenty Regional Council, 1995) and Tauranga District Council (1996) that underpinned them.

Statutory consultees After a considerable delay in consulting with iwi over planning matters, the council offered them some funds for writing iwi resource management plans. One group, Ngaiterangi, took up the offer and produced a plan in 1995. The resulting document was helpful not only to staff writing the plan, but also for building positive relationships. Unfortunately, these efforts were undermined by political resistance to the idea of any sort of partnership or formal role for tangata whenua in council's formal decision-making processes.

Important in developing the 'notified' district plan were the responses of MfE, DoC and EBOP to the 'draft' plan. They were 'especially helpful in improving its structure'. MfE staff commented in depth on the shortcomings of the draft plan and especially its similarity to the district schemes written under the *Town and Country Planning Act* (1977). They emphasized the need for an effects-based approach that better highlighted policies in a restructured cascade of issues, objectives, policies, methods and anticipated environmental outcomes. According to a planning committee councillor, 'the draft plan was poorly written' and MfE staff 'worked closely with the plan-makers in rewriting the plan'. The district planner found this to be a 'difficult, but worthwhile activity'. The revision concentrated on reorganizing and representing the district plan in terms of the mandate for sustainable management set out in the RMA.

In parallel with the revision of the draft plan, the council did further work on Bethlehem, an urban growth area. The 'proposed' plan was publicly notified in January 1997 (a newly negotiated deadline) (Tauranga District Council, 1997b; 1997c; 1997d), opening the period for formal lodgement of submissions, and was followed by a variation dealing with Bethlehem later that same year.

Plan-Making Style and Process

The plan-making process at TDC could be construed as 'muddling through', but such an assessment would be too superficial. The sequence of reporting on technical matters, focus-group discussion of key topics and strategies, and subsequent discussion of chapters in the 'draft' plan shows that the plan-preparation process was iterative. The technical reports flowed from issue identification and were fed into plan writing and rewriting as work progressed.

There was considerable emphasis on technical content, a sign that the rational approach was adopted. The district planner commissioned reports whenever he could, to the extent that resources, circumstances and politics permitted. Further, consultation was threaded through the process, revealing issues that required addressing through further technical reports, including some commissioned in response to feedback on the draft plan. The result was a process that wove together both technical research and stakeholder participation — two strands of the 'incremental' model of planning described in Chapter 2 (see Figure 2.2).

Effects of Managerialism on Planning

While pressures to deal with urban expansion meant that it was important for the council to advance the district plan speedily, progress was not as rapid as hoped. One reason for not meeting the deadline for releasing the 'draft' plan was delays in preparing services structure plans and iwi consultation (Tauranga District Council, 1995: 51).

Indeed, the council's Technical Services Department (TSD) was slow to perform many of the asset management planning functions for which it was responsible in supporting the plan-makers. Its contribution was important because it affected the provisions for urban growth, including financial contributions, a vital part of the plan to get right and one which was going to be controversial. Memoranda between the district planner and TSD, as well as other records, show that internal co-ordination of effort was difficult to achieve. This arose in part because of the impact of successive internal re-organizations on staff numbers that affected morale and a shortage of engineers who were suitably skilled in policy development. In effect, institutional reforms, particularly the separation of service delivery from regulatory and policy-making activities, reduced access to technical advice and resulted in a loss of 'institutional memory'. This loss impacted later in the planning process, when shortcomings in the process for designating the council's own public works were discovered with some re-notification of designations needed.

Another reason for the delay in notifying the plan was further development of the 'silo mentality' wrought by managerialism. On this occasion, the policy development team was separated from resource consent processing. This flowed from the council's strategic decision recorded in the *1991-92 Annual Plan* (Tauranga District Council, 1991b), but not fully implemented until much later. This separation was formalized in October 1996, when the plan-making team became part of the Strategic Planning Department and resource consent staff became part of Environmental Services, but had been a practical reality for some years. This separation was detrimental to plan-making as the new emphasis on customer service (a quantifiable task) demanded a high commitment of time and energy by the consent processing team. Planners had little time for giving feedback on policy or for participation in activities for which costs were non-recoverable.

Given these managerial changes and separations of function, the comments of a planning committee councillor on its development of the district plan are somewhat ironic: 'restructuring took place (in 1996) to get desired outcomes and because people in departments were not talking to each other and this was reflected politically, too, across committees.'

Another likely reason for the delays in notifying the plan was the difficulty of linking the writing of the district plan to the preparation of the new strategic plan. The council had decided to produce a strategic plan fairly early on, but work on this did not start in earnest until 1996 when the new department was created. Hence, some strategic issues were not addressed in a timely fashion for the district plan programme. Council fully appreciated the place of the district plan 'as an important means of implementing council's sustainable management strategies, [but it was] not the only means' (Tauranga District Council, 1994). Clearly, the council envisaged that additional methods for promoting sustainable management would be adopted, but the strategic plan, which co-ordinates all of the council's activities, was not actually finalized until some two years after the district plan was notified. The senior planner observed that 'because of its direct impact on regulating growth and development, the district plan became a "de facto" strategic plan in terms of influencing city form, character and environment'.

Quality of Plan Produced from Process

The 1997 *Tauranga Proposed District Plan* (Tauranga District Council, 1997b; 1997c; 1997d) was the highest-scoring of the 34 notified district plans that we evaluated (see Figure 7.2). It gained a score of 54.8 out of a possible 80 (or 68.5 per cent), indicating that it was very good though not excellent. The TDP was coded before it was amended as a result of submissions, so the results suggest a fundamental strength in this plan that is lacking in many others.

Planning Approach

The comparatively high score for this plan shows that the writers thought deeply about making a plan that was effects-based and true to the mandate. This reflected the advice given by the Ministry for the Environment, not often seen elsewhere.

The Tauranga District plan was in three main parts. Part I (Tauranga District Council, 1997b) was the policy statement, which is organized around five themes: Amenity Values, Natural Resources, Heritage, Hazards and Physical Resources. Within each theme, there are objectives, policies, methods, reasons and anticipated results. Part II (Tauranga District Council, 1997c) contained all of the rules for each 'activity area' and for particular resources, such as Heritage. It provided for the conventional categories of activities — that is, permitted, controlled, restricted discretionary, discretionary, non-complying and prohibited. A feature of the plan was that it had a wide range of permitted, controlled and restricted discretionary activities, which have implications for cost effective implementation and particularly management of urban growth. Part III (Tauranga District Council, 1997d) contained the planning maps, showing zonings and the areas where special policies and rules apply to the protection and management of various resources and assets — for example, the coastal environment and the airport.

Evaluation of the Plan

One of the key measures of plan quality is 'clarity of purpose', and this plan scored well in that regard. A competent and concise introductory chapter sets the scene for what was to come, covering resource management principles, the council's responsibilities under the RMA and the key strategic themes in the plan. The second chapter provided a policy overview of the five distinct themes noted above, around which the plan was organized. The coherence of the whole plan flowed from a clear understanding of the purpose of sustainable management, which is reinforced by its organization into the five policy themes. This plan was characterized by a simple conceptual framework that was systematically implemented and conveyed in plain language.

This plan also scored well relative to other plans in terms of the 'interpretation and application of the mandate'. Further, alternative methods were fully outlined in a separate analysis, as required by section 32 of the RMA.

As the Tauranga District plan was prepared after the EBOP Regional Policy Statement (RPS) (Bay of Plenty Regional Council, 1996), *Regional Coastal*

Environment Plan (RCEP) (Bay of Plenty Regional Council, 1995) and the *Western Bay of Plenty District Plan* (Western Bay of Plenty District Council, 1996) were all publicly notified, it was reasonable to expect a high score for 'integration', a plan-quality criterion that was measured on two dimensions: integration with other plans, and the handling of cross-boundary issues. In fact, 'integration with other plans' was adequately done to the extent that all relevant plans were mentioned. However, more than a limited explanation of the relationship between the Tauranga District plan and other published plans was seldom provided. Consequently, some opportunities to make more explicit provision for integrated management have been missed. On the other hand, while references to the RPS also had limited direct explanations, it was clear that the plan as a whole had been written with the values expressed in national and regional policies and plans in mind, especially in the coastal environment, which we elaborate on in the final section. Curiously, there was a full explanation of the relationship between the council's own strategic plan and the district plan, yet the strategic plan was not completed until two years after the district plan was notified.

A clear explanation of 'cross-boundary issues' indicates that there was good consultation with adjoining districts and the relevant region. Like other plans, the Tauranga plan did only the minimum required by statute — that was, to set out a process for addressing cross-boundary issues. It did not, however, identify substantive issues and state co-ordinated responses to them.

The 'organization and presentation' of a plan are important to ensure that it is user-friendly. The Tauranga District plan had many good features in this regard. A radical difference between this and other plans was that the rules section was in a separate volume from the issues, objectives and policies. There was considerable debate within the council about the wisdom of this split, but the lead plan-maker reasoned that most members of the public would primarily want to know only the rules which apply to their land, so that they can design a proposal which fully complies. The issues, objectives and policies become important when a proposal requires consent as a discretionary or non-complying activity, a less common occurrence under this plan, and therefore the separation benefits more people than it disadvantages.

A weakness in the organization of the plan was the lack of cross-referencing. Given the separation of rules from the issues, objectives and policies, and the several classes of maps, good cross-referencing was essential to enable the reader to identify all of the relevant plan provisions.

A seemingly small but important advantage of the plan was the choice of landscape format and three columns for text. This was pleasing visually and provides flexibility in layout, especially when inserting tables and diagrams, of which the plan made good use in presenting information and rules. Careful attention was paid to the quality of paper, choice of typeface, editing and layout, all of which contributed to ease of use. Most important, the logical and systematic organization of the plan's contents was reinforced by the way it was presented.

Policy Integration for Coastal Hazards

While our other case studies focused on significant natural areas, outstanding landscape values or Māori interests, this one highlights integrated management as manifested through the hierarchy of policies and plans which the RMA provides for the coastal environment.

We delved into plan content in an attempt to find out whether or not practice had improved since the previous planning regime. We compared provisions in the Tauranga District plan relating to open coast hazards with those in the district schemes prepared under previous legislation, to see if the new policy framework was more robust and the coastal environment better managed (see Figure 11.2). We also traced policy from higher-level policy statements and plans through to district plans, to find out whether or not more integrated environmental policy had been achieved, both horizontally and vertically.

Only for the coastal environment is there a *compulsory* national policy statement and regional plan under the RMA. This hierarchy was not present under the previous planning regime (Figure 11.5). Our Tauranga case provided an opportunity to see how these other policies and plans came to bear upon the making of the Tauranga District plan. Does the national coastal policy send clear signals for guiding regional and local plans, and do regional policy statements and regional coastal plans help guide district plans? These questions, it will be recalled, were raised in Chapters 3 and 4 when discussing the roles of central and regional government in facilitating plan-making in local councils.

Addressing these questions meant making comparisons and drawing conclusions as to the vertical and horizontal consistency between various planning documents in relation to their policies and methods, and as to the extent to which local councils had been creative in their choice of methods for fulfilling national and regional directions. Comparisons were not easily made, because at any one time the various documents could be at different phases of the statutory process. Nevertheless, we attempted to do so because policy integration is an important measure of the effectiveness of the new planning regime.

We focused our evaluation primarily on that section of coast lying between Mt Maunganui and Papamoa (Figure 11.1). Our evaluation of planning for coastal hazards was in three parts. First, we compared the district schemes prepared under the previous planning regime for the adjacent jurisdictions of Mt Maunganui Borough Council and Tauranga County Council. Second, under the RMA we studied the vertical alignment in the hierarchy of policies and plans for managing coastal hazards, and its implications for the TDP. Finally, we made a horizontal comparison of district plans prepared under the RMA for the adjacent jurisdictions of Tauranga District Council and Western Bay of Plenty District Council, where, for the latter, policies affecting the Omokoroa coast were examined.

Town and Country Planning Approach to Coastal Hazards

Although a popular place to live, parts of the Mt Maunganui–Papamoa coastline are prone to erosion or inundation. Successive councils under the old planning regime grappled with the issue of how best to deal with the strong demand for coastal living while managing the risk of coastal hazards. With respect to hazards, use of the *Town and Country Planning Act* (1977) by councils aimed more at avoiding any council legal liability for losses, rather than concern for the environment, life and property.

Prior to local government reform in 1989, two local authorities had jurisdiction over this open coastline. Mt Maunganui Borough was responsible for the part which extended from the Mount itself along Ocean Beach Road to Papamoa West, and Tauranga County had jurisdiction over the area from central Papamoa eastwards to the Kaituna River. The boundary separating the two is indicated in Figure 11.1.

Mt Maunganui Borough Council   The district scheme of Mt Maunganui Borough Council (Tauranga District Council, 1990) characterized the planning problem as the management of urban growth along the coast in an area of known risk from natural hazards. The coastal hazard area was identified through scientific studies, first initiated in 1980, and hazard lines were shown on planning maps for areas yet to be urbanized. As with previous district schemes, there were provisions in the fourth scheme review for the foreshore areas to be acquired for public reserves at the time of subdivision as a means of achieving public access to the beach, conservation of the coastal vegetation and hazard avoidance.

The fourth review of the district scheme designated the extension of the existing foreshore reserve, with boundaries fixed 'by the physical characteristics of the foredunes and not by an arbitrary line [and for] subdivision of the land, the precise boundary will be defined' (Tauranga District Council, 1990: 51). Seaward of Papamoa Beach Road, which runs parallel to the coastline, the designated foreshore reserve was acquired at the time of subdivision. A structure plan was prepared to guide future urban growth in the Te Maunga area and it too showed the proposed foreshore reserve as part of an overall reserve acquisition strategy (Figure 11.3).

For new development, Mt Maunganui Borough's coastal hazard policy was simple and fairly effective. However, as the council did not always succeed in acquiring the whole of the foreshore that was subject to coastal erosion, housing continued to be erected in areas that were potentially at risk, creating a policy problem for subsequent planners. No attempt was made to deal with coastal hazards in those parts of the borough that were long-settled.

While public acquisition may have been a blunt instrument, it had merit as it avoided risk to private property by taking land into public ownership. The planning approach was thus driven partly by questions of liability, although the district scheme did not emphasize risk from either natural hazards or financial liability.

Tauranga County Council   The third review of the county scheme (Tauranga District Council, 1991a) relied on a similar approach to that of Mt Maunganui Borough, with similar outcomes. Scientific studies of coastal hazards were commissioned for Papamoa and an area susceptible to hazards was identified, being some 75-80 metres from mean high-water mark. Structure plans guided urban growth and there were clear policies for acquiring public reserves. Between Pacific View Road and Allan Place, the coastal hazard line lay well within the area acquired for public reserve, whereas in Papamoa East (the Taylor Road–Karewa Parade stretch) the foreshore reserve (Figure 11.3) was much narrower and houses lay within the hazard lines (Figure 11.4). The Tauranga County district scheme did not have a clear, easily administered set of ordinances (rules) for controlling subdivision and development.

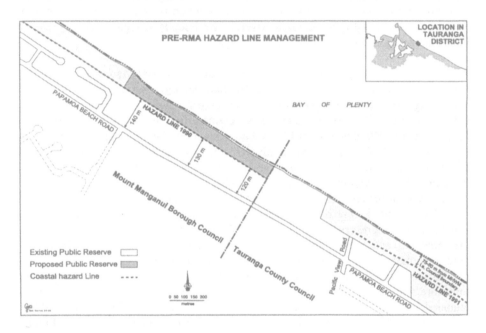

Figure 11.3   Pre-RMA hazard line management along Papamoa Beach (The hazard lines are shown for Mount Maunganui Borough Council (MMBC) and Tauranga County Council (TCC) in relation to existing and proposed public reserves)

Source: Tauranga District Council, 1990 and 1991a

The RMA Approach to Coastal Hazards

Local government reform resulted in the area subject to urban expansion along the Mt Maunganui–Papamoa coastline coming under a single jurisdiction for the first time. Simultaneously, the RMA put in place a more rigorous planning system and the *Building Act* (1991) changed the legal liability of councils. The RMA requires councils to control the actual or potential effects of the use, development or protection of land for the purpose of avoiding or mitigating natural hazards (section 30).

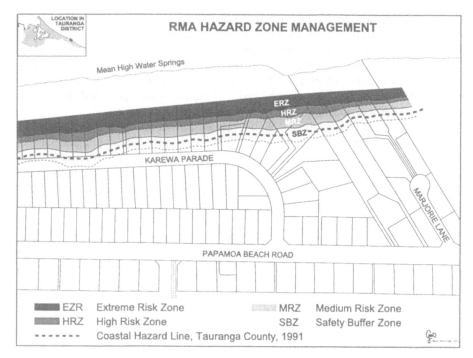

Figure 11.4   Eastern Papamoa showing four risk zones of the Coastal Hazard Erosion Policy Areas of TDC, 1997 and the Coastal Hazard Line of Tauranga County Council, 1991

Source: Tauranga District Council, 1991a: 22; 1997d: CH6

Approach of Tauranga District Council   The Tauranga District plan identified the environmental issue as:

> Use and development in areas prone to subsidence or flooding, or otherwise at risk from hazard events such as coastal erosion or inundation, can expose people and physical resources to unacceptable risk or lead to an accelerated loss of natural resources (Tauranga District Council, 1997b: A71).

The objective on coastal hazards was:

> To avoid coastal erosion and inundation arising from the subdivision, use and development of coastal land and to avoid, remedy or mitigate damage to land, structures and the environment arising from coastal erosion and inundation (Tauranga District Council, 1997b: A72).

A series of policies on foredune protection, subdivision and building control set out how the plan would achieve this objective. Methods of implementation were divided into two parts: rules governing subdivision and alternative methods to rules, such as the provision of information (Land Information Memoranda and

Project Information Memoranda) and the maintenance of a database of known hazards. For coastal hazards, the rules section of the plan defined 'Coastal Hazard Erosion Policy Areas' and 'Coastal Protection Areas' within which subdivision, building, vegetation clearance and certain activities were limited discretionary activities in all cases except in the Extreme Risk Zone, where these activities and stormwater discharges were prohibited. Criteria were provided by which applications for limited discretionary consents would be evaluated.

The 'Coastal Hazard Erosion Policy Areas' (CHEPA) were identified on the planning maps at a scale of 1:2500 in three locations: Mt Maunganui Beach, Omanu and Papamoa. Subdivision and development had already occurred in these areas. There were four risk zones; extreme, high, medium and a safety buffer zone (see Figure 11.4). Rule 17.2.1.2(c) provided that any reassessment of the CHEPA must comply with the methodology presented in the *Regional Coastal Environment Plan* (Tauranga District Council, 1997c: 73).

The 'Coastal Protection Area' encompassed the undeveloped open coastline from Papamoa East to the Kaituna River mouth. The hazard line adopted was 74 metres from the toe of the foredune, or the actual area likely to be at risk from coastal erosion or inundation within 100 years of the lodging of an application for resource consent or a plan change in relation to the land. This area was zoned 'rural' and was not intended to be urbanized within 20 years. A 'Coastal Protection Area' was also denoted on publicly owned land seaward of Maranui Street and Papamoa Beach Road.

The anticipated environmental results for these policies included:

- avoidance of intensive urban development on land subject to natural hazards;
- reduced risk to property from natural hazards;
- maintenance of a natural protective buffer area between the open coastline and development so avoiding the need for hazard protection works;
- choice for owners in using the coastal area so long as it remains prudent;
- avoidance of development close to the foredune that could compromise the stability of the coast or which itself could be subject to risk from erosion or inundation;
- reduction in the net risk of coastal hazards over time; and
- maintenance and enhancement of the natural character of the district's open coastline (Tauranga District Council, 1997b: A77).

The council stated that it was committed to 'monitoring coastal change through systematic data collection and recording programmes and individual site hazard assessments' (Tauranga District Council, 1997b: A101).

Comparisons Then and Now

While there were stylistic differences between the Mt Maunganui Borough and Tauranga County councils' district schemes, there was reasonable consistency in their policy direction and therefore the management of this cross-boundary coastal hazards issue. Thus, the Tauranga District Council inherited similar policy approaches from both previous councils, but not regulatory instruments. North of Papamoa Road, large areas of foreshore were in public ownership, but with stretches of housing located within identified risk areas.

Unlike the earlier district schemes, which controlled development only, in the TDP the issue with coastal hazards was couched in terms of risk management, but not so much of the council's liability as of the real risk to the environment, life and property arising from the adverse effects of erosion and inundation. The discipline of thinking imposed by the RMA and changes in the *Building Act* (1991) regarding liability had provided a much better framework for policy development than under the previous planning regime.

Importantly, the areas subject to coastal hazards in the Tauranga District plan were defined more accurately than in previous plans, with more subtle gradations of risk. Experience since 1980, including the benefit of further in-depth studies (some in conjunction with EBOP), had provided better quality data for planning. There was also greater appreciation of the technical uncertainty surrounding this topic and a realization that processes must be in place to constantly upgrade information.

As a consequence of access to good risk-assessment data, the policy response could be more comprehensive. A wider range of methods was identified in the notified TDP, and they were better targeted to the environmental issue than in previous plans. For example, whereas the Mt Maunganui district scheme offered one approach — public acquisition — the proposed Tauranga plan had six district plan methods and eight other methods (Tauranga District Council, 1997b: 74).

Nevertheless, the Tauranga District plan built on the strategic approach taken by both Mt Maunganui Borough and Tauranga County. For example, it used structure plans to guide urban growth, restrictions on development within areas known to be at risk, and set up a process for further refining the hazard areas at the time of subdivision. Public acquisition was still a means of achieving protection, especially where this would achieve several objectives at once — mitigate the risk of adverse effects from coastal hazards and protect heritage sites or natural ecosystems. Thus, the new policy had evolved from the old, placing emphasis on a mix of private landowner and public sector mitigation.

Another major difference arose because the RMA, unlike the *Town and Country Planning Act* (1977), provided for district plans to specify prohibited activities. Where there were known areas of extreme risk, a council could prohibit subdivision, earthworks or building that would have adverse effects, whereas under the old regime, the only available option was public acquisition. This technique was also of assistance in managing the council's exposure to legal liability, although existing use rights might slow down the managed retreat.

Given the more rigorous planning system under the RMA and the *Building Act*, the legal liability of councils had changed. As a result, the coastal hazard policy for the Mt Maunganui–Papamoa coastal strip had improved in several ways. The Tauranga District Council could:

- commission a scientific study for the whole length of the foreshore that provides a better platform on which to base district plan provisions than the unco-ordinated timing and location of earlier studies;
- focus on managing the risk of adverse effects on the environment, rather than simply options for avoiding legal liability; and
- use the variety of regulatory and non-regulatory methods encouraged by the RMA to gain a better fit between the level of risk assessed and the policy response, thereby helping to avoid the problem of either under- or over-acquisition of land.

Does the Hierarchy Framework Guide Local Planning?

There is both national and regional policy guidance for coastal planning locally through the *New Zealand Coastal Policy Statement* (NZCPS, Department of Conservation, 1994) regional policy statement and regional coastal plan. How well does this hierarchical framework guide the content of district plans, and is it better than under the old planning regime? As we saw earlier, especially in Chapter 4, the new planning regime also encouraged partnerships and co-operation between a regional council and local councils in its area. How effective was this for dealing with coastal hazards?

We examined influences from the mandated NZCPS, *Regional Policy Statement* of the Bay of Plenty Regional Council and the *Regional Coastal Environment Plan* in relation to the district plans of TDC and Western Bay of Plenty District Council (WBOPDC). The relationship between old and new planning regimes is schematized in Figure 11.5, while the relevant policies of TDC and WBOPDC are in the boxed insert.

New Zealand Coastal Policy Statement

The *New Zealand Coastal Policy Statement* (1994) set out the principles and priorities that guide policy development in the coastal environment and fleshed out interpretation of section 6(a) of the RMA, which required:

> The preservation of the natural character of the coastal environment (including the coastal marine area), wetlands, and lakes and rivers and their margins, and the protection of them from inappropriate subdivision, use, and development.

With respect to natural hazards, Policy 3.4 of the NZCPS required that the policy statements and plans of local government: identify areas where natural hazards exist; recognize the possibility of sea level rise; maintain the ability of natural features, such as beach dunes, to protect land uses; recognize that some natural features may in future migrate inland as the result of dynamic coastal processes; and locate and design new subdivision, use and development so that the need for hazard protection works is avoided. In particular, Policy 3.4 provided for avoiding or mitigating hazardous effects, and caution that coastal hazard protection works be permitted only where they are the best practicable option. Importantly, Policy 3.3.1 stated that:

> Because there is a relative lack of information about coastal processes and the effects of activities, a precautionary approach should be adopted towards proposed activities, particularly those whose effects are as yet unknown or little understood (Department of Conservation, 1994: 9).

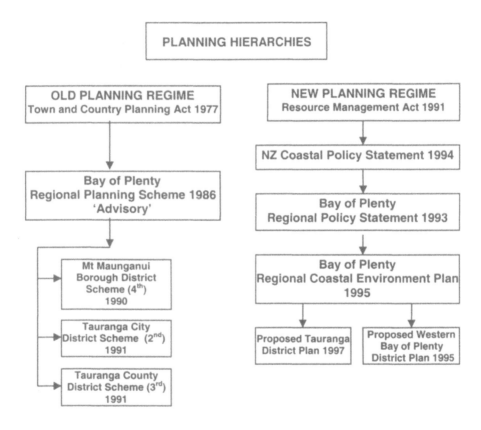

Figure 11.5   Comparison of documents prepared under the old and new planning regimes

EBOP: Policy Statement and Coastal Environment Plan

EBOP's *Regional Policy Statement* (RPS) (Bay of Plenty Regional Council, 1993) was a relatively weak link in the hierarchy in terms of implementation. It referred only once to the NZCPS, probably because both documents were being drafted at the same time, and was not an influence on the two district plans we examined.

Tauranga District Council Proposed District Plan (1997)

Four policies distinguish between levels of risk and also protect the integrity of natural defences to coastal hazards.

*Subdivision, use and development in the coastal environment should not be carried out in areas prone to **extreme** risk from coastal erosion. Uses should be restricted to those that maintain and enhance natural protection against natural hazards provided by the active foredune area (Policy 6.1.4.2: 72).*

*Subdivision, use and development in the coastal environment should not be carried out in areas prone to **high** risk from coastal erosion unless it can be reasonably demonstrated the activity will not accelerate, worsen or result in further erosion or inundation of the land or adjoining land and that the risk is accepted by the landowner (Policy 6.1.4.3: 72).*

*Subdivision, use and development in the coastal environment should not be carried out in areas prone to **medium** risk from coastal erosion unless it can be demonstrated the activity will not accelerate, worsen or result in further erosion or inundation of the land or neighbouring land or buildings and that the risk is accepted by the landowner (Policy 6.1.4.4: 72).*

Subdivision, use and development should not compromise the integrity of natural defences to coastal hazards, the natural character of the coastal environment (particularly in areas where little development has occurred), the relationship of Māori and their culture and traditions to the coast, or public access to the coast (Policy 6.1.4.8: 72).

Proposed Western Bay of Plenty District Plan (1996)

Three policies do not distinguish between levels of risk and only provide for part of the action needed to protect the integrity of natural defences to coastal hazards.

Enable the development or redevelopment of land already subdivided or otherwise developed for urban purposes in areas now known to be at risk from natural hazard only where any likely adverse effects can be avoided or appropriately mitigated (Policy 12.2.2.3: 76).

Enable coastal ecosystems in currently undeveloped areas to migrate inland as a result of dynamic coastal processes (including sea level rise as predicted by recognized national or international agencies) (Policy 12.2.2.6: 77).

Encourage the conservation and enhancement of natural features such as sand dunes and wetlands which have the capacity to protect existing developed land (Policy 12.2.2.7: 77).

The *Regional Coastal Environment Plan* (RCEP) compensated for the shortcomings of the RPS by providing very clear guidance to the local councils. The RCEP was concrete and directive. It identified coastal hazards as an issue of regional significance and recognized the part played by human activities in creating risks. The RCEP adapted the values of the RPS to a particular environment, and applied the principles and priorities of the NZCPS, in some cases importing the precise wording. This served to emphasize policy consistency — see, for example, RCEP Policy 11.2.3(a) and NZCPS Policy 3.3.3.

Because the Tauranga District plan was prepared after the RCEP was well advanced, it was strengthened by the clarity of the region's policy direction. By comparison, the adjacent *Western Bay of Plenty District Plan* was prepared at the same time as the RCEP, and did not have the same rigour. The differences in quality between the two district plans confirm that establishing a clear regional framework is a benefit to district planning. Further, all lower-level policies and plans benefited from the national policy statement because the values that applied to the coastal environment had been thoroughly debated, leaving regional and local councils to focus on policy refinement and methods of implementation.

The proposed Tauranga District plan achieved a good degree of consistency with both the RCEP and the NZCPS in terms of the values that underpinned policy development and the actual approach. Although the RPS was a weak link in the policy hierarchy, this seemed to have had little effect on the quality of policy in the Tauranga District plan, owing to the more concrete influence of the RCEP. There were also some links between the Tauranga District plan and the WBOP proposed district plan, which in turn was not inconsistent with the values and approach of RCEP, although policy was less well executed in the WBOP plan. In summary, the framework provided by the RMA did result in more effective policies at the district level compared with the former regime.

Co-operation on the Coast

Consistency in policy alignment is more likely to be attained when accompanied by co-operation between jurisdictions. Faced with pressure for the intensification of development along its coast, TDC sought expert advice from consultants on how to delineate the coastal hazard zone and options for managing land uses in them. It also sought support from EBOP, which was itself under pressure to produce the regional coastal plan and had turned to experts for advice.

These needs led to joint workshops with consultants and an interchange of information that broadened consideration of coastal hazards for both jurisdictions. In the resulting scientific documents, technical uncertainty was recognized and the precautionary principle adopted (Warrick et al., 1993; Kay et al., 1994; Gibb, 1996). Where the risks were evaluated, policies were well targeted and a range of methods provided that emphasized avoidance and mitigation of adverse effects. Where good data was available, more precise rules were provided in the plan, such as for prohibited activities and assessment criteria for restricted discretionary activities. Where the risks were unknown, there was a requirement for further research before subdivision, use or development could occur. The plans put in place good processes for ensuring that this was done in advance of any development.

Co-operation with EBOP was evident at several stages in the development of the Tauranga plan, from sharing the costs of coastal studies to policy writing. It was also apparent in the neat fit between the provisions of the two plans, notwithstanding their stylistic differences. The respective functions of the two councils were clearly delineated in the RCEP, and so provisions of the TDP were suitably complementary to those. Permeating both documents were the values, principles and priorities of the NZCPS. A less well-developed partnership between EBOP and WBOPDC resulted in a weaker district plan, with no degrees of risk, fewer methods and all proposals for development being assessed as discretionary activities — that is, ad hoc.

Co-operation on the coast helped to reinforce the rational approach to planning adopted by Tauranga District Council. It sought improved scientific information over several years and kept refining its planning provisions accordingly. Co-operation on the coast also helped to highlight how the hierarchy of plans facilitates policy integration at a local level.

Lessons and Epilogue

Climate and coast combine to make the Tauranga District attractive for business and residents. A fast-growing urban environment in a salubrious coastal setting together with a demanding RMA mandate did, however, provide a serious challenge for planners, as did the council's warm embrace of New Right managerialism, which resulted in several restructurings and the progressive separation of functions. Also challenging was the need to reduce risks to development from hazards along the coast. Some key lessons can be drawn from TDC's planning efforts.

Lessons

The creation of a silo system of management reduced the ability of staff from various units within the council to contribute effectively to the plan-making effort. While plan preparation required the integration of staff across significant council units, organizational restructuring reinforced the separation of functions and a sense of competition among the units. It also caused a loss of institutional memory, as experienced staff moved and new staff with limited skills took their place. These changes contributed to the delays in getting the Tauranga District plan to notification. Tauranga was fortunate in having the same lead plan-writer throughout, otherwise the situation could have been worse.

Councils should reflect long and hard about the purpose of a significant structural change and trace through the range of effects, both good and bad, that it can have on operations. It seems unlikely that several structural changes in as many years, as happened in Tauranga, provides enough time to assess whether a change has yielded gains in the efficiency and effectiveness of the organization's operations ahead of further change.

Tauranga provides a more positive lesson from the results we obtained for its coastal planning. We found a consistent philosophy from top (national) to bottom (local) of the policy hierarchy, and collaboration over scientific studies and policy options was fruitful for the Tauranga District plan. Whether the coastal hazard policy for Mt Maunganui–Papamoa would have evolved to the current level of sophistication without the earlier government reforms will never be known. Certainly, there was nothing in the prior planning regime to prevent either commissioning a comprehensive scientific study of the whole coastline, a rigorous focus on risk management or greater inter-organizational co-operation. It is clear, however, that the new planning regime has delivered improved policy, albeit evolutionary rather than radical in its origins. What it is not possible to judge is whether such wholesale reform, with its attendant upheavals, was necessary in order to achieve these evolutionary gains. Still, the more rigorous framework required by the RMA has produced more targeted district plan policies and methods for coastal hazard management in the Tauranga District plan. It will be some years, however, before the effectiveness of these policies can be judged by their results.

Epilogue

There were over 800 submissions to the notified Tauranga District plan, which raised over 3000 submission points. In addition, there were three variations and these added to the hearings process as they 'caught up' with the main plan. The hearing process was a major drain on council's planning and administration resources, with part-time staff being contracted to help with the organization, report distribution, submitter and councillor co-ordination, and recording of the hearings sessions.

A special district plan committee was set up with the power to hear and make decisions on all submissions. Three staff were fully engaged in reporting to the committee through the year-long hearings and decision making process. A good database system was essential.

All decisions were released over a month period between October and November 1998, and 55 references were received. The strategy of trying to negotiate/mediate outcomes with appellants was reasonably successful in settling the majority of appeals, but over a long timeframe. The inability of the Environment Court to really push for solutions (outside the hearing arena) meant there was less pressure on some to work for settlement. Mediation facilitated by court commissioners was the best approach to resolving difficult appeals. After nearly four and a half years, there were still 10 appeals outstanding that were awaiting court hearing time. One such appeal involved the CHEPA (Coastal Hazard Erosion Policy Area).

Papamoa residents had presented submissions opposing the CHEPA to the hearings on the Tauranga District plan, but their arguments did not sway the council so that the CHEPA and its accompanying rules were confirmed. An appeal to the Environment Court followed, seeking the seaward re-location of the CHEPA (so as to lie clear of the properties concerned).

In a 'strenuous defended hearing' involving three separate sessions in court spread over eight months, an array of expert witnesses presented complex technical evidence on the assessment of coastal hazard risk at Papamoa. This was augmented by evidence from local landowners and residents. The nub of the argument was whether Papamoa Beach was in a state of dynamic equilibrium, as claimed by the council, or an accreting beach with a plentiful supply of sand as deposed by the witnesses for the appellant. In contention was the degree of risk to which properties along the beach are realistically subject during the twenty-first century.

The court found that '... the beach is presently in a state of dynamic equilibrium (and) is not only subject to erosive cutback when major storm events occur, but to continual dune line change, depending on wave and wind patterns over different seasons within a year and over successive years ...'. It went on to say that 'predicting the extent of coastal erosion and the associated hazard risk at Papamoa over the next 100 years is very difficult. The marked differences between the witnesses on one side and those on the other underline the point (but) we agree with those witnesses who considered that a lack of field data stands in the way of any safe conclusion that the CHEPA should be modified to the extent of re-siting the hazard zone seaward of beachfront properties' (Wayne Keith Skinner v TDC RMA 1666/98, A163, 2002: 21-2).

Having affirmed the broad approach (the CHEPA) and the location of the extreme, high, and medium risk zones, the court then said that the safety zone had the effect of placing a 'restriction on the properties affected beyond the extent necessary to ensure sufficient and appropriate recognition of coastal hazard risk to those properties during the 100-year forecasting period'. The safety zone was therefore deleted because it was derived from a methodology that 'plied caution with yet more caution' (ibid: 24-5).

This interim decision confirmed the validity of the council's planning approach and showed the relevance of the national and coastal policy framework to the court's decision-making. In particular, the court bore in mind the precautionary principle recognized in the NZCPS when determining how to deal with coastal hazard risk in light of scientific uncertainty. Co-operation between EBOP and TDC paid off — no wedge was driven between the two councils — and the broad approach to risk management set out in the RCEP proved robust. The co-ordination of data collection and analysis was a key element of this inter-agency co-operation.

Mr Skinner, a resident, appealed to the High Court against the Environment Court's decision, but this appeal was struck out because, among other reasons, the grounds did not disclose any errors in law (Wayne Keith Skinner v TDC RMA 1666/98, A163, 2002). The council thus succeeded in defending its planning approach to coastal hazard management at Papamoa.

Clearly, the council invested a lot of time and legal resources in the appeal process overall, including the CHEPA case, and in seeking the court's approval under section 17, First Schedule of the RMA, to get the majority of the Tauranga District plan operative. This became reality on 29 October 2002, some 12 years after the plan-making process initially started within Tauranga District Council.

Chapter 12

Tasman District: Political Populism

Our study of Tasman District Council explored how it went about preparing its resource management policies and plans as a unitary authority with regional and district functions. The council had set forth a challenging goal: to integrate its district plan and various regional plans into one resource management document. In achieving this goal, would being a unitary authority make it any easier to co-ordinate staff and carry out research for dealing with the nationally and regionally important matters laid out in Part II of the RMA? In a district dominated by publicly administered conservation land and where Māori were one of the largest landowners, it was expected that issues to do with natural and cultural heritage would come to the fore. How well did Tasman District Council handle these issues and particularly its relationship with Māori when preparing its integrated resource management plan?

Physical and Cultural Setting

The Tasman District is located at the top of the South Island (Figure 12.1). At 9786 square kilometres, it is one of the largest district land areas in the country. Within it are diverse biophysical conditions. The topography is mostly mountainous, descending to hills and plains on the northern coast. The climate is temperate, with high rainfall to the west and rain shadow to the east. The district's virgin vegetation reflected this physical diversity, with dominantly montane forests (softwood–hardwood–beech) in the mountains and lowland forests (softwood–hardwood) on hills and alluvial plains. Fern and scrub occupied lower hills and wetland parts of the coastal plains. The uniqueness of its environment led to public approval for enclosing 59 per cent of the district's area within three national parks, Nelson Lakes, Abel Tasman, and Kahurangi — almost double the 23 per cent of New Zealand's total land area in national parks (Figure 12.1).

There are five types of managed ecosystems in the district. Plantation timber production and extensive pastoral farming have replaced lowland scrub and forests on the free-draining gravel-lain lower hills, while horticulture and arable cropping have replaced the scrub, fern and forests of the alluvial plains. In the shelter of Golden Bay, aquaculture is expanding in importance, based on mussels and scallops (Figure 12.1).

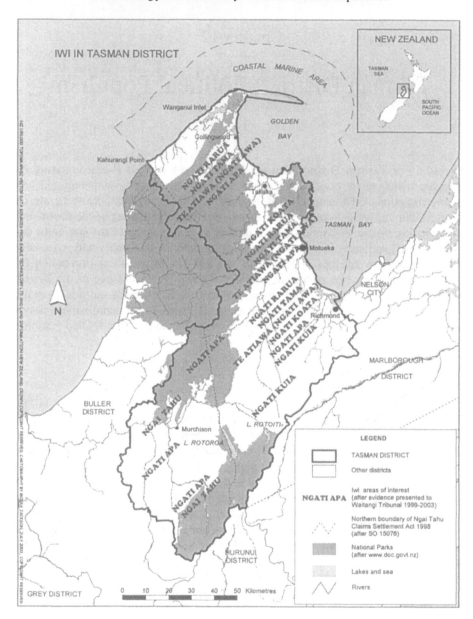

Figure 12.1   Tasman District Council: location of iwi and main settlements

Source: Moira Jackson and Associates Ltd.

Figure 12.2 Northern Whanganui Inlet: an area of cultural significance to Māori, but open to aquaculture and other development if not protected

Source: Department of Conservation, photographer Ron Davidson

There were only three urban communities in the district in 1996, and their populations were small — Richmond (7818), Motueka (6330), and Takaka (1215) — and located near the coast. An extensive road network (2008 kilometres) linked 31 small, dispersed settlements. The attractive environment had enticed alternative lifestylers and retirees, with strong environmental values, and tourists. About 130,000 international and 230,000 national visitors came to the district in 1996, and tourism was increasing rapidly. The district's 1996 population of 37,973 had grown 12 per cent in five years, and there were 2156 businesses employing 10,488 people.

In 1996, Tasman District Council (TasDC or the council) had 17,180 rateable properties returning it an income of $12.45 million. Although Māori accounted for only 7.1 per cent of the total population in the district, they were council's second-biggest landowner after the Crown, and by far the largest ratepayer. The Crown lands, which covered about 5800 of the 9786 square kilometres of the district's land area, were managed by DoC and not rateable by the council.

About 1506 square kilometres of land in the district (or 15 per cent of the total area) was either Māori freehold or general land owned by Māori trusts and incorporations. Of this Māori land, 1260 square kilometres were under the control of the Wakatu Incorporation, representing four of the eight iwi in the district (Figure 12.1). Its lands were valued at $60 million. However, it must be emphasized that most of Wakatu Incorporation's hereditary land titles have been encumbered by Crown-appointed administrators with perpetually-renewable leases to non-Māori tenants.

Apart from the few titles for which the statutorily-imposed and protected leases had been broken, most of the hereditary Māori lands in the region could not be occupied (or worked) by the actual owners who were (and still are) reduced to being little more than passive rent-collectors.

Key Stakeholder Interests

As in any rural district, but perhaps more so in those with high heritage values, development and environment are destined to clash when environmental planning for the future occurs. Tasman District Council was found to be no exception.

Landowning Developers

On the one hand, the main industries developing the district's natural resources defended their interests through relatively strong lobbying of local and central government. Pre-eminent were Federated Farmers and the forestry corporations, with the horticultural and recreational fishing associations more active at local level. There were clashes of interest among these groups (plantation forestry versus pastoral farming), especially over use of marginal lands and the effects of afforestation on surface water reduction (Tasman District Council, 1992).

Environmental Protectionists

On the other hand, various environmental groups were concerned about the effects of resource development on the district's valued ecosystems and landscapes — matters pertinent to sections 6 and 7 of the RMA. Pre-eminent among these groups were the local branch of the New Zealand Forest and Bird Society, Econet (an umbrella for dozens of diverse environmental groups in the district), and the Friends of Nelson Haven and Tasman Bay Inc (focusing on coastal issues).

Iwi Interests

Occupying a unique position were the eight iwi that had resource development and conservation interests — matters of cultural significance pertinent to sections 6, 7 and 8 of the RMA. They included: Ngati Koata, Ngati Rarua, Te Ati Awa, Ngati

Awa,[1] Ngati Tama, Ngati Kuia, Ngai Tahu, and Ngati Apa. All but the last two iwi were within the Te Upoko o Te Ika regional grouping of tribes (Figure 12.1). The residue of Māori reserved lands was owned by the Wakatu Incorporation, which includes the first four iwi named, if not the fifth. In addition to wanting to protect their cultural heritage through the plan, Māori aimed to regain control of at least some historically alienated resources, and this would be in conflict with Pākehā (European) landholders who benefited from them.

Around 1841 when the New Zealand Company first arrived in the Nelson-Tasman region, most of the four thousand or so Māori living in the district were located in the lower parts of catchments, on floodplains and coastal terraces, near river mouths and estuaries. According to Commissioner William Spain's Award of 1845, The New Zealand Company was deemed to have purchased 151,000 acres from local Māori. He further decreed that of the 151,000 acres, 10 per cent (i.e. 15,100 acres) was to be retained by the Māori vendors, along with their pa, cultivations, burial places and wāhi tapu; in all about 18,000–19,000 acres.

The Company and the Crown ignored Spain's Award, leaving local Māori with only about 6,500 acres of reserved lands, large tracts of which were whittled away without compensation by further Government actions. Of the lands which were reserved, most titles were eventually encumbered by leases to Europeans, with perpetual rights of renewal, rent reviews at 21-year intervals, and rents prescribed at fixed percentages of land values at levels well below prevailing market rates. The injustices of all of these actions and inactions are the subject of claims to the Waitangi Tribunal.

A rising issue at the time of plan preparation was tradable water permits. Māori were concerned that their tenants might get control of the resources instead of themselves as the owners of the freehold interest in the lands. Another issue stemmed from some landowners and 'cowboy contractors' hiding the bones of ancestors uncovered during earthworks so as to avoid dealing with the cultural issues implicit in such discoveries. Council's own engineering works had similarly offended customary values on at least one occasion when a registered urupā was destroyed at Wainui Bay during road realignment. These incidents led to iwi pushing the council for a protocol to deal effectively with the various issues, including a requirement for resource consent applicants to consult with iwi over wāhi tapu (sacred) sites held in a 'silent' register.

Institutional and Organizational Arrangements

In 1989, the Nelson Marlborough Catchment Board and Regional Water Board and five borough and county councils were replaced by the Nelson–Marlborough Regional Council, within which were three district councils (Tasman District, Nelson City and Marlborough District).

---

[1] Te Atiawa and Ngati Awa were synonymous until the Waitangi Tribunal process recently saw a separate small claimant group emerge calling itself Ngati Awa (pers. comm., Moira Jackson, 2003).

Local agitation against regionalism resulted in the regional council being axed in mid-1992 and the three district councils each taking on the dual functions of unitary authorities for policy formulation, regulation, advocacy, service delivery, and asset management. There was some talk of Nelson City (population 40,914) and Tasman District combining together as one unitary authority. This did not happen, although the two councils do work together closely on a number of matters, including joint provision of some services.

Both Nelson and Tasman were Agenda 21 councils under the trial programme of the MfE, which serviced these unitary authorities from its regional headquarters in Christchurch. DoC provided its input into their plan-making and advocacy through its conservancy office in Nelson. The Ministry of Agriculture and Fisheries (MAF) and the Ministry of Forestry also operated out of Nelson.

The resource development interests of the eight iwi in the district were held under the umbrella of corporations and trusts, and matters related to statutory planning were delegated to iwi representatives with appropriate skills. People from each iwi were designated to look after the plan-making interests and for reviewing all resource consent applications, not only those deemed pertinent by the council.

Plan-Making Organization

While amalgamation resulted in some loss of regional staff, the internal organization of the council as a unitary authority remained relatively stable throughout the 1990s. The political committees and technical departments mirrored one another, these being environment and planning, engineering services, community services, and corporate services. A fifth structural component was proposed in 1998 for education and advocacy. In addition, two areas, Motueka and Golden Bay, each had an area manager, and eleven sub-committees with delegated powers met as required.

Responsibility for plan development lay with the seven-member Committee on Environment and Planning, formed in 1992. Its make-up had altered twice by 1998, but when the *Tasman Resource Management Plan* (TRMP) was notified in mid-1996, four of the seven members had served more than one term on it (Tasman District Council, 1996d). The committee was serviced by the Department of Environment and Planning, which by 1998 had 51 staff responsible for delivering most environmental management functions, including policy development, resource monitoring, consents (region/district), environmental health, and administration. It was in the small policy development section within this department where much of the plan-making was done.

Plan-Making Resources

It is not easy to be specific about the resources used in preparing the *Tasman District Plan*, because it formed part of a combined TRMP which integrated the district plan with several regional plans already notified — this could, of course, be done only in a unitary authority. As well, the *Tasman Regional Policy Statement* (RPS) provided the policy basis for the TRMP (Tasman District Council, 1994).

Initially, councillors gave staff only six months to prepare the RPS and 30 months for the district plan and its integration with the notified regional plans.

Core staff on the RPS totalled '3.5 person equivalents or FTEs, although that was across a large number of staff who were involved in the process'. For the combined TRMP it was '4.6 FTE staff bolstered by 1 person for 18 months and another for 12 months'. There were also 'two contracts with legal experts' amounting to 0.5 FTE staff. Most of the internal staff had planning qualifications, the remainder being used for policy analysis, environmental management and iwi liaison. The combined cost of preparing the RPS and TRMP to public notification was $1.53 million, of which $171,000 was for the RPS (Table 12.1). The average annual cost of the combined TRMP over the four years to notification was $340,000 or about 2.5 per cent of the council's annual income, or just over $18 per ratepayer per year.

Table 12.1 Costs for key items for the Tasman Regional Policy Statement and Tasman District Plan to notification in May 1996

Items	Regional Policy Statement	District Plan
Staff	130,000	948,000
External consultants	25,000	262,000
External legal advice	-	11,000
Information	-	30,000
Public consultation	-	38,000
Publishing	6,000	221,000
Totals	**171,000**	**1,510,000**

Source: Questionnaires, 1996

Chronology of Plan-Making Process

Council Actions to Notification

It was in late 1993 that the council decided to produce a combined plan — one that would integrate the optional regional land and water plans, already prepared, with the statutory coastal and district plans. The council also decided at this time not to produce a draft of the combined plan for public comment, but rather to publicly notify the proposed TRMP ahead of the October 1995 local body elections. In early 1994, the council also foreshadowed that the policies and methods of the RPS would eventually be drawn into the TRMP, thereby making the former redundant as a separate document.

The regional land, coastal, and district plans combined into the proposed TRMP publicly notified on 25 May 1996, but the job was too big in the time available to include water management, lakes and rivers, and discharge sections. These, it was decided, would be added by way of a later variation. As it was, the council not only missed its original pre-election 1995 deadline for public notification of the TRMP, but also two later deadlines — 23 September 1995 and 26 April 1996. The notified documents referred to above are listed in Table 12.2, along with other the major actions taken in implementing this complex planning process (Tasman District Council, 1996d).

Table 12.2   Main activities in the planning process of Tasman District Council, 1992-97

1 July 1992	TasDC commenced as unitary authority and took over plan preparation processes then under way on RPS, coastal plan, land plan, two water plans, and district plan.
3 October 1992	*Notified Moutere Water Management Plan.*
3 November 1992	*Notified Motueka/Riwaka Plains Water Management Plan.*
23 November 1992	*Notified Regional Plan* [land].
Late 1992	TasDC set tasks, identified issues, produced policy papers, carried out intense policy analysis, prepared drafting instructions, and did just enough community consultation to satisfy statutory obligations.
Early 1993	TRMP discussion papers emerged.
	TasDC held hui on marae over iwi issues in RPS.
Late 1993	TasDC decided on combined TRMP (regional plans, coastal plan, and district plan).
	TasDC foreshadowed the RPS and TRMP eventually being combined into one document.
	Iwi consultants employed by TasDC to help prepare tangata whenua parts of RPS.
Early 1994	Framework for TRMP confirmed.
Mid-1994	TasDC wanted TRMP publicly notified ahead of election; set target date of September 1995.
30 July 1994	Notified RPS and again foreshadowed combining it with the TRMP.
Late 1994	Started on natural heritage areas with student doing literature search for TRMP.
	Iwi consultants employed by TasDC to help prepare tangata whenua parts of TRMP.
Early 1995	TasDC sought DoC help with natural heritage areas, but help delayed five months due to staff ill health, poor weather and lack of resources.
August 1995	RPS hearings and decisions incorporated into the statement. Three months consultation programme on TRMP commenced.
19 February 1996	Pre-notification TRMP received by statutory consultees, like DoC, MfE, Ministry of Forestry and MAF.
22 February 1996	DoC wrote to TasDC stating that an unrealistic timetable had been set for public notification (due to notify 26 April) and a great deal had to be done to the plan.
11 March 1996	Mayor replied to DoC that TasDC will notify as planned; called for DoC response to TRMP.
15 March 1996	DoC said too many issues in TRMP to effectively oblige; noted lack of consultation with landowners on natural and cultural heritage matters. Other statutory consultees responded similarly.

Table 12.2 (continued)

April 1996	TRMP at printers while TasDC sent letters to landowners on aspects in proposed plan.
May 1996	TasDC published *Plan Drafting Protocol: Tasman Resource Management Plan.* TasDC published *Record of Action Taken and Documentation Prepared in Fulfilling Duties to Consider Alternatives and Assess Benefits and Costs Under Section 32, RMA.*
Early May 1996	First that farmers knew of rules for natural and cultural heritage areas; wool-shed meetings attended by council showing controversial maps to which rules applied.
25 May 1996	Notified proposed TRMP (regional land plan, coastal plan, and district plan – two water plans to be added later by a variation to the combined plan); proposed four months of submission time extended ten months due to interest group discontent.
June 1996	TasDC appointed Natural Heritage Working Party to advise on contentious parts of TRMP.
5 September 1996	440 people at public meeting in Richmond Town Hall demanded withdrawal of TRMP.
14 September 1996	TasDC established a Mayoral Working Party to consider options for the TRMP.
October 1996	Natural Heritage Working Party reported recommendations to TasDC.
21 October 1996	TasDC resolved: to retain the proposed TRMP (votes 8 to 5); to have prepared a variation that would delete controversial sections in the plan containing rules; and to further extend the public submission period for the plan.
1 February 1997	Proposed Variation No.1 on the proposed TRMP completed.
18 February 1997	Variation No.1 adopted, ten days prior to submissions closing on the TRMP.
28 February 1997	Submissions closed on TRMP.

Clearly, the council was keen to find advantage in vertically integrating the policies and plans for which it had responsibility. How it went about that task is summarized in the next two sections, which are based mainly on a review of council documents and group interviews with staff from the environment and planning, district policy, coastal, heritage, and consents sections of the council. Quotations are taken from comments by participants.

Research and Policy Analysis

The approach for preparing the TRMP was devised in-house and was the 'same as that for the RPS'. It began with building 'protocols for plan preparation tasks, developing an analytical paper for structuring issues and options, and preparing a think-piece on drafting the plan (such as on how to structure an effects-based plan)'. The team then dealt with plan matters 'issue by issue'. Issue identification was based more on past experience, construed to meet the demands of the RMA, than on newly commissioned research, because neither time nor finances enabled the team to carry out detailed research to the extent desired. Where new issues required attention, such as for iwi interests, information was gained from existing inventories, consulting iwi kaumatua (tribal elders) and employing iwi consultants.

'The framework for the plan was confirmed by early 1994' and the 'policy analysis process was focused into the period mid-1994 to late 1995', during which a series of policy papers on key issues was prepared (Table 12.3) (Tasman District Council, 1995c).

Table 12.3 Issues and options policy papers for preparing the TRMP, 1993-95

Date	Title
April 1993	Regional Coastal Policy Paper No. 1: Port Developments, Small-craft
June 1993	Regional Coastal Policy Paper No. 2: Aquaculture
Sept 1993	Regional Coastal Policy Paper No. 5: Coastal Margins
April 1994	Regional Coastal Policy Paper No. 4: Coastal Conservation
No date	Regional Coastal Policy Paper No. 3: Coastal Water Quality Management
No date	Coastal Water Quality Assessment and Classification
Aug 1994	Reserves and Open Spaces - Strategic Issues and Options
Sept 1994	Agrichemical Discharges to Land, Water and Air
Dec 1994	Classification System for Productive Land in Tasman District
Dec 1994	Rural Cross-Boundary Effects
Feb 1995	Urban Cross-Boundary Effects
Feb 1995	Settlements Issues and Options
April 1995	Designations and Public Works
April 1995	Landscape and Visual Effects
April 1995	Signage Effects
May 1995	Land Disturbance Effects of Mineral Extraction and Land Re-contouring Activities
May 1995	Esplanade Management (Riparian Land Management)
May 1995	Investigation and Assessment of Surface Water Reduction Effects of Afforestation
June 1995	Heritage Policy Paper No. 1 : Natural and Built Heritage
June 1995	Investigation and Assessment of Noise Emissions
June 1995	Assessment of Earthquake and Slope Instability Hazards in Tasman District
June 1995	Transport Effects
July 1995	Hard Rock Quarry Aggregate Source Areas Investigation and Assessment
Aug 1995	Archaeological and Māori Sites Investigation and Assessment
No date	Rural Subdivision Control (Environmental Planning and Assessment for TasDC)

Combining plans into the TRMP meant that many issues and options already publicly addressed by the council in its regional plans were revisited. Other public discussion documents were also emerging at this time, such as the council's *Environmental Monitoring Strategy for Tasman District* (Tasman District Council, 1995b). All of them were directed at drafting the TRMP, using instructions from policy decisions previously made. Finally, all relevant background documents were then 'pulled into a section 32 statement' (Tasman District Council, 1996e), and the plan drafting protocol affirmed (Tasman District Council, 1996c).

Consultation and Consensus Building Actions

Given the statutory requirement to consult iwi, council complied with respect to the RPS, regional plans, and the TRMP pushed by proactive iwi. While an iwi liaison officer was proposed to help the council with this process, none was appointed. Nor were funds made available to assist iwi to participate in it. Nevertheless, iwi were consulted extensively and on marae throughout much of the policy and plan preparation processes, and in general they felt the council had done reasonably well.

Building on consultative processes for earlier regional plans and the RPS, the council consulted other stakeholders on the TRMP. In late March 1995, mid-way through preparing the policy documents, the council wrote to various interest groups explaining that it was working towards the combined TRMP and that consultation would occur over coming months. Through public meetings, special issues and interest group meetings, and street-side meetings, the council would 'explain recommendations in policy papers and discuss issues related to them'. This was to allow 'any policy issues that are flagged as being incompatible with general public opinion [to] be taken back by councillors'. The letter also stated:

> The fact that council is, at the same time, asking for further submissions on the Proposed Regional Policy Statement, and is also processing some significant [regional] Plan changes, does not make it easy for people to get involved and have an overall understanding of the processes and where they are heading ... Accordingly, we understand if you prefer to await public notification of the [combined] Plan (in September) before getting further involved (Tasman District Council, 30 Mar. 1995).

The public was invited to comment on twelve of the 25 policy papers (Table 12.3) then ready for review, which were available in libraries and council offices or by purchase at prices ranging from $5 to $30 per copy — causing a Federated Farmers representative to lament, 'You had to buy your opportunity to be involved'.

Issues raised in the policy papers were discussed by councillors and staff in a travelling road-show around the region's communities during the first half of 1995, as had happened for the earlier RPS and proposed regional plans. Public responses were then synthesized by council staff into a slim volume for review by councillors.

Consequently, staff did not start drafting the TRMP until mid-1995. Councillors were disappointed that the second pre-election deadline of September

1995 for notifying the plan would not be met, and heavily pressured staff to meet a new deadline of April 1996. After the TRMP was substantially completed in late February 1996, there was, according to staff, 'statutory consultation, a little interest group consultation, legal audit, and some testing of rules'. The plan was publicly notified near the end of May 1996.

Evaluation of Plan Quality Produced From Process

When assessed according to our plan quality criteria, the RPS (1993) and TRMP (1996) were good. They scored 47 out of 80 (58.8 per cent) and 48.1 out of 80 (60.2 per cent), respectively. However, when compared with policies and plans in councils elsewhere in the country, they were excellent (second out of 16) and very good (fifth out of 34), respectively. (See Figures 6.2 and 7.2 in previous chapters.) Evaluating components of the policy statement and combined plan shows both comparable strengths and weaknesses, and some significant differences in criteria that underscore measures of quality.

Overall Evaluation

For both the proposed RPS and TRMP, the clarity of purpose criterion was sounder than that for application and interpretation of key provisions of the RMA. Nevertheless, sustainable management and effects-based planning were stressed as driving forces in both documents. Unlike the RPS, the TRMP had no early overview that signalled the anticipated environmental outcomes to be achieved through implementation. Apart from information and monitoring (section 35) and its purpose (section 59), the policy statement was vague on other sections of the Act (e.g. sections 6, 7, 8, 30 and 62). Words like 'The Act requires ...' were used often, but the actual section was not specified, nor interpreted for the circumstances of the district. Except for most matters of national and other importance (sections 6 and 7), the plan was vague also in explaining how key provisions of the RMA (sections 8, 32, 35, 67 and 75) would be implemented. For the all-important functions of regional and district councils (sections 30 and 31) the TRMP gave no explanation at all.

When compared with other councils in the country for which policy statements and plans were analysed, the treatment of 'cross-boundary issues' for the district in the RPS and TRMP was very good, because there was genuine understanding of the need for collaboration with neighbouring councils. This did not, however, carry over to the integration of other internal and external policies and plans. This lack of integration was a problem noted by several people making submissions on the plan. It is surprising, given the dual functions of the council and its broadened perspective as a unitary authority, but perhaps inevitable because other sections of the TRMP were not released at the same time as the district or coastal plans.

Topic Evaluation

Generalising across selected topics (iwi interests, natural hazards, significant natural areas, and the coast), it was clear that the strong policy analysis approach adopted by staff when preparing the RPS and TRMP had resulted in clear identification of issues and strong internal consistency within each document. Thus, for each topic chapter, there were clear linkages in the cascade of issues, objectives, policies, methods and indicators for monitoring results.

However, both the RPS and TRMP had a weak fact base, although they did contain strong descriptive analysis in the introductions to topic chapters and to a lesser extent in the principal reasons and explanations. The lack of factual information in the documents might suggest that the council relied more on institutional memory for dealing with many of the identified issues than on sound research data. However, separate documents containing factual information, including a section 32 analysis, were available for review, though not well referenced in the plan. Also weak were provisions for monitoring, especially in terms of quantitative indicators of use in determining the extent to which desired environmental outcomes were to be achieved.

Overall, the organization and presentation of each document were in marked contrast — the TRMP being the easier to read and understand. The blocked and shaded headings in the plan helped readability. Of the ten items considered important for document accessibility, the plan had five (including table of contents, glossary of terms, cross-referencing of issues, clear illustrations, and a user's guide). The user's guide in Chapter 1 explained how to work with the plan in order to find what type of consent was required and what rules applied. Even so, the plan was not easy to use, because if the user had a particular activity in mind, it was hard to know which sections of the plan needed to be considered. In addition, matters on which the council reserved discretion did not give certainty as to whether an applicant's proposal would be acceptable.

In general, the environmental effects-based TRMP adopted a regulatory approach when dealing with important natural and cultural heritage issues, and disaster potential from hazardous events. These were, however, supported by other methods, such as the provision of information and education. It is not practical here to provide details about the various topics reviewed, so instead we focus on one as illustration.

Iwi interests   Three issues of importance to iwi were clearly identified in the chapter on tangata whenua (local Māori by ancestry) interests in the proposed RPS: iwi/council relationships; cultural values in environmental management; and iwi commercial interests. In the TRMP, the emphasis was on cultural values in environmental management within a chapter on natural and cultural heritage. We focus on sites of significance to Māori.

Although the RPS specified issues and policies on sites of significance to Māori, it did not contain any objectives, a shortcoming in terms of internal consistency. In other words, scores could not be given for links between objectives and issues, objectives and policies, and results and objectives in the cascade.

In contrast, the issues and objectives in the TRMP were clearly linked in the cascade and carried through to policies, methods, and monitoring (Annex 12.1). The two issues and related objectives of relevance to iwi in the plan were that:

1.   heritage resources are finite and need protecting, and
2.   a 'new method [be developed] for areas where the presence of unknown archaeological sites, or sites of significance to Māori, are likely'.

Policies in the TRMP (Chapter 10) therefore aimed to protect specific sites (known or unknown) within specified areas (of known occupation), by requiring a resource consent application for any activity within these areas (Policy 10.1.4), and ensuring that subdivision of land did not damage or separate such sites (Policy 10.2.3). For significant sites on coastal or river margins, reserves and strips would be created to protect them (Policy 10.1.5).

In the special area rules section of the TRMP (Chapter 18), one of the topics for which a strict regime was adopted was controlling activities that might affect sites of archaeological interest falling within cultural heritage areas. The main reason for the rules was given as follows:

> The Cultural Heritage Area covers those areas of the District where there are dense concentrations of sites of archaeological interest and for which rules were developed that were consistent with a landowner's obligations under the Historic Places Act in the event that an archaeological site is to be disturbed while undertaking land use or development. It is better that the prospect of disturbance be investigated before the work actually begins (Tasman District Council, 1996d: Section 18.2, 18/51).

In effect, the rules were aimed mainly at protecting known and unknown sites of significance to Māori which could be destroyed by future development. The strict regime would require a resource consent for any building, other structure, land disturbance or tree planting within the cultural heritage area. What is more, the landowner would be required to consult with the tangata whenua before lodging an application for resource consent.

In the TRMP, the means for monitoring these policies and methods were four-fold:

- liaising with other relevant agencies to advocate further investigation of significant sites;
- developing a programme of public education and advocacy for the protection of sites;
- providing free and subsidized advice from council staff; and
- giving financial assistance from the council to help identify and record significant sites and areas.

Performance of policies and methods was to be monitored by assessing the effects of the setting of heritage sites or areas as a result of subdivision or building (10.2.40).

Community Reactions to the Plan

There was a dramatic public reaction to the TRMP. Some interest groups, especially farmers, wanted it withdrawn because they had not been properly consulted and did not agree with rules that affected or might affect them. Others, who disagreed with some aspects in the plan, simply wanted their say through the normal submissions process.

A planning consultant acting for a large number of property owners and several stakeholder groups disaffected with the notified TRMP explained it to us this way:

> Council did do a lot of consulting early on, but it was more to do with broader issues and principles. Property owners were not consulted over what mattered most, the methods and rules that affect them. They felt that good objectives were compromised by some poor rules.

Why Had This Happened?

In its indecent rush to notify, the council had allowed only two months (February–March 1996) for the statutory consultees to review and comment on its TRMP. The statutory consultees implored the council to delay notification so that it could do more work on the plan, including: better research on key issues; field testing policy provisions, such as special area rules; more comprehensive consultation with landowners over methods and rules; and publication of a draft plan as a preliminary to statutory notification. They warned the council that without adequate consultation there would be a public backlash on a wide range of natural and cultural heritage matters. For example, DoC and Ministry of Forestry staff wrote to the council more than once during the two month statutory consultation period, urging that it delay notification in order to rectify errors and consult with landowners over rules on natural heritage areas identified on the planning maps. They had much earlier warned that methods used for identifying significant natural areas were questionable and much of the data old and unreliable.

Councillors thus came to realize that their travelling road-show around the district a year earlier for discussing heritage matters outlined in discussion papers had been insufficient, and eventually agreed to delay notification of the plan. They delayed it by just one month so that they could feed key aspects of the plan back to their constituents, but not all councillors chose to do so.

Thus, the first that most landowners knew about provisions for protecting heritage values was in wool-shed meetings with council personnel in early May 1996, where maps to which rules applied were shown to farmers and others. This was in spite of a policy paper, with rules and other provisions, having been published for public comment the year before. As news of the provisions spread around the district, discontent escalated. Landowners were incensed at errors on maps and rules for protecting indigenous vegetation on private land, but they were also unhappy with proposals for protecting Māori heritage, landscape priorities, surface water yield, and rural site developments.

In trying to damp community dissatisfaction with the TRMP, the council established a Natural Heritage Working Party in June 1996 to advise it on how better to identify significant natural areas (Tasman District Council, 1996f). It failed to set up an equivalent for cultural heritage. All this reaction excited further interest in the plan and broadened the opposition to it. An environmentalist said that it was like 'Chinese whispers, with Murchison farmers taking the lead to have the plan canned. But foresters, horticulturalists and other landowners were all unsettled by the plan, and a fax war started which was picked up by the local paper. Councillors received threats by fax'. Fact and fiction merged in a heady mix. It was a winter of discontent.

On 5 September, around 440 citizens, mostly from the Murchison and Golden Bay areas, marched on Richmond Town Hall to express their anger directly to the councillors and plan-makers. One of the consultants said it was the most intense meeting he had ever endured.

> Racial hatred was evident and acknowledgement of Māori interests in the plan let it surface ... Rabid comments from some red-necked ratepayers were frightening ... and made hair on the back of my head stand on end ... At one point, a back-country farmer held aloft an artifact shouting the best thing for it was to flush it down the toilet.

The five or six iwi representatives present were not exactly fearful for their safety, but two later acknowledged that they were glad they were all big men and had sat together for mutual support during the meeting. The local newspaper reported:

> Mr Hurley (Pangatotara ratepayer) said that withdrawing the plan is the only option, and would avoid further confrontation and possible litigation. 'A responsible decision has to be made — put out a draft plan, restore some credibility and get back ratepayer confidence... Only 10 or so people wanted to keep the plan (*Nelson Mail*, 6 Sept).

Mr Hurley was elected Mayor of TasDC in the following local body elections of October 1998.

How Did Council React?

A week after the Town Hall boil-over, the council established a Mayoral Task Force to deal with the clamour. Even though environmentalists had a representative on the task force, neither Econet nor DoC was invited. This was in spite of DoC having been heavily involved throughout the plan preparation process (see Table 12.3) and having expressed its concerns about the TRMP in writing to the council. The iwi representative got there only after his associate 'did a haka'. It is not surprising that some members of the task force thought that it was stacked with too many strident landowners from the protest march and with council staff, and that in consequence members were soon being asked to vote on the removal of sections from the plan. Two councillors on the task force who wanted the TRMP thrown out admitted they had not read it, yet they were among those who had pressed for its premature notification. Some members put great pressure on TasDC staff and even ridiculed them in the media.

On the other hand, environmentalists and DoC were included in the Working Party on Natural Heritage, which had a more enduring presence than the Mayoral Task Force. A leading environmentalist in the district reported that the Working Party did good work. Its report to the council concluded that the provisions as written could not be legally defended, more robust identification and assessment methods were required, a plan variation should be used to delete notations from the planning maps (unless landowners wanted them retained), and consequential amendments made. After proper study, significant natural areas should then be reintroduced into the plan following consultation with affected landowners (Tasman District Council, 1996f).

On 21 October 1996, soon after receiving the working party report, and four months before submissions on the TRMP were to close, the council resolved by eight votes to five to:

- retain the proposed TRMP;
- prepare a plan variation aimed at deleting controversial sections containing rules, for reasons to be given at a later date; and
- extend the submission period on the TRMP from four to ten months in deference to public concerns.

The council adopted Variation No. 1 for the TRMP on 18 February 1997, with a submission period of six weeks (Tasman District Council, 1997a).[2] The variation was extensive and included deleting provisions for natural and cultural heritage, landscape priority, surface water yield protection, rural site development, and flood hazard. Of these topics, flood hazard was the only urban issue.

Because it was adopted just ten days before submissions on the TRMP closed and while a regional plan was also under public review, Variation No.1 caused great confusion for people who were then finalizing their submissions on the plan. It also caused a lot of extra work for already overburdened staff.

The Way Forward

In proposing the variation, the council devised a two-year strategy to deal with the TRMP that included three complex sets of parallel activity:

1. on the TRMP — further submissions, reports, meetings, hearings, decisions and appeals;
2. on variations to the TRMP, especially on natural and cultural heritage — submissions, technical investigations, reports, decisions, hearings, and appeals;

---

[2] Recall from earlier chapters that a 'variation' is a legal means for changing a plan from that proposed, and must be achieved through the public submissions process.

3. on water resources for the TRMP — investigations, policy papers, plan drafting, consultation, notification, submissions, and so on.

By 2000, the council had held 20 hearings and adopted six plan variations. It was about two-thirds of the way through the hearing process, and had yet to consider natural hazards, significant natural areas, and cultural heritage. Completing this long and costly process was made more difficult by having a newly elected council (October 1998) with no previous councillors from the Environment and Planning Committee, which had been responsible for preparing the TRMP to notification. The changed membership of the 1998-2001 council reflected community reactions to planning; several previous councillors were voted out because of their support for the draft TRMP and several new councillors were elected because of their stated intent to dispose of much of the TRMP.

Cultural Heritage and Iwi Interests

What, then, happened to cultural heritage and iwi interests through this complex planning process? It will be recalled that the policies and methods for cultural heritage with respect to Māori interests were to apply to all developments within a large area of the district. In the absence of a lengthy period of consultation with and a strong education campaign for key stakeholders, especially farmers, these were bound to be highly controversial in a district where land issues between Māori and Pākehā had simmered since 1841.

A sound protocol had been developed for including iwi interests in plan preparation by having councillors meet on marae and employing two iwi planners to help develop the RPS and TRMP. However, several councillors who had obvious antipathy towards Māori issues did not attend any of the hui or meetings with iwi representatives; these members were often the most vociferous in their support of anti-TRMP and anti-Māori heritage/cultural values in the community. When the plan was notified, however, the section on cultural heritage areas, including iwi interests, was a major focus for discontent.

Other landowners, especially farmers, complained bitterly that iwi had been consulted throughout the plan preparation process about matters of cultural heritage, but they had not. This, however, ignores the fact that a draft policy paper on the matter was released for public comment a year before the TRMP was notified, and comments on it incorporated into a revised paper that was adopted by the council's Environment and Planning Committee in August 1995. This too was made public, and contained rules and other provisions for protecting cultural heritage (see Table 12.3). Meetings were held with the representative of Federated Farmers throughout this process.

Plan Variation for Iwi Interests

Forsaking its protocol for consultation with iwi, the council responded to its disaffected citizens by excluding the cultural heritage areas and area rules in its

Variation No.1 for this reason:

> The council has become aware of a number of problems with the administration of the Cultural Heritage Area rules and considers that further investigation, consultation and review of the area and area rules is necessary in 1997. The Council wishes to remove legal obligations on land owners under the area rules while these tasks are carried out ... deletion of the above Area and area rules [does not affect] authorization ... under the Historic Places Act ... (Tasman District Council, Variation No.1, 18 February 1997a).

It had become clear early in the administration of the proposed provisions for iwi interests, that there were problems for the council, private developers and iwi. First, the cultural heritage areas covered a large part of the district, especially along the coastal margins where much of the original Māori settlements had been located. Second, the rules in the plan covered all development activities in the cultural heritage areas, and even minor alterations to existing buildings would need resource consent. This was in fact a drafting error due to staff misinterpreting what iwi had recommended, and it was to be rectified in the submission process. Nevertheless, that error was itself a trigger for considerable anti-Māori sentiment in the community. The rule was intended by Māori to apply only to new green-field developments, and not small-scale uses like extensions to existing buildings. Third, the obligations on property developers in relation to sites of significance to have an archaeological survey done and to consult with the tangata whenua of the area affected by the activity, would be a drain on the resources of all parties (Tasman District Council, November 1996b).

The overall problem with this section of the TRMP was not with the general approach to protecting cultural heritage. Rather, it was that not enough research and testing had been carried out before notification, coupled with a failure to prepare administrative procedures and guidelines for assisting parties seeking information and consents. Beyond these deficiencies was the disaffection of some Pākehā with things Māori, which may or may not have lessened with appropriate consultation between parties in the affected areas. Be this as it may, the council did not consult iwi on these perceived problems and how they might be solved.

Iwi Response

The council's action in adopting Variation No.1 placed its Māori consultants in an invidious position. Because of the historical Māori distrust of Pākehā governance, it had taken a great deal of effort by them to convince kaumatua (elders) from the various iwi that they should join the consultative process for preparing the TRMP. As one said, 'It was hard to get them on board; really hard work!' Having done so to the point of what seemed to be mutual satisfaction over issues of iwi interest, the council then ditched them in response to vociferous protests from some landowners. The council had consulted iwi well up to the time of plan notification, and then ignored them over its variation. The iwi consultant continued, 'We jumped through all the non-Māori hoops and got a non-Māori result by those non-

Māori who did not follow the non-Māori process. They stamped their feet loudest to get their way'.

Just because cultural values were seemingly protected in the TRMP did not mean that iwi were entirely happy with it. In their submissions on the plan, they objected to rules on several cultural and resource development matters (Tasman District Council, 1997b). For example, papakainga was defined too narrowly, restricting it to marae-based land instead of all Māori land (see Glossary). On development, they objected to zoning Marahau land for farming when it has conservation and tourist potential — not surprising, given that Māori were the owners through Wakatu Incorporation of more than 280 acres at Marahau in five large blocks, some of them in prime beach-front locations. They also opposed rules on site development, and so on.

The main concern of iwi was not, however, with the content of the plan, but with the process by which it came to be publicly dealt with. Submissions on the plan clearly show that iwi did not want either the rules or notations on planning maps withdrawn by a plan variation. To do so was seen as a betrayal of the democratic process of submissions, cross-submissions, and appeals. The Ngati Tama Manawhenua ki Te Tau Ihu Trust said in its submission (emphasis in original):

> ... the decision to remove Section 18 ... was taken without consultation with tangata whenua, representatives of whom had been involved in work which led to [its] drafting ... In its peremptory decision ... Council gave in to unseemly mob rule, exacerbated by certain statements and actions of a small number of councillors who had not participated to any noticeable extent in the public consultation processes which had preceded drafting the [TRMP] and who chose subsequently to distance themselves from it. We know which Councillors and staff attended the several meetings and hui to discuss Māori heritage and cultural issues ... and we know which Councillors acted most vigorously to torpedo those sections of the proposed [TRMP].
>
> We believe that Council's decision to accede to pressure from a very vocal minority was an abdication of democratic principles.... Every time there is a clamour at the gate is Council going to give in to whoever shouts loudest or makes the most outrageous threats? We find the implications of Council's behaviour quite frightening – not for the actual decision that was taken, but for *the process* which Council allowed itself to follow.
>
> By all means throw out Sections 18.1–18.5 if *at the end* of the submission and cross-submission period the overwhelming *sensible* public input identifies *serious* shortcomings and/or – better still – more appropriate alternatives recommended. But do not [do so] because a few people choose to air their prejudices ... as in that infamous public meeting last year (Ngati Tama Manawhenua Ki Te Tau Ihu Trust, 27 February 1997).

These strong words show how deeply the Tasman District Council had offended iwi by its actions. How imperative was it for the council to proceed in the manner stated in this iwi submission in the face of stormy protests from some other stakeholders?

Public Submissions

A review of submissions on the cultural heritage section (TRMP, Chapter 10) showed that 57 remedies sought in submissions by all parties on its issues, objectives and policies supported the section (75 per cent) (Tasman District Council, 1997b). For fifteen others (19 per cent), qualified support was given. Only four remedies (6 per cent) sought, opposed this section. This pattern was loosely reflected for the three policies that included matters of interest to Māori.

However, when the special rules for cultural heritage areas in Chapter 18 were considered, this pattern was dramatically skewed by what can be explained only as a 'write-in campaign', which wanted the council to 'Remove all references to Cultural Heritage Areas from Chapters 10 and 18.2, and all Area maps'. This phrase is repeated word-for-word 170 times in the summary of decisions requested, and makes up 76 per cent of the 222 remedies sought on rules. Another 28 remedies (14 per cent) also wanted the cultural heritage provisions or rules deleted from the plan, with a further five (2 per cent) wanting this done until such time as the methods and rules had been revised following consultation with affected parties. Only seven submissions (3 per cent) could be regarded as supporting the rules for cultural heritage areas.

The number of negative responses to the rules may well have been encouraged by the council's own actions after notifying its plan. First, in June 1996 it wrote to people in the cultural heritage areas outlining a less onerous method of compliance with rules in the plan (Tasman District Council, 1996a). Second, in October 1996 the council signalled that a variation to the notified plan regarding the methods and rules was imminent and that the deadline for submissions on the plan itself would be extended by four months, thereby allowing opposition to mount (Tasman District Council, 1996b).

Unique or Ubiquitous?

It is worth knowing whether the public response to cultural heritage and iwi interests was unique or part of a general pattern. Our evaluation shows that the patterns of public response on natural hazards and natural heritage, including significant natural areas, were similar to cultural heritage, but stronger. Remedies sought on issues, objectives and policies for natural hazards and significant natural areas were 96 per cent and 95 per cent supportive, respectively. For rules, both topics gained more than 50 per cent support, except for flooding, which was just under that democratically significant number. As for cultural heritage areas, one might well question why these rules and notations were removed by Variation No.1, instead of allowing the hearings process to run its course.

Inside Looking Out at the Outside Looking In

It is clear from our narrative and analysis that Tasman District Council produced a good plan which squarely addressed the main environmental and cultural issues warranting attention under the RMA. This can be attributed to the dedication and skill of its technical staff and helpful input from many stakeholders. It is equally clear that community acceptability of the TRMP was badly compromised by truncated research and, especially, consultation processes. This can be attributed largely to the questionable behaviour of too many councillors.

Obviously, developing a combined plan was a new experience, and a great deal was being learnt as the process unfolded. Many lessons can be gained from this by drawing directly on the candid views of those most directly responsible for, and affected by, it — staff, councillors and key stakeholders, including iwi. These views are summarized next.

The Plan-Makers

Looking back, the planning staff believed that the plan-making 'approach [to notification] was comprehensive, well structured, and analytically robust', but some elements were 'underestimated at key points in the process'. Lead staff saw the wider planning team as being too 'small' in number; too 'uneven' in quality, 'unable to keep up with the complexity of issues addressed due to other work commitments', and needing 'better integration'. Indeed, 'consultants were employed to do specific tasks, because council staff did not have enough analytical skills', and while 'some consultants were very good, others were problematic'. Staff from elsewhere within the council helped, but the plan-makers found them to be 'stressed and daunted by the process', especially by having to work under very tight time constraints. Staff also thought that 'more time and money and better staff selections would have greatly reduced many of the problems'.

Staff believed that the 'compressed time-lines meant that not only was there no draft plan' for public comment, but also the 'plan-making period for the TRMP was six to nine months shorter' than it ought to have been. In that extended period of time, they 'could have dealt with the problems known to them prior to notification'. They also noted that the Government did not make allowance in the RMA for the more complex combined plans in setting the statutory deadlines. Nevertheless, staff did accept the deadlines, and so deflated the expectations of councillors and community when they could not be met.

The staff also thought that, 'while very trusting of staff, many councillors sleep-walked through the process, unaware of the mounting conflicts' that would eventually lead to 'the need for damage control'. Although the public had been given good opportunity to comment on policy documents, with 'outcomes synthesized into a slim volume for councillors to review', that mode of consultation 'did not warrant the effort because methods and rules were needed to sharpen up [public and council] responses ... Tough public comments on the notified plan led to more investigations and meetings with stakeholders'. Parts of the plan were 'withdrawn ... to maintain good community relations', even though some segments, like 'cultural heritage, were leading-edge planning'. On natural areas, staff noted that 'we were in such a hurry, we made mistakes and it stoked the fires of concern ...We misread the complexity and contentiousness of this issue earlier [and] got behind in providing necessary information ... This stretched all concerned ... It was a mistake to include the information in the TRMP'.

The Councillors

A senior member of the planning committee said that, 'in retrospect, we said that the plan had high priority, yet failed to fund it properly. ... Had we been less political ... as committees over getting [our cut of] resources and keeping rates down ... more funds would have gone into the combined plan. ... We ought to have forgotten about sealing the roads, and put adequate funds into planning'. The 'consequence of inadequate funding', ironically, was that the plan preparation process 'has taken too long and has been too costly'.

Extraordinary though it may seem, one councillor confessed, in a self-remonstrative way, to there being a suspicion that staff wanted delays in notifying the plan in order 'to protect their jobs', since it was believed that 'fewer planning staff would be needed' after that. This view, which will have helped drive deadlines, reveals as much naïveté over what is required to publicly review notified plans as it does over preparation of them.

A key councillor on the committee was disappointed at the lack of public response to the advertised policy papers among the community, but thought that overall the committee had learnt a great deal about the district from the plan-making process. Of particular note was a lot of education about Māori interests in the community. The councillor also noted that more than 80 per cent of the plan was initiated by staff, and that councillors did not become much involved until the submission process got under way. He would 'vary the plan-making process little' if doing it again, which is a disconcerting comment given the dramatic public reaction to the notified TRMP.

Iwi

Iwi consultants regarded the staff and councillors with whom they dealt as good, but resented the few councillors who attended none, or very few, of the consultative iwi meetings, but then proceeded to sabotage iwi efforts. They acknowledged that staff were stretched to do too much too soon, yet said they were good under pressure. But Māori, too, were stretched by having to meet the demands of the RMA across three unitary authorities in their rohe (tribal area). Unlike councillors and council staff, iwi were not recompensed for their time and efforts which must have accumulated to thousands of person-hours. Each iwi contributed from their own coffers to a small fund that partly supported six part-time iwi representatives who kept their fingers on the pulse of resource activities in the region and matters relating to the RMA, including resource consents and archaeological sites.

Some Other Stakeholders

The representative from the 100-strong Friends of Nelson Haven and Tasman Bay felt that the council consulted well on coastal matters because they talked, listened, and mostly acted appropriately, unlike some other councils they had dealt with.

The Tasman District Horticulturalists, representing nine associations, accepted the need for effects-based regulations, but were critical of the council for not applying proposed rules in a practical context and discussing them with those potentially affected by them.

A representative from Weyerhauser NZ Ltd (formerly Fletcher Challenge Forests Ltd) noted that the long journey on the land plan showed that 'between start and finish there was a large swing in approach [which] resulted in terrible rules [and] outcomes that seemed conceptually at odds with earlier consultation and concepts'. Consultation had 'started well, but then faded'. He was critical of the 'confusion caused by council through its submission processes'.

A spokesperson for the Forest and Bird Society and Econet said that 'public consultation was not good, especially in the later stages', and they had major concerns about the structure and content of the plan.

From Federated Farmers came the view that:

> the RMA is sound, but needs good advocacy ... The council has to educate and nurture landowners [but] this reaction to the TRMP has set the process back 10 years ... Council did consult farmers well on some issues, like riparian strips, but not on other key issues, like significant natural areas. Specific staff must share blame for problems over consultation. We cannot work with them. Much effort by Federated Farmers, DoC, Ministry of Forestry, and others on landscape priorities seems to have been ignored, until the crisis hit.

The Ministry of Forestry representative said that 'dribbling out 27 issue papers early in the consultation process for informal discussion was ineffective, although the quality of some issue/policy papers was quite good'. After a 'long period of non-contact, the consultation draft of the TRMP, which was incomplete, arrived for review [which was] expected to be done within three weeks. When the notified plan appeared, it had a lot more in it, but it was of poor quality.' He believed that 'TasDC was trying to do too much with too few resources, and staff had a hard time trying to produce an effects-based plan'.

Better Plan-Making in a Unitary Authority?

Clearly, only in a dual-function unitary authority was it possible to integrate regional and district documents into one combined plan. The Tasman District Council chose that option, and used its RPS as the policy basis for the TRMP. Elsewhere, district councils were required only to produce plans that were not inconsistent with the RPS prepared by the relevant regional council. Nevertheless, staff at Tasman District Council found it very difficult to complete the task of combining all plans in the time allocated by the councillors. Thus, some of the regional plans had to be integrated into the TRMP by variation after its notification.

Agencies such as DoC and forestry companies appreciated the 'one-stop shop' provided by a unitary authority. It meant that they did not have to deal with physically and functionally separated regional and district agencies when addressing RMA matters.

Iwi had mixed views, and while they also found it useful to be arguing for their interests with a single set of planners and councillors for both region and district, there were problems for most of the iwi whose rohe (tribal boundaries) extend across at least two other unitary authorities — Nelson City and Marlborough District Council. It was inefficient and confusing when the Resource Management Plans (RMP) of each unitary authority contained different criteria, rules and processes. Iwi also experienced many instances where it was only the planning staff of the council who seemed to understand who the iwi were, who the iwi representatives were, what iwi views were about certain issues, what the Treaty and Treaty principles in the RMA require/demand, and what the council's own RMP said about such matters. Iwi often found that many senior staff of other divisions of council had little or no idea of such matters. This situation was no great improvement from the days prior to the abolition of the regional council. For iwi, the unitary authority model made little difference to their aspirations.

On the other hand, environmental groups maintained that the system of separate regional and district councils allowed for checks and balances, which tended to be lost in unitary authorities. They saw internal conflicts arising quite a lot in the council, even though the legislation requires they be avoided. More specifically, the representative of the Friends of Nelson Haven and Tasman Bay had read all regional coastal plans for the country, and thought that regional councils could focus on the task of developing a coastal plan better than could dual-function unitary authorities, such as TasDC. His members were wary of combined plans, because they were not seen to be as effective as a separate coastal plan. For this group, the TRMP lacked clarity in defining regional and district functions, and lacked consistency with various plans — in spite of TasDC being a unitary authority. We also found in our evaluation that the TRMP, while good on cross-boundary issues, was not nearly as good as we supposed on integrating other plans. This flaw was likely due to the rush to publish and notify the TRMP.

Almost all stakeholders we interviewed in the region thought that unitary authorities in general, and Tasman District Council in particular, tended to be under-resourced for carrying out their dual-function tasks. Ironically, the savings that would accrue from combining as a unitary authority was one of the main arguments from the local ratepayers for creating them. What is more, it is clear to us that most regional and district councils in New Zealand, especially the smaller ones, were also under-resourced for carrying out their plan-making functions.

Tasman being a unitary authority did not avoid other shortcomings we saw in other councils. For example, Tasman's councillors proposed unrealistic deadlines for completing the TRMP and would not agree to a draft plan being made available for stakeholders to comment on. The staff did not carry out effective research early enough in the plan preparation process, failed to test important rules to see if they would work, and were unwilling or unable to carry out, ahead of publicly notifying the TRMP, stakeholder consultation on methods and rules that were likely to be controversial. As in other councils, farmers complained about iwi consultation and reacted badly to methods for protecting matters of national importance.

Nevertheless, within the council, the planning staff approved of the unitary structure because it reduced the sense of hierarchy (from region to district), and enabled staff from one department, like engineers concerned with flood problems, to collaborate with planners from another department in solving problems, rather than having to make submissions to a regional council. On the other hand, some staff, especially from the former regional council, felt that their interests had simply been grafted onto those of the district council. Consequently, the organizational structure had changed little since amalgamation in 1992, resulting in their interests in resource information appearing to be of lesser importance than other aspects of environmental planning and management. Nahkies (1998) was commissioned by the Department of Environment and Planning to review its structures. He found a lack of integration across structures and that some staff who could have contributed to the final preparation and review of the TRMP did not.

The local politicians generally supported the unitary authority structure, because they saw little point in having two separate councils for carrying out integrated resource management in the area. It was, after all, seen as a means of keeping down costs to the ratepayers.

Lessons and Epilogue

Tasman District Council showed enthusiasm for the unitary authority system and its perceived efficiency and effectiveness, and its staff considerable foresight in working towards an integrated approach to the sustainable management of resources in this area. As a unitary authority, council could 'integrate policy' with respect to regional and district functions and with regard to the ecologically sustainable use, development and protection of its natural and physical resources. What is more, it could work towards an 'integrated process' for marshalling resources within and beyond its organization that aimed at developing a combined plan — an 'integrated product'. Finally, as a unitary authority, it could aspire towards an 'integrated outcome' through adopting an 'integrated state of the environment management strategy' (Bush-King, 1997; Frieder, 1997: 37-41).

Unfortunately, councillors did not grasp the magnitude of the task of preparing a combined plan, and too few resources were made available for its completion. Consequently, councillors and staff were placed under great stress. We are moved to say that 'never was so much done by so few in such a short time with so little'. What are the long-term costs of this behaviour?

Lessons

This case revealed unrealistic political expectations in plan-making, along with blatant political populism. On the one hand, the processes for identifying issues and policies and for discussing them with interested parties was of a standard that ought to have produced a high-quality plan. On the other hand, unrealistic deadlines set by councillors, coupled with inadequate resources for producing so many plans in tandem, resulted in a combined plan of lesser quality.

In the end, these factors drastically compromised the ability to develop methods and rules that were not only technically feasible, but also publicly acceptable. Failure to achieve the latter was due largely to a truncated consultation process that left farmers believing they had been ignored in comparison with iwi. While this affected planning provisions across many of the district's natural and cultural resources, it was profound with regard to iwi interests. This served to foster the legacy of distrust among iwi and farmers, to the cost of all.

Councillors are elected to govern according to the laws of the land, including the RMA and its clauses regarding Treaty matters and Māori interests — regardless of personal beliefs and prejudices. Forging a meaningful relationship with iwi requires the senior political figures in both council and iwi agreeing on how best to proceed with respect to mutual interests. A good partnership fostered by key councillors and their staff and iwi leaders and their representatives is easily damaged by the 'white-anting tactics' of the prejudiced few.

Epilogue

The effort and resources required to bring the TRMP together have been substantial. Recently, senior staff reported that trying to give effect to the theme of integrated resource management and bring together a combined plan representing all unitary functions has been an intellectual and logistical challenge. This has been met by progressing the plan in parts, which has created difficulties for members of the public in keeping up with the associated workload demands. The council's attempts to protect iwi issues in relation to cultural heritage sites have been problematic because of public response and fears, but the importance of the issue has not diminished. Appeals were lodged by iwi to the deletion of the cultural heritage maps from the plan in 1997, but the matter has still not proceeded to a hearing by 2003. Mediation had taken place and a programme of work to better define areas of cultural sensitivity had been undertaken.

Iwi concerns over the use of coastal space have emerged through a complicated and lengthy Environment Court process determining the plan provisions for aquaculture. How iwi aspirations will be provided for is still subject to ongoing discussion. This aspect alone has been further complicated by other measures, such as further legislative reforms, Waitangi Tribunal claims and other legal proceedings. These external factors greatly influence any local authority's ability to address iwi issues through its resource management and planning framework.

An iwi representative put it this way: 'The coastal plan has been so controversial — especially in the ways it has or has not dealt with aquaculture — that it has taken more than two years in the Environment Court and probably more than $4 million in legal fees. Iwi are in the thick of this squabble, trying to balance and protect their many commercial interests in quota fisheries and aquaculture, alongside their traditional/customary interests and their Treaty Article III rights in recreational fishing'.

Annex 12.1

Planning for sites of significance to Māori through the notified Tasman Resource Management Plan of 25 May 1996
(relevant excerpts from the plan)

Issue

Inventories of heritage resources are incomplete. A significant issue is the likely presence of archaeological sites which are currently unknown, but for which protection is desirable (including) sites of significance to Māori.

Objective 1

Protection of heritage resources of significant value to the district.

Policies

1.3 To recognize and protect ... archaeological sites, or sites of significance to Māori which are included in the Historic Places Trust register.
1.4 To protect archaeological sites, or sites of significance to Māori within specified areas whether there are known or unknown sites, by requiring a resource consent application for any activity within these areas.
1.5 To protect archaeological sites, or sites of significance to Māori in coastal margins or river and lake margins to the greatest extent possible by means of creation of esplanade reserves or esplanade strips.
(Other policies protecting natural heritage indirectly have significance for Māori.)

Methods of Implementation

(a) Regulatory: rules in the plan to achieve the above policies, except for activities on sites in residential, business, commercial and industrial zones.
(b) Investigations and Monitoring: Liaise with other relevant agencies to advocate further investigation is undertaken in respect of the location, nature and significance of archaeological sites, or sites of significance to Māori.
(c) Education and Advocacy: (i) Implement a programme of public education and advocate for the protection of significant archaeological sites and sites of significance to Māori; (iv) provide free or subsidized technical advice from council staff regarding archaeological sites, sites of significance to Māori ...
(Other policies indirectly supporting archaeological sites of significance to Māori.)

(e) Financial Incentives: (i) Financial assistance to protect or enhance natural and built heritage features and values and archaeological sites, and sites of significance to Māori in the district ...(iii) Rates relief on covenanted land or for heritage resources or sites which are registered either by the Historic Places Trust or in the plan.

Objective 2

Protection of relationship heritage resource may have with surrounding land.

Policies

2.3 To control the subdivision of land within the Cultural Heritage Area to ensure that there is no damage or destruction of archaeological sites, or sites of significance to Māori as part of the subdivision process; and to ensure that these sites are not unnecessarily or unreasonably separated.

Method of Implementation

Regulatory rules in the Cultural Heritage Area to protect ... sites significant to Māori.

Conclusion

A Decade On: Unfulfilled Expectations

New Zealand's brave new world under the RMA has not eventuated. Instead, early enthusiasm for the new planning regime has been eroded by controversy over plans, coupled with persistent public discontent about their administration. This discontent has been manifested particularly in the business sector's campaign for reduced compliance costs, but is by no means confined to those with vested interests. Indeed, the RMA is blamed for all sorts of things, including things that have little to do with it, so much it has become a 'blob in the middle of the New Zealand psyche'. For most of the 1990s, the Government starved its departments of the resources needed to implement the new mandate, especially for capacity-building at regional and local levels. Consequently, the credibility of the planning system was undermined, the planning profession demoralized and, we must assume, the environment suffered as a result. Here, we explain why the great expectations for the RMA were not realized.

Radical Changes in Planning Governance and Law

The Government engineered 'system change' in the late 1980s, in order to create the intergovernmental structure and partnerships necessary for implementing its new RMA 'mandate'. This included a new central administration (MfE, DoC, and PCE) and amalgamations to increase the 'capacity' of regional and local councils to comply (Chapter 1).

There were great expectations for the success of the RMA. National policies and standards would emerge through MfE and DoC that would help guide local government in the preparation and implementation of policy statements and plans (Chapter 2). Councils would be 'induced' into preparing innovative plans for managing environmental effects. These would be more permissive than their predecessors, but would require more rigorous research and policy analysis. Implementation through resource consents would ensure quality environments resulted.

Our research into planning and governance under the RMA shows that there is cause for concern for our environmental futures. The quality of notified regional policy statements and district plans that we evaluated ranged from only good to poor, with most falling below a pass mark (Chapters 6-7). While the statutory hearings process tended to improve 'community acceptance' of notified plans, this was achieved at the expense of their 'environmental fit' (Chapters 9-12). In other words, good effects-based plans got watered down to meet the demands of some property owners, while many councils avoided public backlash by producing activities-based plans of old.

While there were good gains on paper for Māori because of their inclusion in the plan preparation process as statutory consultees, competing political ideologies in Government (social democrats versus liberal democrats) resulted in a lack of clarity on partnership and that affected the commitment of councils to Māori interests (Chapter 5).

These results suggest that current plans alone are unlikely to yield the quality of environmental outcomes sought by the RMA. This is not so much the fault of the RMA, although some key provisions in the Act still need to be clarified. Nor is it so much the performance of hard-pressed planners in the councils. Rather, these results can be sheeted home primarily to failings in the intergovernmental system (Chapters 3-5). In other words, while shortcomings in professional practice go some way towards explaining the mediocre quality of plans, it is shortcomings in governance at each level in the intergovernmental partnership that prevented better outcomes being achieved.

Governance Shortcomings at the Centre

Every system has a goal, but the problem with the RMA is that the goal is not clear. Uncertainty about the meaning of Part II of the RMA — purpose and principles — caused difficulties for plan-makers charged with developing a framework for achieving sustainable management of natural and physical resources. The problem was one of 'mandate design'. This was exacerbated by a lack of funding by Government for capacity-building, such as for directing councils on matters of national importance and environmental standards. This was a problem of 'implementation'.

Mandate Design

Limiting sustainable development The exclusion of non-renewable resources from the RMA meant that councils could take only a piecemeal approach to sustainability. As well, by being narrowly construed, the RMA mandate avoided meeting head-on the challenge of integrating environmental protection and economic development in regional and local planning. This enabled a shift in focus away from the politically less palatable regulatory controls on land use activities and towards planning for environmental outcomes. The emphasis in the RMA was therefore on 'environmental sustainability' rather than 'sustainable development' as it is understood internationally.

This approach was an attempt by the government to reconcile competing political imperatives. On the one hand, the green movement was pushing for greater environmental protection; on the other, the New Right agenda required a smaller role for government and a greater role for the market in allocating resources (see Introduction). The internalizing of environmental externalities through resource consents would, it was argued, ensure that resources were wisely used for the benefit of present and future generations, obviating the need to define 'sustainable development' more fully in the statute. Government's failure to reconcile these opposing political demands and to not provide national direction left a legacy of poorly targeted district plans and helped to create widespread disenchantment with the RMA.

Meaning of sustainable management The lack of direction for plan-makers was also manifested in lengthy arguments over the meaning of 'sustainable management' in the Act's purpose (section 5), especially its conditioning sub-clauses (see Appendix 1).

Councils had trouble understanding what was required of them under section 5 and were confused further by messages coming from the Minister for the Environment (Simon Upton) who used the 'bully pulpit' to promote the RMA as a statute setting out environmental bottom lines, thus emphasizing one element of the purpose of the Act — avoiding, remedying or mitigating any adverse effects — at the expense of a more holistic view of sustainable management (Chapters 3 and 9).

When preparing plans, many councils either ducked the task of articulating sustainable management by choosing to write district plans that were based largely on their earlier activities-based plans, or regurgitated key phrases from the Act to avoid 'getting it wrong'. To avoid confrontation, many councils negotiated a resolution to disputes over plan content rather than defend their policies in the Environment Court (e.g. Chapters 9 and 12). Only when a proposal for using or developing resources required a specific consent was the inadequacy of the plan revealed in its lack of a rigorous framework for assessing environmental effects. Many plans were therefore enabling, not so much because they fulfilled the intention of the RMA, but simply because they were weak.

Māori partnerships Failure to clarify the nature of the partnership between the Crown and Māori and between the Crown and local government in relation to Māori interests had serious implications for developing environmental plans (Chapter 5). Our findings show that nearly half of the respondents from district councils did not understand the provisions in Sections 6 and 8 of the RMA in this regard (Chapter 7). By and large, councils and Māori were left to work things out for themselves, with wide variations resulting. A few councils took the initiative and worked with Māori, sometimes funding their participation in plan-making, but, for the majority, uncertainty was a licence to ignore or to pay only lip-service to their obligations. Not knowing what it means to be in partnership and lack of capacity-building for Māori among all three tiers of government undermined implementation of the RMA, resulting in poor-quality plans and much litigation as Māori turned to the courts to articulate their rights (Chapter 5). Where sound plans led to good gains for Māori, vocal minorities ensured they were rolled back during the statutory hearings process (Chapter 12). The upshot was that Māori expectations of an enhanced role in environmental management under the RMA were not fulfilled.

Clarity of mandate provisions   We expected the design of the planning mandate to have an important influence on the quality of regional and local plans. When key legislative provisions in the mandate were clearly understood, the capability of regional and local councils to plan and the quality of their planning documents was correspondingly higher. Indeed, the strength of the statistical evidence for the clarity of mandated goals was strong (see Chapter 8).

However, while clarity of the mandate was an important predictor of plan quality, findings presented in Chapters 6 and 7 indicated that many key provisions in the RMA were not clearly understood by most plan-makers. The lack of clarity caused confusion for councils, thereby contributing to the generally poor quality of plans. In spite of many amendments to the RMA throughout the decade, key provisions, such as its purpose and principles (sections 5-8), remained largely unchanged. The government seemed to expect that regional and district councils would satisfactorily work out for themselves the intentions of the mandate. The uncertainty that flowed from this expectation enabled councils to develop plans that reflected the spectrum of political ideologies. This suggests that central government should not have left interpretation of the mandate entirely to local governments. Rather, higher levels of government ought to have made the purpose of the environmental legislation clear and assisted councils in its interpretation by providing leadership, policy direction and adequate support.

Implementation Efforts

A devolved and co-operative mandate assumes that not only are local governments free to devise the best means, through plans (and other methods), for reaching the goals in the mandate, but also that they have the capacity to create high-quality plans. When local capacity is lacking, the challenge for central government is to provide leadership in building it in ways that facilitate without being too intrusive.

In most respects, our findings support the theory. Indeed, the strength of statistical evidence presented in Chapter 8 indicated that the influence of local capacity-building efforts by central government was strong in helping councils to produce high-quality plans. This was further illustrated by our four case studies (see Chapters 9–12). Where staff in the regional offices of MfE helped willing local councils to develop effects-based plans, the outcomes were good, as we saw for the Tauranga and Far North plans. Where well-intentioned DoC staff failed to provide reliable data and technical assistance, good plans suffered, as seen in the Far North and Tasman cases.

Overall, however, our findings showed that there was a major gap in the intergovernmental capacity-building hierarchy at the central government level. The consequences of this became most evident over matters of national importance and the lack of inducements for helping low-capacity councils to develop their policies and plans.

Matters of national importance   Councils that genuinely tried to define sustainable management by developing the effects-based plans that the Minister sought encountered protracted and costly processes at best and public hostility at worst.

This happened in part because there was a lack of policy direction, methods, and data from Government on the six matters of national importance identified in section 6 of the RMA, including Māori interests.

Apart for the coast, not one national policy statement was produced to provide councils with policy direction. There was also an absence of methods to help councils identify these matters in their areas, many of which were in small low-capacity rural councils. This meant that councils were left to work it out for themselves in 86 different ways around the country — an absurdly time-wasting and costly process economically, socially, and emotionally, as planners in those councils that produced good effects-based plans could attest.

The hands-off approach from the centre that aimed at fostering local innovation at the periphery was well-intentioned, but could only succeed if accompanied by policy direction and well-funded capacity-building activities from the centre. Whether we were examining outstanding landscape values in Queenstown Lakes District, significant natural areas in the Far North District, or iwi interests in Tasman District, results were much the same. Having been left in a policy vacuum and with limited support from the centre, these councils struggled when dealing with matters of national importance, and in all three cases there was a great deal of ratepayer discontent over provisions included in the plan, even though as effects-based plans they scored well.

Minimizing central resources   Although its capacity-building role was well-recognized by the Ministry for the Environment (as the main central government implementer of the RMA), it was knee-capped by Government from the outset. This severely limited the help it could give to councils. The capacity-building intention of MfE was clear from its *Resource Law Transition Plan* of 1991, which contained a range of activities to help ensure the efficient and effective transition from old to new resource planning regimes. It did not, however, include proposals for preparing a suite of national policy statements and standards. Modestly costed, the transition plan was rejected by the new National government. Having funded the preparation of the RMA, Government's view was that it was now up to the newly reformed councils to implement it. Naively, too many local politicians agreed, with a 'leave it to us, we can do it' attitude. Others, were, however, disappointed that promised resources from the centre never came.

Denying MfE its transition plan contravened the most basic assumption of Government's co-operative and facilitative mandate — to ensure councils had the capacity to comply with its goals. Indeed, our research showed that larger wealthier councils had better RMA policy statements and plans and implementation processes, than smaller poorer councils — evidence that capacity matters. Since small councils make up the large majority of councils, amalgamations of the late 1980s had not gone far enough. Further reform seemed necessary, but that would not have obviated the need for capacity-building by Government.

The poor start to capacity-building by Government was made all the worse by severe budget cutting, signalled in the 1991 'Mother of All Budgets'. The annual cuts that followed for several years not only pruned what little perceived fat there might have been in the central environmental administration, it cut deep into the bone.   Thus, from the outset, MfE and DoC were much too under-funded for effectively performing their functions, including building capacity in councils under the RMA. Budget-cutting simply made a bad situation worse. By 1995, DoC was gripped by a poverty mentality, yet had nearly four times more money for its RMA work on the coast than MfE had for the rest of the RMA and the whole of the country. By the time policies and plans were tumbling out of councils for MfE to review in 1995, the Resource Management Directorate's staff had almost halved since 1991 to 22 FTE, including its regional offices, and it remained at that level into the twenty first century.

Thus, in the first six years under the RMA when MfE ought to have had the capacity to provide councils with a clear mandate along with policy direction, methods, and data for matters of national importance, it was struggling to cope. For the most part, MfE was forced by Government's mean-fistedness into being a reactive rather than proactive capacity-builder. When it was realized early on that adequate resources would not materialize, the Minister and his advisors exhorted local councils to comply with the RMA through political rhetoric over environmental bottom lines, rather than compromise Government's budget-cutting.

The dilemma faced by councils in relation to Government and the RMA can be epitomised in this exchange (Lewis Carroll, 1947: 80-1):

> "Would you tell me, please, which way I ought to walk from here?"
> "That depends a good deal on where you want to get to", said the Cat.
> "I don't much care where------" said Alice.
> "Then it doesn't matter which way you walk", said the Cat.
> "------ so long as I get *somewhere*", Alice added as an explanation.
> "Oh, you're sure to do that", said the Cat, "if you only walk long enough!".

Governance Shortcomings in the Regions

Theoretically, the main reasons for having a regional tier of governance are to identify significant regional issues that transcend the boundaries of local councils, fashion regional plans that promote integrated management across environmental media, provide technical assistance and advice to local councils (i.e. capacity-building), and support consensus-building to ensure regionally responsible decisions by them. Under New Zealand's devolved mandate, these responsibilities called for a co-operative partnership between a regional council and the local councils within its area so that integrated planning would occur.

We found a major gap in the intergovernmental capacity-building hierarchy at regional level that further underscored governance problems under the RMA. Initially, the partnership roles of the 14 regional councils (including one unitary council) were fostered by financial 'inducements' from Government to help prepare the regional policy statements and coastal plans (including iwi input into them) that would give policy direction for local councils (Chapter 4). Significantly, the level of grant was based on the ability of a regional council to perform: poor councils got more than wealthy ones. Unfortunately, the grants ceased after the two year statutory deadline had been reached for policy statements and coastal plans, and were never extended to the 74 local councils.

In spite of this regional council capacity-building effort of Government, our statistical analysis of data showed that they had limited influence in enhancing the capacity of local councils and the quality of their plans (Chapter 8). Planning in regional and local councils through the 1990s operated largely independently of each other, with only weak inter-organizational relations and little integration of policies in plans — except perhaps for some metropolitan areas like Auckland and Wellington, the coast (e.g. Chapter 11), and unitary authorities (Chapter 12). This might suggest that capability (commitment and capacity) is a more important factor than institutional form (regional council and unitary authority) in determining integrated environmental management within regions. (See also Parliamentary Commissioner for the Environment, 1999.)

Our evidence suggests that turf protection together with lack of capacity, problems with professional practice, and unhelpful statutory deadlines all contributed to the disconnection between regional and district councils (Chapters 6-12). These findings are of great concern and have important practical implications for achieving the goals of the RMA — integrated management for quality environments. They further illustrate problems of RMA governance.

Shortcomings in Planning Practice

While the contextual conditions created by poor governance in the intergovernmental system go a long way towards explaining why the quality of regional policy statements and district plans was only fair to poor, planning practice within councils also played a part. The RMA set up a rational-adaptive approach to planning that required plan-makers to carry out rigorous research and policy analysis that would include state of the environment reporting and ongoing monitoring, coupled with comprehensive consultation and public participation (Chapter 2). The former (rigour) would fill a gap evident under the old town and country planning regime, while the latter (participation) would extend its provisions. We assumed that effective implementation of the rational-adaptive approach would be influenced by such factors as plan quality, organizational capability, and institutional arrangements within councils.

Plan Quality

Our theorizing about the content of plans specified eight criteria or principles for evaluating plan quality (Chapter 2, Table 2.2). We maintained that plans incorporating these criteria would be of higher quality than plans that did not. We thus assumed that high-quality plans would be more effective than lesser-quality plans in guiding councils towards achieving the goal of protecting and enhancing the environment through the sustainable management of natural and physical resources.

Lack of rigour   We found that, generally, plan-makers did not have a good understanding of the basic principles of plan quality. Thus, by-and-large we found that councils were producing planning documents that needed considerable improvement. The main weaknesses stemmed from the lack of data on the state of the environment. The 'fact base' in plans received the lowest score of the eight criteria, indicating the lack of time and skills for carrying out the necessary research and therefore an absence of analytical rationales for defining and prioritising issues. It also partly explains the generally lacklustre scores for the criteria of 'issue identification' and 'monitoring'.

Another weakness that had a pervasive effect on plan quality was revealed by the poor scores achieved for 'interpretation of the mandate'. Local plans in particular did not provide clear explanations of how the goal of sustainable management applies to local physical and social conditions. Unclear provisions in Part II of the RMA made it difficult for plan-makers to interpret them in the circumstances of their district. Consequently, it was hard to distil any 'vision' of the future or sense of what constitutes a sustainably managed environment from their plans and policy statements. Overall, regional policy statements scored better in this regard, reflecting early Government pressure on regional councils to concentrate on the physical environment (i.e. the effects of development on air, water and land) and not be distracted by reference to social and economic circumstances. A small number of regions did, however, take a broader view of sustainable management, especially those wanting to manage metropolitan growth.

Integrated management of natural and physical resources is a cornerstone of the RMA. We therefore looked for evidence that councils were enacting the policies of higher-order plans and collaborating over cross-boundary issues. Instead, we found that, based on the low score for the 'integration' criterion, plans showed a lack of inter-organizational co-ordination. While consultation processes were identified and other local, regional, and national plans and policies mentioned, local plans lacked clear explanations of how the objectives and policies of these documents would be taken into account.

This result is explained not only by the newness of the RMA, but also timing issues with regard to higher level policies and plans. Councils tackled plan-writing to suit their own programmes, and therefore it was not uncommon for a district plan to precede the relevant regional policy statement and/or the *New Zealand Coastal Policy Statement*. Competitive behaviour, especially among regions, was also a reason for the lack of co-ordination. The RMA set October 1993 as the deadline for notifying regional policy statements and coastal plans, and beating other regions was the name of the game in many councils, including poorly resourced ones. Old tensions between regions and districts also surfaced, further reducing the likelihood of achieving good 'integration'.

The plan quality criteria of 'internal consistency' and 'clarity of purpose' gained the highest scores, but these were still only fair. With 'internal consistency', we found that the linking of objectives, policies and methods in plans was reasonably well done, reflecting past experience under town and country planning legislation. However, the link between objectives and anticipated environmental results was weak, as were the provisions for monitoring — both new RMA requirements. Writers had neither a clear understanding of what the plan was trying to achieve nor the skills to express this via a rigorous cascade of tightly crafted policies and rules (Chapter 2, Figure 2.3).

Although writing plans was a decades-old requirement, too many plan-writers seemed to have forsaken their 'plan organization and presentation' skills. The generally low scores for this criterion may simply reflect the perplexity plan-makers experienced in developing new effects-based plans and/or the haste with which they were done.

Truncated consultation Prior experience with town and country planning meant that most councils quickly established comprehensive consultation programmes. In spite of a great deal of consultation, however, many councils still encountered stiff resistance from various property owners when the plan was notified, especially if it was effects-based. The key problem was that too much effort had gone into public consultation early in the plan-making process over 'issues' and 'objectives', when only limited research on the environment had been done, and too little effort went into consultation with affected property owners late in the plan-making process over 'methods' and 'rules'. In a number of cases, the latter was impeded by the need to carry out research late in the process to help fill gaps that had emerged as the plan-making unfolded. Councillors pressuring staff to notify the plan before it had been satisfactorily completed and reviewed by affected stakeholders aggravated this. All these problems were evident to a greater or lesser extent in our four case studies (Chapters 9-12). But where councils produced a draft plan well in advance of a notified plan, the risks of adverse public reaction were reduced.

The next generation While 'research' and 'consultation' form essential components of the rational-adaptive approach and are carried out in an iterative fashion, it is essential that their timing and emphasis mesh properly with the sequence of steps in the plan-development process (Figure 2.2). Tightening up the application of this process would go a long way in improving the next generation of RMA policies and plans. This prospect would be enhanced if their fuzziness of purpose was sharpened and they are informed by state of the environment monitoring. Only then may they serve as a focal point for co-ordinating decision-making and helping diverse stakeholders reach consensus about desired environmental outcomes (Chapman, Ericksen and Crawford, 2003).

Organizational Capability

The theory we outlined in Part I assumed that the capability (commitment and capacity) of regional and district councils has an important influence on not only how well the plan-making process is carried out, but also ultimately plan quality (Godschalk et al., 1999). Our findings supported the theory. Indeed, the strength of evidence from our statistical analysis (Chapter 8) and case studies (Chapters 9-12) is striking. When capability was strong, the quality of plans, and thus the effectiveness with which they guide councils to achieve sustainable environmental outcomes, was significantly greater. This was especially true given that our data on local capability, context, and plan quality came from independent sources. Overall, however, our findings reveal troubling gaps in commitment and capacity throughout the planning process at regional and local levels.

Council commitment   This was mixed across and within councils. Where there was good political ownership of the plan, we saw higher levels of organizational commitment to both the planning process and the goal of sustainable management. Too often, however, commitment to the plan-making process by elected officials was uneven and their leadership weak. Councillors often did not make the effort to understand the mandate, let alone the plan. They set unrealistic deadlines, often rushing to notify the plan ahead of local elections. They mistrusted planning staff and were not sympathetic when deadlines were not met.

Most planners were committed to the task, but struggled to meet the multiple demands of the rational-adaptive approach to plan-making, as reflected in the lack of research and rigour in policy writing. Too often the plan-makers were not supported by councillors in the face of opposition to notified plans by organized stakeholder groups, and thus became demoralized. The lack of compensatory support from central government or regional councils was discouraging, especially where good effects-based plans had been produced.

Commitment to iwi interests by councils was very mixed. Even where relatively good, it may be explained more by the statutory requirement to take iwi interests into account than by new-found regard for their inclusion. Planners were, however, much more attuned to this need than were councillors. The limited resources allocated to consultation with Māori reflect the lack of political support given to planners, although a few councils did very well in this regard.

Council capacity   Perhaps a truism, but 'capacity' is critical for making good plans. This is evident in Figure C.1, which locates councils in a matrix that relates 'plan quality' to council 'capacity to plan'. It shows the importance of capacity for creating a good plan, but some councils did not fit this general rule (e.g. Hutt City).

The cost-cutting in central Government was mirrored in local government. Thus, too often inadequate resources were made available by local councils for developing their policies and plans. For example, over one-quarter of local councils employed less that one FTE staff for preparing the district plan, and another one quarter between one and two FTEs — wholly inadequate staff levels.

While capacity is uneven across councils, we also found that it was inefficiently used within councils at different stages of the planning process. Many planners and elected officials reported being daunted by unknowns in effects-based planning. They devoted inadequate time at the outset to strategic thinking about the interpretation of the mandate and to project management, and often failed accurately to assess the scope and type of work needed at different stages of the plan-making process. Frequently, estimates of staff, time and budget needs were unrealistic, as illustrated to a greater or lesser extent in all four case-study councils.

		Council capacity		
		High	Medium	Low
Plan Quality	High	Tauranga Christchurch Waitakere	Queenstown Lakes Tasman Masterton Far North	
	Medium	Palmerston North Matamata–Piako Wellington Dunedin Hurunui Clutha	Gore Tararua Rotorua South Taranaki Waikato Southland Kapiti Coast	Rangitikei Horowhenua Otorohanga Timaru
	Low	Hutt City Stratford	Waimate	Kaipara South Waikato Kawerau Papakura

Figure C.1   District council plan quality scores and capacity to plan

Institutional Arrangements

Local institutional arrangements were expected to have a significant influence on planning. Arrangements that foster well-organized agencies were assumed to enhance communication, help provide a common set of facts to decision-makers, reduce the likelihood of conflict and duplication of effort, and lessen the chances of mistrust and misunderstanding developing among local agencies, stakeholder groups and citizens. These activities are important contributors to how well local governments are able proactively to foster innovation and change through planning, and produce high-quality plans.

We evaluated two aspects of local institutional arrangements: those within councils, and those between Māori and councils.

Arrangements in councils   With respect to institutional arrangements within councils, our findings supported the theory. Local case studies and interviews with key informants revealed that the government reforms of the 1980s, together with increased managerialism in the early 1990s, compromised local efforts to create 'effective' institutional arrangements for planning. Because the reforms emphasized reduced spending, down-sized bureaucracy and increased efficiency, several obstacles were created. In many cases, councils separated policy (and planning) from regulatory and/or service delivery functions in the hope of improving transparency and accountability. This caused a loss of co-ordination and resulted in limited feedback from the regulatory and service delivery sections of councils to the policy sections. Moreover, the policy sections, where plan preparation occurs, suffered from loss of specialized staff. Where staff members responsible for service delivery did participate in plan-writing, there were more opportunities for tailoring policy solutions to suit local circumstances.

Councils were also under pressure by Government to monitor and audit their own performance. Auditing was made easier when tasks were streamlined along narrow functional lines of organizational responsibility. This required setting up departments or business units in councils to deal with those aspects of resource management that were visible, measurable and politically feasible, such as meeting the prescribed deadlines for processing applications for consents. As well, plan-making and policy development were not well served in resource allocation and priority setting because the results were seen as long-range and somewhat diffuse; not up-front and immediate.

Many councils, as exemplified by our Tauranga case, experienced multiple restructuring, suggesting the changes were poorly conceived. The implications are that local elected officials and staff did not carefully consider the effects of initial change, and did not allow enough time to judge the effectiveness of the changes. Some changes were aimed more at shedding staff or simply pursuing the fashionable trend for reform than at creating well-organized and effective institutions. Because the transaction costs of organizational change are high (in time, resources and staffing), multiple organizational changes were costly, wasteful, demoralizing to staff and detrimental to planning.

On the other hand, the reforms did promote 'efficient' organizational arrangements, such as: more business-like systems and attitudes towards customer service; improved processing of resource consents and monitoring of consent compliance; and ability to charge applicants for the costs of processing consents. Annual reporting to the MfE (as part of the ministry's ongoing monitoring of the RMA) on the numbers of consents handled and performance in relation to statutory deadlines reinforced this focus on efficient procedures.

Māori and local government   Our theory suggested that organizational arrangements designed to integrate iwi interests into local planning initiatives would be an important factor in creating high-quality plans. Our statistical evidence showed that working with Māori early in the planning process has a positive influence on how well plans advance their interests. As discussed in Chapter 7, however, only 43 per cent of planners in district councils believed that this would work. Case studies revealed, that although gains accrued to iwi from their being consulted, as required by the RMA, many Māori felt that their involvement in policy decisions affecting their interests was insufficient. Some iwi who engaged in the plan-making process in good faith felt betrayed when changes in plans forced by more powerful stakeholder groups undercut their contributions, as happened in Tasman District (see Chapter 12).

High Price Paid for Risky Reforms

The reforms that were undertaken in New Zealand in the late 1980s and early 1990s were far-reaching attempts to devolve responsibilities from national to local government, instil co-operation, and promote market solutions in the management of natural and physical resources. These reforms attracted international attention due to their innovations. Given the almost revolutionary nature of the change in governance and planning, however, public officials and stakeholders at all levels in New Zealand have been on a steep learning curve in their efforts to realize the promise of the reforms. Of particular concern is the harsh and persistent public criticism of the RMA, a symptom of widespread disillusionment. In a system based on communities taking responsibility for local action to promote sustainability, this disillusionment is a fundamental threat to the RMA's chances of success.

Changing suddenly from activities- to effects-based planning — to say nothing of moving from state welfarism to market liberalism — was a dramatic and risky policy shift that involved all levels of government. In a newly devolved system, where strong guidance from central government was suddenly eschewed in favour of innovation at local level, the risks escalated along with the costs as councils struggled to achieve success.

Perhaps the learning curve needed under such dramatic change was too steep? This book documents the struggles that policy-makers, planners, and others faced in establishing planning for sustainability and making it work under a devolved and co-operative intergovernmental framework. Noteworthy obstacles remain in the way of accomplishing policy goals. The main challenges in New Zealand are to improve the weak quality of local plans, the limited local capability to plan, and the lack of involvement by central government in building local capacity to plan. Māori aspirations for a meaningful role in planning must also be addressed.

An alternative to dramatic change would have been to take a more prudent, measured and evolutionary course that emphasized co-operation and blended experiment with policy learning. This would have enabled higher levels of government to clarify national policy, cultivate the capabilities of local government, disseminate information about successful local plan-making practices, and monitor plan performance and provide feedback. An evolutionary course would also have enabled local governments to respond first to the need for organizational change and capability, then to proceed with plan preparation. Another benefit would have been to allow adequate time to establish strong partnerships with Māori, and to build trust through participation. Regional and local governments would also have gained under an evolutionary course in terms of policy co-ordination, information sharing and pooling of resources for environmental monitoring. Legislative reforms could then have embedded these new innovations and practices.

There are many obstacles to be overcome before high-quality plans emerge from the planning efforts of councils. Our findings suggest that focusing on examples of best practice as a means of improving plan quality by planners within councils will not of itself lead to better plans, improved implementation and thereby quality environmental outcomes.

Weak plans are only a symptom of fundamental problems that must be systematically addressed if better performance is to be achieved. Three sets of organizational factors make a big difference in preparing plans for environmental sustainability: local capacity to create good plans; Government's mandate design; and Government's capacity-building efforts. All of these factors underpin the need for improved governance, and should be the main focus of future action; not simply plan-writing. Addressing them satisfactorily will require greatly improved Government funding.

Signs of Hope?

Since the return of a new Labour government in late 1999, several significant steps have been taken to improve governance and planning. Government has reacted positively to urgings by the Parliamentary Commissioner for the Environment (2002) to develop an integrated framework for sustainable development (Department of Prime Minister and Cabinet, 2003). The Foundation for Research, Science and Technology (FRST) has been directed to fund integrated approaches to research on sustainable development and Māori development (Foundation for Research, Science and Technology, 2002). Government has increased funding for MfE and appointed a new leader who, unlike predecessors, has publicly supported the need for national policy statements and standards (Ministry for the Environment, 2003: 2). While recent changes to the RMA do not address problems with the Part II purpose and principles, the Minister for the Environment (Marian Hobbs) has signalled that national policy statements and standards will be prepared for matters of national importance and promote consistency and quality in RMA plans and processes (Hobbs, 2003). The new *Local Government Act* (LGA, 2002) enhances devolution by giving powers of general competence to councils with respect to relevant statutes, providing they have good community buy-in for the development of long-term council community plans. These new plans are aimed at promoting the environmental, social, cultural and economic well-being of communities as a means for achieving sustainable development.

Many of these actions reflect some of the recommendations that we made to Government in early 2001 (see Annex C.1). As well, the process of feeding findings to key stakeholders as the research unfolded seems to have positively influenced behaviour in relevant agencies. Nevertheless, we do not think that the changes go far enough. In our view there are fundamental problems of governance in New Zealand that require urgent attention if the sustainable management of natural and physical resources under the RMA is to lead to improved environmental outcomes.

From our general observations, this need for improved governance applies across the Government's policy spectrum. Beyond the RMA, Government continued devolving functions to local government in the decade to 2000. It withdrew or reduced services and funding in 13 areas of activity, leaving councils to fill the gap. To this 'unintended devolution' were added 17 areas of activity where functions had been 'formally devolved' to local government. Of these, seven are unfunded (including those under the *Biosecurity Act* [1993]), five are funded (e.g. through local fees, such as the *Building Act* [1991]), and five are partially funded (including the RMA) through local sources (Local Government New Zealand, 2000).

It is not difficult to devolve functions from centre to the periphery; the challenge is in their effective implementation. In our view, New Zealand suffers from poor policy implementation, which confounds sound policy initiatives. Planning under the RMA exemplifies this malaise, and the portents are that it will extend to proposals for long term council community plans under the new LGA.

Large well-resourced councils have driven the quest for integrated community plans, but there is the likelihood that the problems experienced by the many smaller and lesser-resourced councils under RMA will be visited upon them under the LGA. And, as for the RMA, Government has not provided its implementing agency — the Department of Internal Affairs (DIA) — with the resources needed for carrying out the capacity-building in sub-national government that is required for sound implementation. Indeed, the DIA has less than half of the funds that MfE had for implementing the RMA, with no assurance that they will extend beyond a second year of implementation activities in 2004. This wholly inadequate funding may well cause another unwelcome 'blob in the middle of the New Zealand psyche'.

Annex C.1

Summary of recommendations to the Government

Drawing on the PUCM Phase 1 research findings on plan quality and organizational factors influencing the preparation of plans, we made a number of recommendations to Government aimed at improving the implementation of the RMA. Entitled *Plan Quality, Resource Management and Governance* (Ericksen, et al., 2001), it contained a summary of main findings followed by 20 recommendations organized into five interrelated sets. In summary, these sets were:

1. *Improve the national framework for sustainability* by revising policy on sustainable development and management; clarifying key provisions in the RMA for the users; and adopting an integrated set of national policies to give direction to regional and local government.

2. *Build national capability for environmental planning* by strengthening the Ministry for the Environment so that it can better co-ordinate the actions of central government agencies and more effectively collate and transfer information to regional and local councils.

3. *Develop a national programme to build local capability* by continuing the local government reforms; supporting land-owners in protecting nationally important environments; extending the environmental education programme; providing better factual information about the environment; and assessing the implementation of plans and their environmental outcomes.

4. *Develop an integrated state of the environment reporting system* involving all levels of government, and provide regular reports aimed at helping to improve the monitoring of policies and plans.

5. *Improve the quality of regional and local planning documents,* by strengthening both the organizational processes for plan development and the technical aspects of plan-making.

APPENDICES

Key Provisions of the RMA Affecting Local Government Functions

5. Purpose — (1) The purpose of this Act is to promote the sustainable management of natural and physical resources.

(2) In this Act, "sustainable management" means managing the use, development, and protection of natural and physical resources in a way, or at a rate, which enables people and communities to provide for their social, economic, and cultural wellbeing and for their health and safety while:

(a) Sustaining the potential of natural and physical resources (excluding minerals) to meet the reasonably foreseeable needs of future generations;

(b) Safeguarding the life-supporting capacity of air, water, soil and ecosystems; and

(c) Avoiding, remedying, or mitigating any adverse effects of activities on the environment.

6. Matters of national importance — In achieving the purpose of this Act, all persons exercising functions and powers under it, in relation to managing the use, development, and protection of natural and physical resources, shall recognise and provide for the following matters of national importance:

(a) The preservation of the natural character of the coastal environment (including the coastal marine area), wetlands, and lakes and rivers and their margins, and the protection of them from inappropriate subdivision, use, and development;

(b) The protection of outstanding natural features and landscapes from inappropriate subdivision, use, and development;

(c) The protection of areas of significant indigenous vegetation and significant habitats of indigenous fauna;

(d) The maintenance and enhancement of public access to and along the coastal marine area, lakes and rivers;

(e) The relationship of Māori and their culture and traditions with their ancestral lands, water, sites, waahi tapu, and other taonga.

7. Other matters — In achieving the purpose of this Act, all persons exercising functions and powers under it, in relation to managing the use, development, and protection of natural and physical resources, shall have regard to:

(a) Kaitiakitanga;

(b) The efficient use and development of natural and physical resources;

 (c) The maintenance and enhancement of amenity values;

 (d) Intrinsic values of ecosystems;

 (e) Recognition and protection of the heritage values of sites, buildings, places, or areas;

 (f) Maintenance and enhancement of the quality of the environment;

 (g) Any finite characteristics of natural and physical resources;

 (h) The protection of the habitat of trout and salmon.

8. **Treaty of Waitangi** — In achieving the purpose of this Act, all persons exercising functions and powers under it, in relation to managing the use, development, and protection of natural and physical resources, shall take into account the principles of the Treaty of Waitangi (Te Tiriti o Waitangi).

30. **Functions of regional councils under this Act** — (1) Every regional council shall have the following functions for the purpose of giving effect to this Act in its region:

 (a) The establishment, implementation, and review of objectives, policies and methods to achieve integrated management of the natural and physical resources of the region;

 (b) The preparation of objectives and policies in relation to any actual or potential effects of the use, development, or protection of land which are of regional significance;

 (c) The control of the use of land for the purpose of -

  (i) Soil conservation;

  (ii) The maintenance and enhancement of the quality of water in water bodies and coastal water;

  (iii) The maintenance of the quantity of water in water bodies and coastal water;

  (iv) The avoidance or mitigation of natural hazards;

  (v) The prevention or mitigation of any adverse effects of the storage, use, disposal or transportation of hazardous substances.

 (d) In respect of any coastal marine area in the region, the control (in conjunction with the Minister of Conservation) of:

  (i) Land and associated natural and physical resources;

  [(ii) The occupation of space on land of the Crown or land vested in the regional council, that is foreshore or seabed, and the extraction of sand, shingle, shell, or other natural material from that land;]

  (iii) The taking, use, damming, and diversion of water;

  (iv) Discharges of contaminants into or onto land, air, or water and discharges of water into water;

  (v) Any actual or potential effects of the use, development, or protection of land, including the avoidance or mitigation of natural hazards and the prevention or mitigation of any adverse effects of the storage, use, disposal, or transportation of hazardous substances;

  (vi) The emission of noise and the mitigation of the effects of noise;

  (vii) Activities in relation to the surface of water.

(e) The control of the taking, use, damming, and diversion of water, and the control of the quantity, level, and flow of water in any water body, including:
  (i)   The setting of any maximum or minimum levels or flows of water;
  (ii) The control of the range, or rate of change, of levels or flows of water;
  (iii) The control of the taking or use of geothermal energy.
(f) The control of discharges of contaminants into or onto land, air, or water and discharges of water into water;
(g) In relation to any bed of a water body, the control of the introduction or planting of any plant in, on, or under that land, for the purpose of:
  (i)   Soil conservation;
  (ii) The maintenance and enhancement of the quality of water in that water body;
  (iii) The maintenance of the quantity of water in that water body;
  (iv) The avoidance of natural hazards.
(h) Any other functions specified in this Act.
(2) The functions of the regional council and the Minister of Conservation [under subparagraph (i) or subparagraph (ii) or subparagraph (vii) of subsection (1)(d)] do not apply to the control of the harvesting or enhancement of populations of aquatic organisms, where the purpose of that control is to conserve, enhance, protect, allocate or manage any fishery controlled by the Fisheries Act 1983.

31. **Functions of territorial authorities under this Act** — Every territorial authority shall have the following functions for the purpose of giving effect to this Act in its district:
(a) The establishment, implementation, and review of objectives, policies, and methods to achieve integrated management of the effects of the use, development, or protection of land and associated natural and physical resources of the district;
[(b) The control of any actual or potential effects of the use, development, or protection of land, including for the purpose of the avoidance or mitigation of natural hazards and the prevention or mitigation of any adverse effects of the storage, use, disposal, or transportation of hazardous substances;]
(c) The control of subdivision of land;
(d) The control of the emission of noise and the mitigation of the effects of noise;
(e) The control of any actual or potential effects of activities in relation to the surface of water in rivers and lakes;
(f) Any other functions specified in this Act.

Appendix 2

Methodology

The framework for our study given in Figure 1.3 shows the range of factors that we theorized would influence the making of plans and thereby plan quality. Each factor required a method or methods to help uncover its influencing nature. Thus, a range of social science methods was adapted to the specific elements shown in Figure 1.3. For the most part, development of methodology was reasonably straightforward, if somewhat time-consuming. Much more demanding was the need to develop a robust method for evaluating plan quality, there being little in the international literature to guide us. Here, we summarize the main approaches used in our study, and then in more detail the methods and sources of data for it.

Rationale for Data Collection Methods

Our study of plan quality and influencing factors used two main approaches: nationwide overviews; and detailed case studies. It combined five main methods of data collection: three for the nation-wide surveys and two for the case studies. Typically, planning studies take one of two methodological approaches: in-depth case study of a single or small group of local planning programmes; or an overview of the activities of a class of local programmes.

The *case study approach* has the advantage of providing detailed information, but it does not allow generalizations about results. When only a small group of planning programmes is examined, it is difficult to specify the causes of success or failure and to know if an unsuccessful programme would be successful in a different setting. An *overview approach* can provide information on the importance of organizational capabilities to plan, central government capability-building activities, and the contextual setting at a specific point in time, but often lacks an in-depth examination of the dynamics of organizational and planning processes.

Taking advantage of the strengths of both approaches, our study relied on a nationwide survey of regional and district plans and planning programmes (Part III of book), three studies of co-operative governance and partnerships (Part II) and four case studies of local planning programmes (Part IV).

Overview Approach

The nationwide surveys were designed to provide an evaluation of the quality of regional policy statements, regional plans and district plans, and of the influences on plan quality of the clarity of provisions in the RMA mandate, central government capability-building activities, and local organizational capability. The surveys included use of a mailed questionnaire, face-to-face interviews, and a systematic evaluation of the quality of plans using a plan coding protocol.

The first set of data was derived from the mailed questionnaire, which aimed at eliciting factual information about the planning process and resources for plan preparation in councils. The second set of data was derived from open-ended interviews with lead planners, councillors, and consultants in regional and local councils, and of key staff members in central government agencies. The intent of the interviews in councils was to provide nuances of council capabilities to plan, and of the effectiveness of central government agencies in assisting councils to prepare plans. This data was used to give more thorough interpretations of results derived from the questionnaire data. The interviews in central government agencies provided information on the history of government policy since passage of the *Resource Management Act* in 1991, and on the evolving capabilities of key agencies to implement the Act. The intent of this data was to help assess the role of central government agencies in capability-building in councils, and their capabilities for doing so (Chapters 3-5). The third source of data came from application of a plan coding protocol, the elements of which included criteria denoted earlier in Table 2.2. Its development was based on review of the international literature and interviews with nearly 40 practitioners in New Zealand. The prototype was tested in peer review group workshops with a further 80 planning practitioners in Auckland, Wellington and Christchurch. The plan coding protocol yielded information about the quality of all regional policy statements, regional plans, and district plans that had been publicly notified prior to March 1997 (Chapters 6-12).

Case Study Approach

The case study method, the fourth source of data, was used to provide in-depth assessments of: a) co-operative governance and partnerships in central and regional government; and b) planning programmes and plan-making in district (local) councils. This enabled key research questions as well as issues identified through the national surveys to be further explored.

For co-operative governance and partnerships we focused on the role of central government (especially the Ministry for the Environment and Department of Conservation) in building the capabilities of local government and Māori to implement the national resource management mandate, and the role of regional councils in supporting local councils (Chapters 3-5). We drew upon various other sources of information, including the aforementioned plan coding, questionnaires, and interviews. Additional sets of interviews were conducted for analysis of central government agencies, including agency staff and independent policy analysts. Various documents of central government were reviewed, including annual reports and publications of key environmental agencies. We also drew on our previous research (1992-1995) on governance and environmental management, which included extensive interviews in government agencies, thus contributing to the historical or longitudinal dimension of our analysis (May, et al., 1996).

For planning programmes and plan-making in local councils, the case studies focused on the relationship between the steps in the plan-making process and the plan-quality principles or criteria noted in Figure 2.2, as well as external forces affecting the plan preparation effort. The four local councils selected for this part of the study were drawn from the upper quintile of plan quality scores, rather than a random selection across the 34 councils for which we had plan quality scores (Figure 7.2). This was because most plans were of poor quality, and we wanted to provide a more positive story for practitioners and policy makers by highlighting good practice in good plans. Within this constraint, we wanted at least one South Island council (Queenstown-Lakes District), one unitary authority (Tasman District), one coastal council (Tauranga District), and one other North Island council (Far North District) (Chapters 9-12). Data for the local case studies were drawn from several sources. They included: plan coding information, in order to highlight features in the plan that scored well; questionnaires, in order to provide factual information about the council; and interviews that had been done for the nationwide surveys, in order to gain insights from councillors and staff about the plan preparation processes and procedures. Other data came from more focused face-to-face interviews with representatives from key stakeholder groups in the district (especially council and end-users) and from documentary searches within council (such as submissions on policy documents by stakeholders, including government agencies).

Types and Sources of Data

The sources of data for our research are discussed in more detail below.

Plan Coding

In New Zealand, there are 74 district councils (including four unitary authorities) each of which must prepare a district plan, and 12 regional councils, each of which must prepare a regional policy statement and a regional coastal plan. Regional councils may also prepare other regional plans. The four unitary authorities have the dual functions of district and regional councils, and must prepare a regional policy statement, regional coastal plan, and district plan, but can elect to prepare a combined or integrated plan (e.g. Chapter 12).

The source of plan coding data for regional councils was the universe of 16 regional policy statements. The source of data for district councils was a sample of district plans. Because of resource limitations, not all plans from the 55 district councils that had publicly notified plans by March 1997 (the end point for our coding effort) could be analysed. Thus, a two-step sampling strategy for selecting district plans was used.

The first involved a selection of regions to be included in the sample. Regions were selected to assure balanced geographic and demographic coverage. The relative populations in 1996 of the two main islands of the country (South Island, 931,000 and North Island, 2,724,000) suggested that three regions be selected from the South Island and six regions from the North Island. Another criterion for selection required that the regions represent metropolitan and rural areas of each island. A further criterion was the need to include at least one of the four unitary authorities that have dual regional and district functions.

The second step dealt with selection of districts within the nine resulting regions. Up to four districts were randomly selected from each region. For all regions with less than four districts, all districts were selected. This procedure resulted in the selection of 34 district councils, including two unitary authorities.

Evaluation protocols were prepared in order to conduct a content analysis of the selected district plans (34) and regional policy statements (16) (Appendix 3). Each protocol contained questions to be used by coders in assessing each plan. Measures for the questions were nominal ("yes" = item present; "no" = item not present, or present, but not relevant) or ordinal (e.g., "clear" = item present and clearly explained; "vague" = item present and vaguely explained; "no explanation" = item present and not explained; and "absent" = item not present). The questions were designed to assess the extent to which each of the eight plan quality criteria listed in Table 2.2 were being promoted by a plan.

To increase reliability in evaluation ratings, a detailed guide to application of the protocol was prepared. Intercoder-reliability checks were not possible for all plans and policy statements given the amount of time needed to evaluate each plan – between one and three weeks. Thus, three regional policy statements and five district plans were double coded, which served as a partial check. An intercoder reliability score was computed equal to the number of coder agreements for items, divided by the total number of items. An overall reliability-agreement score of 80 per cent was achieved. The literature suggests that a score in the range of 70 to 80 per cent is acceptable (Miles and Huberman, 1984). In addition, the two coders prepared a good practice report at the end of the coding effort as a review and evaluation of their work.

These evaluation ratings were then used to create indices for each of the eight plan quality criteria or principles. Computation for each index consisted of two steps. The first involved summing the scores assigned to each corollary principle under each of the eight main principles (Table 2.2). The second step was to standardize the indices by dividing the sum of scores by the maximum possible and multiplying by 10. Thus, the maximum score for each plan was 80.

Postal Questionnaires

Two additional sources of data were derived from postal questionnaires sent to 32 district councils that prepared district plans, plus two unitary authorities (that prepared combined resource management plans, including regional and district plans), and 12 regional councils, plus the four unitary authorities that prepared regional policy statements. The questionnaire aimed at gaining factual information about the qualifications, skills, and experiences of staff in the plan-making group and the environment in which the plan was being made. The survey was conducted with techniques developed by Dillman (1978). After three follow-up letters and a telephone call to non-respondents, all 32 district councils, 12 regional councils and four unitary authorities responded.

To maximize reliability of data returned by councils, all letters and questionnaires were sent to planning managers responsible for preparing the regional policy statement or district plan. In a cover letter, managers were asked to complete the questionnaire or, where appropriate, to pass it on to the chief plan writer to do so. Of the 34 respondents from councils with district functions, 35 per cent (n = 12) were planning directors, 44 per cent (n = 15) were senior planners, and the remaining 21 per cent (n = 7) were resource managers, supervising officers of consents, and environmental management and regulatory services managers. Of the 16 respondents from councils with regional functions, 11 (69 per cent) were planning directors or managers, and the remaining 5 (31 per cent) were senior staff planners.

Interview Data

An interview schedule was developed to guide the nationwide interviews with the planners, council consultants, and councillors in regional and local councils. These face-to-face interviews were conducted in 1997 and aimed at gaining information about: the strategic nature of the planning process; the planning team; plan preparation; internal factors influencing plan preparation; external factors influencing plan quality; integration, consistency and cross-boundary issues; and devolution. In all, 119 interviews were conducted. This information was used in support of the statistical analysis of plan quality in Part III of the book. It also provided rich detail of value in extending our understanding for the qualitative analyses in the various case studies in Parts II and IV.

For the detailed studies of central government agencies and four district councils, semi-structured interview schedules were tailored for each circumstance. The interviews with staff in central agencies aimed at gaining information about such matters as: the clarity of the RMA mandate, their approach to dealing with plan-making in councils; elements considered important in their review of council policy statements and plans; and capacity-building activities for councils. The interviews in the four district councils aimed at gaining information from selected staff and councillors about: organizational and institutional arrangements for plan-making; the approach to making an effects-based plan; the chronology of main events in developing the plan; the extent of research and consultation that contributed to the 'environmental fit' and 'community fit' of the plan; factors that influenced positively or negatively plan-making; and the reactions of stakeholders to the notified plan. Interviews were held with representatives of key stakeholder groups to gain their views of the plan-making process and outcomes.

Other Documents

For the case studies in particular, pertinent documents held in central, regional, and local government were reviewed. For example, in central government, annual reports and strategic documents were reviewed, in order to trace sub-national capacity building activities in relation to available staffing and funding. In order to understand in more detail the plan development process and organizational factors affecting plan making within the four case study district councils, documents were reviewed relating to policy, research, and consultation for the plan and the organizational processes and procedures affecting its preparation.

Census and Other Data

The final source of data consisted of district and regional census and other information. District and regional population size and median home value were used to represent social and economic conditions. These variables were then used as controls in assessing factors that affect the organizational capacity of regional and district councils to produce quality plans.

Appendix 3

Plan Coding Protocol

**CODING SCHEDULE: CRITERIA FOR EVALUATING PLANS
UNDER THE RESOURCE MANAGEMENT ACT**
(7 July 1996)

COVER PAGE

1. **Title of regional policy statement, regional or district plan**

   _____

2. **Status of the policy statement/plan (at date of coding)**

   _____

3. **Other supporting documents e.g., Section 32 analysis**

   _____

4. **Name of council**

   _____

5. **Name of coder**

   _____

6. **Date of coding**

   _____

7. **Topics in Part II**

   [ ] **Iwi issues**
   [ ] **Natural hazards management**
   [ ] **Other**

A copy of this cover page, with the topic indicated, should be attached to the front of each topic coded.

PLANNING UNDER A COOPERATIVE MANDATE

CODING SCHEDULE: CRITERIA FOR EVALUATING PLANS
UNDER THE RESOURCE MANAGEMENT ACT:

Title of regional policy statement, regional or district plan   **Date of coding**

PART I: OVERALL CRITERIA FOR PLAN EVALUATION

CLARITY OF PURPOSE

Sustainable management, the purpose of policy statements and plans prepared under the RMA, is a process and not an end in and of itself. Therefore, there is a need for policy statements and plans to articulate in a comprehensive overview, preferably early on, the environmental outcomes which the process adopted is designed to achieve. This overview, goal statement or 'vision' must be expressed in terms of effects. To the extent that social, economic and cultural matters are relevant to the management of environmental effects, these may be addressed. A clear statement of purpose is an indicator of both the quality of thinking and analysis with which the plan was prepared and of the consultative process.

C1 Does the overview signal the anticipated environmental results/outcomes to be achieved through the policy statement or plan?

Criteria for scoring

2 (coherent overview of environmental outcomes, with a discussion of how social, cultural and economic matters affect those outcomes)

2 (coherent overview of environmental outcomes; limited mainly to ecosystems, natural resources and amenities)

1 (no overview)

0 (coherent overview of the future development of the region/district, including social, cultural and economic outcomes)

APPLICATION AND INTERPRETATION OF THE RMA

A good plan ought to be based on a sound interpretation of the RMA. It is not sufficient, however, to simply quote the relevant provisions. In a good plan, key provisions of the RMA to be incorporated in a plan or policy ought to be identified and the application and interpretation of these to the particular circumstances of the region or district should be explained. (See Part II of the RMA.)

C2 Does the plan *explain* how the following RMA provisions are applied and/or interpreted in the particular circumstances of the region or district?

Criteria for scoring

2	**Clear**	(provisions are identified, plus explanation of how plan implements provisions).
1	**Vague**	(provisions are identified or mentioned, but little explanation of how plan implements provisions).
0	**No explanation**	(only cites the RMA).

[If the item does not feature, do not code.]

	clear	vague	no explanation
Matters of national importance (ss 6 (a)-(d); 7 (b)-(h))	2	1	0
Treaty of Waitangi (ss 6 (e), 7 (a) & 8)	2	1	0
Duties to consider alternatives; assess benefits and costs (s 32)	2	1	0
Duty to gather information, monitor and keep records (s 35 (1) & (2))	2	1	0
For regional policy statement Functions of a region (s30 (1) (a & b))	2	1	0
Purpose of reg policy stmt (sec 59)	2	1	0
Contents of reg policy stmt (sec 62)	2	1	0
For regional plans Functions of a region (s30 (1) (a & b))	2	1	0
Purpose of reg plan (sec 63)	2	1	0
Contents of reg plan (sec 67)	2	1	0
For district plans Functions of a district (s31 (a))	2	1	0
Purpose of district plan (s72)	2	1	0
Contents of district plan (s75)	2	1	0

INTEGRATION WITH OTHER PLANS

A good plan should be integrated with other policy instruments, both internally and externally. Internal integration improves the ability to take self-directed actions in defining alternatives. External co-ordination enables extra-local interests to be incorporated into the proposed actions of the plan.

C3 For each plan or policy instrument mentioned in the plan being coded, how clear is the explanation of its relationship to the district plan/regional plan/regional policy statement being coded?

Criteria for scoring

2 **Full explanation**   (full explanation of how other plan is related to plan being coded.)

1 **Limited explanation** (limited explanation of how other plan is related to plan being coded.)

0 **No explanation**

[Only score a plan which is mentioned]

	Full explanation	Limited explanation	No explanation
own strategic plan	2	1	0
own annual plan	2	1	0
own vision statement	2	1	0
own district plan sections	2	1	0
iwi resource mgt plan	2	1	0
other district plans	2	1	0
reserve mgt plans	2	1	0
NZ coastal policy stmt	2	1	0
regional policy stmt	2	1	0
regional plan: coastal	2	1	0
reg. energy strategy	2	1	0
reg. land transport strategy	2	1	0
reg. land transport plan	2	1	0
reg. waste mgt strategy	2	1	0
reg. tourism strategy	2	1	0
reg. air quality mgt plan	2	1	0
reg. vegetation clearance plan	2	1	0
reg. catchment plan	2	1	0
reg. water allocation plan	2	1	0
reg. discharge to water plan	2	1	0
reg. water quality plan	2	1	0
river & lake bed use	2	1	0
water conservation orders	2	1	0
conservation mgt strategy	2	1	0
conservation mgt plan	2	1	0
district waste mgt plan	2	1	0
heritage strategy	2	1	0
heritage orders	2	1	0
pest management strategy	2	1	0
other	2	1	0

C4 Extent to which cross-boundary issues are clearly explained.

A clear explanation of cross-boundary issues, including substantive matters as well as the procedures to be followed in dealing with them, indicates that there has been good consultation with adjoining regions and districts and also indicates a degree of integration.

2 Clear (fully-developed explanation of how plan deals with the issues.)

1 Vague (issues are not well-defined; moderate to well-developed explanation of how plan deals with the issues)

0 No explanation

ORGANIZATION AND PRESENTATION

A good plan must be readable and accessible for both lay and professional people so that they can readily find out what they need to know, particularly its implications in regard to specific properties.

C5 Which of the following items are used in the plan for improving its accessibility (i.e. organization and quality of presentation) for users, especially the public?

Yes The feature is present

No The feature is absent

[If item is not applicable, do not code.]

Yes	No	
1	0	detailed table of contents (not just list of chapters)
1	0	detailed index to locate specific rules, policies, etc. (not just page refs.)
1	0	glossary of terms and/or definitions
1	0	user's guide
1	0	"executive summary"
1	0	cross-referencing of issues/objectives/policies and rules
1	0	clear illustrations such as maps, diagrams and pictures
1	0	(for RPS) spatial information available in a manner useful for preparing regional and district plans.
1	0	Where plan provisions affect individual properties, these properties can be readily identified and related to the relevant provisions e.g., by maps, GIS, etc.
1	0	indicates that supporting documentation, such as explanatory pamphlets and videos is available.

PART II: CRITERIA FOR EVALUATING EACH PLAN TOPIC

NAME OF TOPIC:

A good plan should have a high quality fact base, issues linked to effects, be internally consistent from issues to rules, and contain provisions to monitor its effectiveness in achieving its objectives. These four elements of a good plan are required by the RMA. Each element is addressed in turn. NB: For this part, two or more topics will be selected which occur in all policy statements and plans.

QUALITY OF FACT BASE

C6   Make a general assessment of the degree to which the plan topic incorporates and explains the use of factual data.

Criteria for scoring

Yes   (incorporated; explanation of relevance of facts to issue, objectives and policies)

No   (incorporated; little or no explanation of relevance to issue, objectives and policies)

[If item does not feature, do not code.]

	Yes	No
maps that display information in formats that are relevant and meaningful;	1	0
facts presented in relevant and meaningful formats;	1	0
methods used for deriving facts are cited (e.g., by a survey of housing or noise)	1	0
issues that are prioritised based on analytical methods (e.g., need-based ranking of traffic congestion problems)	1	0
benefit/cost of main alternatives	1	0
background information is sourced/referenced	1	0

IDENTIFICATION OF ISSUES

The clear identification of issues is an important measure of plan quality. Not only must the issues be based on effects but they should also be explained in a well-reasoned manner in terms of the effects-based orientation of the RMA.

C7 For each issue identified, how well is it explained in terms of management of effects?

Criteria for scoring

2	**clear**	(issue is defined with well-developed explanation)
1	**vague**	(issue is defined but poorly-developed explanation)
0	**no explanation**	(issues are identified, but no explanation)

	clear	**vague**	**no explanation**
Issue 1	2	1	0
Issue 2	2	1	0
Issue 3	2	1	0
Issue n	2	1	0

ASSESS INTERNAL CONSISTENCY OF POLICY STATEMENT/PLAN

*The internal consistency of plans can be assessed by looking at the strength of the cascade of links from issues, objectives, policies, methods, anticipated results through to indicators. *Rules are also important, but for reasons of efficiency, will be tested in case studies.*

C8 Assess internal linkages and consistency.

Criteria for scoring

2	**clear**	(relationship between the two items is strong)
1	**vague**	(relationship is weakly developed)
0	**absent**	

Name of Issue	Objectives	Policies	Methods	Anticipated Results	Indicators	Rules*
1.___	___	___	___	___	___	___
2.___	___	___	___	___	___	___
n.___	___	___	___	___	___	___

PROVISIONS FOR MONITORING

Good planning takes a systems approach and therefore the provisions in a plan for monitoring both its effectiveness and the environmental outcomes, which complete the feedback loop, are important measures of plan quality.

C9   Does the plan topic contain provisions for monitoring the suitability and effectiveness of its objectives and policies for its region or district?

> **Yes**   1
> **No**   0

C10   To what degree of detail does the plan topic contain provisions for monitoring environmental results?

> **2   detailed**   (identifies specific indicators to be monitored and relevant data bases)
> **1   vague**   (identifies the need to monitor in words, but does not indicate data base or types of indicators)
> **0   absent**   (monitoring systems not identified)

C11   Does the plan topic identify the key agencies responsible for monitoring the indicators of environmental results?

> **Yes**   1
> **No**   0

PART III: COMMENTS AND OBSERVATIONS

Coders: Copy extracts of model plan elements, label and file together with hard copy of the notes

MODEL PLAN ELEMENTS

C12   Does the plan contain innovative features of value as a model for other councils?   If yes, then indicate which sections (by page #s) and briefly explain why they represent model approaches.

GENERAL CHARACTERIZATION OF THE PLAN

C13   Briefly characterize the plan; its quality of thinking; its strengths and weaknesses.

DEFINITION OF SUSTAINABLE MANAGEMENT

C14   Briefly discuss how the policy statement or plan interprets sustainable management as an organizing theme?   Is there an explanation of how this will be carried out or is it necessary to read the whole plan to understand the approach taken?

References Cited

Preface

Bührs, T. and Bartlett, R.V. (1993), *Environmental Policy in New Zealand*, Oxford University Press, Oxford, England.

Boston, J., Martin, J., Pallot, J., and Walsh, P. (1996), *Public Management in New Zealand: The New Zealand Model*, Oxford University Press, Auckland.

Ericksen, N., Crawford, J., Berke, P. and Dixon, J. (2001), *Resource Management, Plan Quality, and Governance: A Report to Government*, International Global Change Institute, University of Waikato, Hamilton.

Harris, P. and Twiname, L. (1998), *First Knights: An Investigation of the New Zealand Business Roundtable*, Howling at the Moon Publishing Ltd., Auckland.

Kelsey, J. (1995), *The New Zealand Experiment: A World Model for Structural Adjustment?* Auckland University Press with Bridget Williams Books, Auckland.

May, P., Burby, R., Ericksen, N., Handmer, J., Dixon, J., Michaels, S. and Smith, D.I. (1996), *Environmental Management and Governance: Intergovernmental Approaches to Hazards and Sustainability*, Roultledge Press, London.

Memon, P.A. (1993), *Keeping New Zealand Green: Recent Environmental Reforms*, University of Otago Press, Dunedin.

Memon, P.A. and Perkins, H. (eds) (2000), *Environmental Planning and Management in New Zealand*, Revised Edition, Dunmore Press, Palmerston North.

Pressman, J. and Wildavsky, A. (1973), *Implementation*, University of California Press, Berkely, California.

Sumits, A.P. and Morrison, J.I. (2001), *Creating a Framework for Sustainability in California: Lessons Learned from the New Zealand Experience*, Report prepared for Pacific Institute for Studies in Development, Environment and Security, Oakland, California.

Glossary

Boast, R.P. (1989), *The Treaty of Waitangi: a Framework for Resource Management Law*, New Zealand Planning Council and Victoria University of Wellington Law Review, Wellington.

King, M. (1988), *Being Pakeha. An encounter with New Zealand and the Māori Renaissance*, Hodder and Stoughton, Auckland.

Parliamentary Commissioner for the Environment (1998), *Kaitiakitanga and Local Government: Tangata Whenua Participation in Environmental Management.* Parliamentary Commissioner for the Environment, Wellington.

Introduction: From Rio to RMA – Great Expectations

Bassett, M. (1987), 'Reform of Local and Regional Government', Statement accompanying the *Economic Statement* of 17 December 1987, New Zealand Government, Wellington.

Boston, J., Martin, J., Pallot, J. and Walsh, P. (eds) (1991), *Reshaping the State: New Zealand's Bureaucratic Revolution*, Oxford University Press, Auckland.

Bührs, T. and Bartlett, R.V. (1993), *Environmental Policy in New Zealand*, Oxford University Press, Oxford, England.

Elwood, B.G.C. (1988), 'Local Government Reform', in J. Martin and J. Harper (eds), *Devolution and Accountability*, Studies in Public Administration no. 34, GP Books, Wellington, pp. 107-111.

Ericksen, N.J. (1990), 'New Zealand Water Planning and Management: Evolution or Revolution', in B. Mitchell (ed) *Integrated Water Management: International Experiences and Perspectives*, Belhaven Press, London and New York, pp. 45-87.

Harris, P. and Twiname, L. (1998), *First Knights: An Investigation of the New Zealand Business Roundtable*, Howling at the Moon Publishing Ltd., Auckland.

Hill, K. (2002), 'Interview with Bill English and Mark Gosche on Auckland Transport and Roading', *National Radio*, 8 February 2002.

McCaskill, L. (1973), *Hold this Land: A History of Soil Conservation in New Zealand*, Local Government Commission, Wellington.

Martin, J. (1991), 'Devolution and decentralisation', in J. Boston, J. Martin, J. Pallot, and P. Walsh. (eds), *Reshaping the State: New Zealand's Bureaucratic Revolution*, Oxford University Press, Auckland, New York and Melbourne, pp. 268-296.

Martin, J. and Harper, J. (eds) (1988), *Devolution and Accountability*, New Zealand Institute of Public Administration, Wellington.

McKinlay, P. (ed) (1990), *Redistribution of Power? Devolution in New Zealand*, Victoria University Press for the Institute of Policy Studies, Wellington.

May, P., Burby, R., Ericksen, N., Handmer, J., Dixon, J., Michaels, S. and Smith, D.I. (1996), *Environmental Management and Governance: Intergovernmental Approaches to Hazards and Sustainability*, Roultledge Press, London.

Ministry for the Environment (1987), *Corporate Plan: 1 April 1987 – 31 March 1988*, Ministry for the Environment, Wellington.

Ministry for the Environment (1988), *People, Environment, and Decision Making: the Government's Proposals for Resource Management Law Reform*, Ministry for the Environment, Wellington.

Ministry for the Environment (1990), *Discussion Paper on the Resource Management Bill. No. 69 (December 1990)*. Ministry for the Environment, Wellington.

Ministry for the Environment (1991), Report of the Review Group on the Resource Management Bill, No. 71 (February 1991), Ministry for the Environment, Wellington.

Ministry for the Environment (1992), *The Context: Sustainable Development as a Backdrop for the Resource Management Act*, Sustainable Management of Resources Information Sheet Number One, June 1992, Ministry for the Environment, Wellington.

Murphy, M. and McKinlay, P. (1998), *Devolution, Partnership or Ad Hocism? The Relationship between Central and Local Government*, Report Funded by State Services Commission, Wellington.

New Zealand Government (1991), *New Zealand Report to the United Nations Conference on Environment and Development: Forging the Links*, Ministry for the Environment and Ministry for External Relations and Trade (December 1991), Wellington.

Palmer, G. (1988), 'Political Perspectives', in J. Martin and J. Harper (eds), *Devolution and Accountability*, New Zealand Institute of Public Administration, Wellington, pp. 1-7.

Powell, P. (1978), *Who Killed the Clutha?*, John McIndoe, Dunedin.

Sandrey, R and Reynolds, R. (1990), *Farming without Subsidies,* Government Print, Wellington.

Searle, G. (1975), *Rush to Destruction*, A.H. and A.W. Reed, Wellington, Sydney, London.

Wilson, R. (1982), *From Manapouri to Aramoana: The Battle for New Zealand's Environment*, Earthworks, Auckland.

Chapter 1: Planning Mandates: From Theory to Practice

Berke, P. (1995), 'Evaluating Environmental Plan Quality: the Case of Planning for Sustainable Development in New Zealand', *Journal of Environmental Planning and Management*, Vol. 37(2), pp. 155-69.

Berke, P., Crawford, J., Dixon, J. and Ericksen, N. (1999), 'Do Co-operative Environmental Planning Mandates Produce Good Plans? Empirical Results from the New Zealand Experience', *Environment and Planning B: Planning and Design*, Vol. 26, pp. 643-64.

Berke, P., Dixon, J. and Ericksen, N. (1997), 'Coercive and Cooperative Intergovernmental Mandates: A Comparative Analysis of Florida and New Zealand Environmental Plans', *Environment and Planning B: Planning and Design*, Vol 24, pp. 451-68.

Berke, P. and French, S. (1994), 'The Influence of State Planning Mandates on Local Plan Quality', *Journal of Planning Education and Research*, Vol. 13(4), pp. 237-50.

Berke, P., Roegnik, D., Kaiser, E. and Burby, R. (1996), 'Enhancing Plan Quality: Evaluating the Role of State Planning Mandates for Natural Hazard Mitigation', *Journal of Environmental Planning and Management*, Vol. 39, pp. 79-96.

Burby, R. and Dalton, L. (1994), 'Plans Can Matter! The Role of Land Use Plans and State Planning Mandates in Limiting Development of Hazardous Areas', *Public Administration Review*, Vol. 53, pp. 229-38.

Burby, R. and May P., with Berke, P., Dalton, L., French, S. and Kaiser, E. (1997), *Making Governments Plan: State Experiments in Managing Land Use,* Johns Hopkins University Press, Baltimore.

Claridge, M. and Kerr, S. (1998), *A Case Study in Devolution: the Problem of Preserving Kiwi Habitat in the Far North*, Treasury Working Paper 98/7a, Wellington.

Cullingworth, J. B. (1994), 'Alternative Planning Systems: Is There Anything to Learn from Abroad?', *Journal of the American Planning Association*, Vol. 60(2), pp. 162-72.

Dalton, L. and Burby, R. (1994), 'Mandates, Plans and Planners: Building Local Commitment to Development Management', *Journal of the American Planning Association*, Vol. 60, pp. 444-61.

DeGrove, J. (1992), *The New Frontier for Land Use Policy: Planning and Growth Management in the United States*, Lincoln Inst of Land Use Policy, Cambridge, MA.

DeGrove, J. (1994), 'Following in Oregon's Footsteps: The Impact of Oregon's Growth Management Strategy on Other States', in C. Abbott, D. Howe and S. Adler (eds) *Planning the Oregon Way: A Twenty-Year Evaluation*, Oregon State University Press, Corvallis, Oregon.

Deyle, R. and Smith, C. (1998), 'Local Government Compliance with State Planning Mandates: the Effects of State Implementation in Florida', *Journal of the American Planning Association*, Vol. 64(4), pp. 457-69.

Earnest New and Associates Ltd. (1993), *Origins of the Southland Sustainable Management Group: Background Paper No. 12*, Vol. 1 (Papers 1-7) and Vol. 2 (Papers 8-13), Southland Sustainable Management Group, Invercargill.

Ericksen, N.J. (1994), 'Collaborative Research on Cooperative Natural Hazard Mandates Under the Resource Management Act', in J.W. Selsky, R.K. Morgan and A. Memon (eds) *Environment and Resource Management in New Zealand*, No 5(6), Environmental Policy and Management Research Centre, University of Otago, pp. 161-70

Faludi, A. (1987), *A Decision Centred View of Environmental Planning*, Pergamon Press, Oxford, England.

Fishman, R. (1978), 'The State of the Art in Local Planning', in R. Fishman (ed) *Housing for All Under Law: New Directions in Housing, Land Use and Planning Law*, Report to the American Bar Association, Advisory Committee on Housing and Urban Growth. Ballinger, Massachusetts.

Googin, M., Bowman, A., Lester, J. and O'Tolle, L. (1990), *Implementation Theory and Practice: Toward a Third Generation*, Scott, Foresman, Glenview, Illinois.
Healy, R. and Rosenberg, J. (1979), *Land Use and the United States*, Second Edition, Johns Hopkins University Press, Baltimore.
Innes, J. (1992), 'Group Processes and the Social Construction of Growth Management,' *Journal of the American Planning Association*, Vol. 62(4), pp. 460-72.
Kerr, S., Claridge, M. and Milicich, D. (1998), *Devolution and the New Zealand Resource Management Act*, Treasury Working Paper 98/7, Wellington.
McDonnell, L. M. and Elmore, R. F. (1987), 'Getting the Job Done: Alternative Policy Instruments', *Educational Evaluation and Policy Analysis*, Vol. 9(2), pp. 133-52.
May, P.J. (1993), 'Mandate Design and Implementation: Enhancing Implementation Efforts and Shaping Regulatory Styles', *Journal of Policy Analysis and Management*, Fall.
May, P., Burby, R., Ericksen, N., Handmer, J., Dixon, J., Michaels, S. and Smith, D. (1996), *Environmental Management and Governance: Intergovernmental Approaches to Hazards and Sustainability*, Roultledge Press, London.
May, P.J. and J.W. Handmer (1992), 'Regulatory Policy Design: Co-operative Versus Deterrent Mandates', *Australian Journal of Public Administration*, Vol. 51(1), March 1992, pp. 43-53.
Murphy, M. and McKinlay, P. (1998), *Devolution, Partnership or Ad Hocism? The Relationship between Central and Local Government*, Report Funded by State Services Commission, Wellington.
Pressman, J. and Wildavsky, A. (1973), *Implementation*, University of California Press, Berkely, California.
Talen, E. (1996), 'Do Plans Get Implemented? A Review of Evaluation in Planning', *Journal of Planning Education and Research*, Vol. 10(3), pp. 248-59.
Younis, T. (ed) (1990), *Implementation in Public Policy*, Gower Press, Brookfield, Vermont.

Chapter 2: Making Plans: From Theory to Practice

Baer, W.C. (1997), 'General Plan Evaluation Criteria: An Approach to Making Better Plans', *Journal of the American Planning Association*, Vol. 63(3), pp. 329-44.
Berke, P. (1994), 'Evaluating Environmental Plan Quality: the Case of Planning for Sustainable Development in New Zealand', *Journal of Environmental Planning and Management*, Vol. 37(2), pp. 155-69.
Berke, P., Crawford, J., Dixon, J. and Ericksen, N. (1999), 'Do Co-operative Environmental Planning Mandates Produce Good Plans? Empirical Results from the New Zealand Experience', *Environment and Planning B: Planning and Design*, Vol. 26, pp. 643-64,
Berke, P. and French, S. (1994), 'The Influence of State Planning Mandates on Local Plan Quality', *Journal of Planning Education and Research*, Vol. 13(4), pp. 237-50.
Berke, P., Roegnik, D., Kaiser, E. and Burby, R. (1996), 'Enhancing Plan Quality: Evaluating the Role of State Planning Mandates for Natural Hazard Mitigation', *Journal of Environmental Planning and Management*, Vol. 39, pp. 79-96.
Braybrooke, D. and Linbloom, C. (1963), *A Strategy of Decision*, Free Press of Glencoe, New York.
Burby, R. and May P., with Berke, P., Dalton, L., French, S. and Kaiser, E. (1997), *Making Governments Plan: State Experiments in Managing Land Use*, Johns Hopkins University Press, Baltimore.
Dalton, L. and Burby, R. (1994), 'Mandates, Plans and Planners: Building Local Commitment to Development Management,' *Journal of the American Planning Association*, Vol. 60, pp. 444-61.

Dixon, J., Ericksen, N., Crawford, J., and Berke, P. (1997), 'Planning Under a Cooperative Mandate: New Plans for New Zealand', *Journal of Environmental Planning and Management*, Vol. 40 (5), pp. 603-614.

Healy, P. (1993), 'The Communicative Work of Development Plans', *Environment and Planning B: Planning and Design*, Vol. 20, 83-104.

Kaiser, E. and Godschalk, D. (1995), 'Twentieth Century Land Use Planning: a Stalwart Family Tree,' *Journal of the American Planning Association*, Vol. 61(3), pp. 365-85.

Kaiser, E., Godschalk, D. and Chapin, S. (1995), *Urban Land Use Planning*, Fourth Edition, University of Illinois Press, Chicago, Illinois.

Kent, T.J. (1990), *The Urban General Plan*, Chandler, San Francisco.

Wenger, D., James, T. and Faupel, C. (1980), *Disaster Relief and Emergency Planning*, Irvington, New York.

Chapter 3: Central Government: Walking the Talk

Bassett, M. (1989), 'Reforming Local and Regional Government', *Local Authority Management*, Vol. 15(2), pp. 7-10.

Boston, J., Martin, J., Pallot, J., and Walsh, P. (1996), *Public Management in New Zealand: The New Zealand Model*, Oxford University Press, Auckland.

Bangura, Y. (2000), *Public Sector Restructuring: The Institutional and Social Effects of Fiscal, managerial, and Capacity-Building Reforms*, Occasional Paper 3, United Nations Research Institute for Social Development, Paris.

Bush, G. (1995), *Local Government and Politics in New Zealand*, Second Edition, Auckland University Press, Auckland.

Bushnell, P. and Scott, G. (1988), 'An Economic Perspective', in Martin, J. and Harper, J. (eds), *Devolution and Accountability*, Proceedings of the 1988 Conference of the New Zealand Institute of Public Administration', Studies in Public Administration, No. 34. Wellington: New Zealand Institute of Public Administration, pp. 19-36.

Commission of Inquiry (1995), *Report: Commission of Inquiry into the Collapse of a Viewing Platform at Cave Creek Near Punakaiki on the West Coast*, The Department of Internal Affairs, Wellington.

Department of Conservation (1990-1999), *Annual Reports to Parliament*, Department of Conservation, Wellington.

Department of Conservation (1990), *Draft New Zealand Coastal Policy Statement*, August 1990, Department of Conservation, Wellington.

Department of Conservation (1992), *Proposed New Zealand Coastal Policy Statement*, September 1992, Department of Conservation, Wellington.

Department of Conservation (1994a), *Draft Guide to Regional Coastal Plan Submission Preparation*, August 1994, Department of Conservation, Wellington.

Department of Conservation (1994b), *New Zealand Coastal Policy Statement*, 5 May 1994, Department of Conservation, Wellington.

Department of Conservation (1995), *RPS: Issues of Concern to the Department of Conservation and Possible Hearings and Appeals*, prepared by Mairi Jorgensen, Department of Conservation, Wellington.

Department of Conservation (1998a), *Strategic Business Plan 1998-2002: Restoring the Dawn Chorus*, Department of Conservation, Wellington.

Department of Conservation (1998b), *Working with Communities on National Conservation Issues: A Strategy for Involvement in the RMA*, Draft of 16 April 1998, Department of Conservation, Wellington.

Dormer, A. (1994), *The RMA 1991: The Transition and Business*, A Report prepared for the Business Roundtable, Wellington.

Froude, V. (1997), *Territorial Authorities Addressing Section 6(c) of the RMA for Terrestrial Ecosystems and Wetlands: Advantages and Disadvantages of Various Approaches*, paper prepared for the Ministry for the Environment and distributed at the workshop: 'Significant Natural Area Requirements,' convened by Local Government New Zealand, 26 September 1997. Wellington: Ministry for the Environment.

Government of New Zealand (1997), *The 1997 Budget*, New Zealand Government, Wellington.

Halligan, J. (1997), 'New Public Sector Models: Reform in Australia and New Zealand', in J.E. Lane (ed), *Public Sector Reform: Rationale, Trends, and Problems*, Sage Publications, London.

Harris, P. and Twiname, L. (1998), *First Knights: An Investigation of the New Zealand Business Roundtable*, Howling at the Moon Publishing Ltd, Auckland.

Hutchings, J. (1999), 'Will the Proposed Amendments to the RMA Give Rise to Efficiency Gains?', *Resource Management News*, Resource Management Law Association, September 1999.

Kerr, S., Claridge, M., and Milicich, D. (1998), *Devolution and the New Zealand RMA*, Treasury Working Paper 98/7, Department of the Treasury, Wellington.

Local Government New Zealand (1997), *Significant Natural Area Requirements*, Proceedings of a Workshop Convened by Local Government New Zealand, 26 September 1997, Local Government New Zealand, Wellington.

Local Government New Zealand (2000), *Devolution – Fact or Fiction?* A report on the extent of new functions and costs acquired by local government since reform in 1989, Local Government New Zealand, Wellington.

Local Government New Zealand and Department of Conservation (1996), *Protocol Between Local Government New Zealand and the Department of Conservation Concerning Responsibilities, Roles and Processes for Policy and Plan Preparation Under the RMA*, 18 December 1996, Department of Conservation, Wellington.

McDermott, P.J. (2000), 'The Costs of Environmental Planning: Implications of the RMA, in A. Memon and H. Perkins (eds), *Environmental Planning and Management in New Zealand.* Revised Edition, Dunmore Press, Palmerston North, pp. 48-63.

McShane, O. (1998), *Land Use Control Under the RMA: A "Think Piece"*, a report commissioned by the Minister for the Environment, with critiques by: K. Tremaine, B. Nixon, and G. Salmon, Wellington.

Martin, J. (1991), 'Devolution and Decentralisation', in J. Boston, J. Martin, J. Pallot and P. Walsh (eds) *Reshaping the State: New Zealand's Bureaucratic Revolution*, Oxford University Press Auckland, pp. 268-96.

Martin, J., and Harper, J. (eds) (1988), *Devolution and Accountability*, Proceedings of the 1988 Conference of the New Zealand Institute of Public Administration, Studies in Public Administration, No. 34, Wellington, p. 19-36.

May, P., Burby, R., Ericksen, N., Handmer, J., Dixon, J., Michaels, S. and Smith D. (1996), *Environmental Management and Governance: Intergovernmental Approaches to Hazards and Sustainability*, Routledge, London/New York.

Ministry for the Environment (1987), *Corporate Plan: 1 April 1987 – 31 March 1988*, Ministry for the Environment, Wellington.

Ministry for the Environment (1988-2000), *Annual Reports to Parliament*, Ministry for the Environment, Wellington.

Ministry for the Environment (1988a), *People, Environment, and Decision Making: the Government's Proposals for Resource Management Law Reform* (December), Ministry for the Environment, Wellington.

Ministry for the Environment (1988b), *Ecological Principles in Resource Management*, No. 34 (July 1988), Ministry for the Environment, Wellington.

Ministry for the Environment (1990), *Discussion Paper on the Resource Management Bill*, No. 69 (December 1990), Ministry for the Environment, Wellington.

Ministry for the Environment (1991a), Report of the Review Group on the Resource Management Bill, No. 71 (February 1991), Ministry for the Environment, Wellington.

Ministry for the Environment (1991b), Resource Law Transition, Unpublished Proposal for funding implementing the Resource Management Act, Ministry for the Environment, Wellington.

Ministry for the Environment (1991c), *Resource Management: Guide to the Act*, No. 75 (August 1991), Ministry for the Environment, Wellington.

Ministry for the Environment (1991d), *Managing Our Future*, No. 77 (August 1991), Ministry for the Environment, Wellington.

Ministry for the Environment (1991e), *Consultation with Tangata Whenua*, No. 78 (September 1991), Ministry for the Environment, Wellington.

Ministry for the Environment (1991f), *Regional Policy Statements and Plans*, No. 80 (September 1991), Ministry for the Environment, Wellington.

Ministry for the Environment (1991g), *Resource Management: Guidelines for Subdivision*, No. 82 (October 1991), Ministry for the Environment, Wellington.

Ministry for the Environment (1991h), *Resource Management: Transitional Provisions*, No. 83 (November 1991), Ministry for the Environment, Wellington.

Ministry for the Environment (1991i), *Resource Management: Guidelines for District Plans*, No. 84 (November 1991), Ministry for the Environment, Wellington.

Ministry for the Environment (1992a), *Kia Matiratira: A Guide for Maori*, No. 89 (June 1992), Ministry for the Environment, Wellington.

Ministry for the Environment (1992b), *Scoping of Environmental Effects*, No. 90 (June 1992), Ministry for the Environment, Wellington.

Ministry for the Environment (1992c), *Hinengaro Bay: a Fictional Case Study*, No. 95, August 1992, Ministry for the Environment, Wellington.

Ministry for the Environment (1993a), *Taking into Account the Principles of the Treaty of Waitangi: Ideals for the Implementation of Section 8*. No. 105 (January 1993). Ministry for the Environment, Wellington.

Ministry for the Environment (1993b), *RMA: Section 24 Monitoring Report*, No. 107 (February 1993), Ministry for the Environment, Wellington.

Ministry for the Environment (1993c), *Section 32 – A Guide to Good Practice*, No. 121, August 1993, Ministry for the Environment, Wellington.

Ministry for the Environment (1993d), *Section 24 Monitoring Report*, No. 123, August 1993, Ministry for the Environment, Wellington.

Ministry for the Environment (1994a), *Resource Consents and Good Practice*, No. 128 (February 1994), Ministry for the Environment, Wellington.

Ministry for the Environment (1994b), *Time Frames for Processing Resource Consents*, No. 130 (March 1994), Ministry for the Environment, Wellington.

Ministry for the Environment (1994c), *Section 24 Monitoring Report*, No. 131 (March 1994), Ministry for the Environment, Wellington.

Ministry for the Environment (1994d), *Investment Certainty Under the RMA*, No. 133 (March 1994), Ministry for the Environment, Wellington.

Ministry for the Environment (1994e), *Waste and the RMA*, No. 134 (April 1994), Ministry for the Environment, Wellington.

Ministry for the Environment (1994f), *Resource Rentals for the Occupation of Coastal Space: A Discussion Document*, No. 136 (June 1994), Ministry for the Environment, Wellington.

Ministry for the Environment (1994g), *Enabling Sustainable Communities: A Strategic Policy Paper*, No. 137 (June 1994), Ministry for the Environment, Wellington.

Ministry for the Environment (1994h), *Issues, Objectives, Policies, Methods and Results Under the RMA*, No. 140 (July 1994), Ministry for the Environment, Wellington.

Ministry for the Environment (1994I), *Case Law on Consultation: Working Paper 2*, No. 152 (November 1994), Ministry for the Environment, Wellington.

Ministry for the Environment (1994j), *Coastal Rentals Under the RMA*, No. 153 (October 1994), Ministry for the Environment, Wellington.

Ministry for the Environment (1994k), *A Guideline to Administrative Charging Under Section 36 of the RMA*, No. 154 (November 1994), Ministry for the Environment, Wellington.

Ministry for the Environment (1995a), *Marine Pollution Regulations Under the RMA*, No. 156 (February 1995), Ministry for the Environment, Wellington.

Ministry for the Environment (1995b), *Proposed Taranaki Power Station – Air Discharge Effects: Report of the Board of Inquiry*, No. 158 (February 1995), Ministry for the Environment, Wellington.

Ministry for the Environment (1995c), *Principles and Priorities for the Development of Environmental Standards and Guidelines*, No. 160 (May 1995), Ministry for the Environment, Wellington.

Ministry for the Environment (1995d), *Odour Management Under the RMA*, No. 165 (June 1995), Ministry for the Environment, Wellington.

Ministry for the Environment (1995e), *Summary of Submissions on Pollution Under the RMA*, No. 172 (September 1995), Ministry for the Environment, Wellington.

Ministry for the Environment (1995f), *Principles and Priorities for the Development of Environmental Standards and Guidelines*, No. 180 (September 1995), Ministry for the Environment, Wellington.

Ministry for the Environment (1995g), *Principles and Processes for Developing Guidelines and Standards: A Position Paper*, No. 181 (October 1995), Ministry for the Environment, Wellington.

Ministry for the Environment (1995h), *Implementing the RMA: Key Messages for Councillors*, No. 189 (December 1995), Ministry for the Environment, Wellington.

Ministry for the Environment (1996a), *Assessment of Environmental Effects*, Working Paper No 6. No. 190 (March 1996), Ministry for the Environment, Wellington.

Ministry for the Environment (1996b), *Awarding of Costs by the Planning Tribunal (Environment Court)*, No. 191 (April 1996), Ministry for the Environment, Wellington.

Ministry for the Environment (1996c), *Health Effects of Bathing at Selected New Zealand Marine Beaches*. No. 192 (December 1996), Ministry for the Environment, Wellington.

Ministry for the Environment (1996d), *Proposals for Regulations Under the RMA to Control Marine Pollution*, No. 203 (June 1996), Ministry for the Environment, Wellington.

Ministry for the Environment (1996e), *Analysis of Questionnaire to Territorial Councils on Monitoring Duties Under Section 35 of the RMA*, No. 205 (May 1996), Ministry for the Environment, Wellington.

Ministry for the Environment (1996f), *Notification Under the RMA*, No. 6. No. 214 (October 1996), Ministry for the Environment, Wellington.

Ministry for the Environment (1996g), *Amenity Values Under the RMA*, No. 7, No. 215 (October 1996), Ministry for the Environment, Wellington.

Ministry for the Environment (1996h), *The Monitoring Guide*, No. 227 (December 1996), Ministry for the Environment, Wellington.

Ministry for the Environment (1996I), *Findings of the Annual Survey of Local Authorities*, No. 228 (December 1996), Ministry for the Environment, Wellington.

Ministry for the Environment (1997a), *A Monitoring Framework Under Section 24*, No. 191 (March 1996), Ministry for the Environment, Wellington.

Ministry for the Environment (1997b), *To Notify or Not to Notify Under the RMA: Background Report*, No. 253, Ministry for the Environment, Wellington.

Ministry for the Environment (1997c), *To Notify or Not to Notify Under the RMA: A Good Practice Guide*, No. 254 (October 1997), Ministry for the Environment, Wellington.

Ministry for the Environment (1997d), *Proposal to Include Limited Notification Procedure for Resource Consents*, No. 256 (December 1997), Ministry for the Environment, Wellington.

Ministry for the Environment (1998a), *Making a Difference: Our Priorities 1998/99*, Ministry for the Environment, Wellington.

Ministry for the Environment (1998b), *New Zealand's Biodiversity Strategy: Our Chance to Turn the Tide*, A Draft Strategy for public consultation, December 1998, Ministry for the Environment, Wellington.

Ministry for the Environment (1998-2001), *Annual Surveys of Local Authorities 1996/97, 1997/98, 1998/99, 1999/2000*, Ministry for the Environment, Wellington.

Ministry for the Environment (2000), *Bio-What? Addressing the Effects of Private Land Management on Indigenous Biodiversity*, Preliminary Report on the Ministerial Advisory Committee, February 2000, Ministry for the Environment, Wellington.

Morris, R. (1994), *The Pricing of Policy Advice. The Proposed Cabinet Benchmark*, Paper presented to AIC Conference on Managing Quality Policy, Wellington.

Nixon, R. (1997), *Working with the RMA*, Paper presented to AIC Conference, 11-12 August 1997.

Organisation for Economic Co-operation and Development (1996), *Environmental Performance Reviews: New Zealand*, Organisation for Economic Co-operation and Development (OECD), Paris.

Palmer, G. (2003), *Whither the Resource Management Act?*, Plenary Address by Sir Geoffrey Palmer to Vision 50/50, New Zealand Planning Institute Conference, 22-24 May, Hamilton.

Parliamentary Commissioner for the Environment (1992), *Proposed Guidelines for Local Authority Consultation with Tangata Whenua*, Parliamentary Commissioner for the Environment, Wellington.

Parliamentary Commissioner for the Environment (1993-1998), *Annual Reports to Parliament*, Parliamentary Commissioner for the Environment, Wellington.

Parliamentary Commissioner for the Environment (1998), *Kaitiakitanga and Local Government: Tangata Whenua Participation in Environmental Management*, Parliamentary Commissioner for the Environment, Wellington, New Zealand.

Parliamentary Commissioner for the Environment (1999), *Local Government Environmental Management: A Study of Models and Outcomes*, Office of the Parliamentary Commissioner for the Environment, Wellington.

Pavletich, H.J., and McShane, O. (1997), *Taking the Heat off the RMA*, Report for Hon Simon Upton, Minister for the Environment, 22 May 1997, Christchurch and Auckland.

Reid, M. (1999), 'The Central-Local Government Relationship: the Need for a Framework', *Political Science*, Vol. 50(2), pp. 165-81.

Richardson, M. with Clifford, P., Cumming, I., Griffin, D. and Caygill, D. (1999), *Taking the Canterbury Communities into the New Millennium: The Role of Local Government*, Report presented at a Forum on the Future of Government, Christchurch, 2-3 June 1999, Christchurch City Council, Christchurch.

State Services Commission (1995), *Review of the Department of Conservation: Under Section 6(b) of the State Sector Act 1998*, State Services Commission, Wellington.

State Services Commission (1988), *Sharing Control: a Policy Study of Responsiveness and Devolution in the Statutory Social Services*, Report to the Steering Group of Permanent Heads from the Task Group on Devolution, State Services Commission, Wellington.

The Officials Co-ordinating Committee on Local Government (1988a), *Reform of Local and Regional Government,* February 1988, The Officials Co-ordinating Committee on Local Government, Wellington.

The Officials Co-ordinating Committee on Local Government (1988b), *Reform of Local and Regional Government: Funding Issues — A Discussion Document,* December 1988, The Officials Co-ordinating Committee on Local Government, Wellington.

The Officials Co-ordinating Committee on Local Government (1988c), *Synopsis of Submissions on Reform of Local and Regional Government,* Prepared by The Bridgeport Group, June 1988, The Officials Co-ordinating Committee on Local Government, Wellington.

The Officials Co-ordinating Committee on Local Government (1989), *Reform of Local and Regional Government: Synopsis of Submissions on Funding Issues — A Discussion Document,* Prepared by Synergy Applied Research Ltd, July 1989, The Officials Co-ordinating Committee on Local Government, Wellington.

Upton, S. (1991), *Resource Management Bill: Third Reading,* Hansard 516 NZPD 3030, 4 July 1991.

Upton, S. (1994), *The Resource Management Act, Section 5: Sustainable Management of Natural and Physical Resources,* Keynote Speech to the Second Annual Conference of the Resource Management Law Association, 7 October 1994, Ministry for the Environment, Wellington.

Upton, S. (1995a), *Purpose and Principle in the Resource Management Act,* The Stace Hammond Grace Lecture, 26 May 1995, University of Waikato. Ministry for the Environment, Wellington.

Upton, S. (1995b), *Problems of Rural Subdivision,* Address to New Zealand Planning Institute Conference, 26 May 1995.

Upton, S. (1997a), *Understanding the Role of Policy Under the RMA,* Speech to the New Zealand Planning Institute Conference at Palmerston North by the Hon Simon Upton, Minister for the Environment, 2 March 1998, Wellington.

Upton, S. (1997b), *The Efficient Use and Development of Natural and Physical Resources,* Address to the 5th Annual Resource Management Law Association Conference, 29 August 1997.

Upton, S. (1998), *The Resource Management Act: Pressures and Problems,* Address to the Auckland Branch of the New Zealand Planning Institute AGM at by the Hon Simon Upton, Minister for the Environment, 18 April 1997.

Upton, S. (1999), *Address to Centre for Advanced Engineering (CAE) Conference: Assessments of Environmental Effects — Information, Evaluation, and Outcomes,* Speech by Hon. Simon Upton, Minister for the Environment, 18 March 1999, Wellington.

Chapter 4: Regional Government: A Non-Partner

Auckland Regional Authority (1983), *Guideline: Comprehensive Catchment Planning,* Auckland Regional Authority, Auckland.

Beanland, R. and Huser, B. (1999), *Integrated Monitoring: A Manual for Practitioners,* Environment Waikato, Hamilton.

Berke, P. (1995), 'Evaluating environmental plan quality: the case of planning for sustainable development in New Zealand', *Journal of Environmental Planning and Management,* Vol. 37(2), pp. 155-69.

Burby, R. and May, P. with Berke, P., Dalton, L., French, S. and Kaiser, E. (1997), *Making Governments Plan: State Experiments in Managing Land Use,* Johns Hopkins University Press, Baltimore.

Court of Appeal (1995a), *Auckland Regional Council vs North Shore City Council*, Declaratory Judgement, CA 29/95, Court of Appeal, Wellington.

Court of Appeal (1995b), *Canterbury Regional Council vs Banks Peninsula District Council and Others*, Declaratory Judgement, CA 29/95, Court of Appeal, Wellington.

Elazar, D.J. (ed.) (1991), *Constitutional Design and Power-Sharing in the Post-Modern Epoch*, Jerusalem Center for Public Affairs/Center for Jewish Community Studies, University Press of America, Jerusalem, Lanham.

Ericksen, N.J. (1986), *Creating Flood Disasters? New Zealand's Need for a New Approach to Urban Flood Hazard*, Water and Soil Miscellaneous Publication No. 77, Wellington.

Ericksen, N.J. (1990), 'New Zealand Water Planning and Management: Evolution or Revolution?', in B. Mitchell (ed.), *Integrated Water Management*, Belhaven Press, London and New York, pp. 45-87.

Ericksen, N., Dixon, J. and Berke, P. (2000), 'Managing Natural Hazards under the Resource Management Act', in A. Memon and H Perkins (eds), *Environmental Planning and Management in New Zealand*, Dunmore Press, Palmerston North, pp. 123-32.

FACIR (1991), *Mandates and Measures Affecting Local Government Fiscal Capacity: Intergovernmental Impact Report*. Florida Advisory Council on Intergovernmental Relations (FACIR).

Goldsmith, M. (1993), 'The Europeanisation of Local Government', *Urban Studies*, Vol. 30(4/5), pp. 683-99.

Gow, L. (1994), *Regional Policy Statements: State of Play*, Speech to New Zealand Local Government Association Conference, 3 June 1994, Ministry for the Environment, Wellington.

Griffiths, G. and Ross, P. (1997), 'Principles of Managing Extreme Events', in M.P. Mosely and C.P. Pearson, *Floods and Droughts: the New Zealand Experience*, The Caxton Press, Christchurch.

Hawkes Bay Regional Council, Taranaki Regional Council, Manawatu-Wanganui Regional Council, Otago Regional Council, and Otago Regional Council (1997), *Regional Policy Statements and Regional Plans: A Guide to Their Purpose, Scope, and Content*, Draft Report, December 1997.

Hinton, S. and Hutchings, J. (1994), 'Regional Councils Debate Responsibilities', *Planning Quarterly*, Vol. 115, pp. 4-5.

Hughes, H. R. (1991), *Environmental Management and Regional Councils*, Office of Parliamentary Commissioner for the Environment, Wellington, 8 pages.

Hutchings, J. (1994), *Interauthority Integration of Approaches and Issues: Regional Approaches*. Paper presented at the New Zealand Planning Institute Conference, 27-30 April, Nelson.

Local Government Commission (1973), *Regions and Districts of New Zealand: Area Adopted to Date for Various Administrative and Research Purposes*, Government Printer, Wellington.

Local Government Commission (1988), *Communique to all Regional Authorities, Territorial Authorities, Special Purpose Authorities*, 28 September 1988, Local Government Commission, Wellington.

Local Government Commission (1995), *Annual Report*, Local Government Commission, Wellington.

McBride, N. (1990), *The Functions and Significance of Economic and Social Programmes in New Zealand's Regional Councils*, Paper for 1990 Regional Government Conference, Whakatane, 3 May 1990, 16 pages.

McDonnell, L. M. and Elmore, R. F. (1987), 'Getting the Job Done: Alternative Policy Instruments', *Educational Evaluation and Policy Analysis*, Vol. 9(2), pp. 133-52.

May, P., Burby, R., Ericksen, N., Handmer, J., Dixon, J., Michaels, S., and Smith, D. (1996), *Environmental Management and Governance: Intergovernmental Approaches to Hazards and Sustainability*, Routledge, London/New York.

Michaels, S. (1996), *Regional Councils in New Zealand as Partners with Local Councils in Mitigating Threats from Natural Hazards*, Paper prepared at Department of Urban and Environmental Policy, Tufts University, Boston.

Ministry for the Environment (1992), *Resource Management Subsidies Criteria 1992/93*, Ministry for the Environment, Wellington.

Ministry for the Environment (1994a), *Environmental 2010 Strategy*, Ministry for the Environment, Wellington.

Ministry for the Environment (1998), *Annual Survey of Local Authorities 1996/97*, Ministry for the Environment, Wellington.

Mitchell, B. (ed) (1990), *Integrated Water Management: International Experiences and Perspectives*, Belhaven Press, New York and London.

Press, D. (1995) 'Environmental Regionalism and the Struggle for California', *Society and Natural Resources*, Vol. 8(4), pp. 289-306.

Scarlet, D., and Matthews, R. (1995), 'The Costs of RMA Processes - Are they Sustainable? A Regional Council Perspective', *Sustainable Management: A Sustainable Ethic?*, Third Annual Conference, Resource Management Law Association of New Zealand Inc., 6-7 October 1995.

Chapter 5: Māori Interests: Elusive Partnership

Baragwanath, Hon. Justice (1997), *The Treaty of Waitangi and the Constitution*, Paper presented at New Zealand Law Society Seminar on Treaty of Waitangi Issues: The Last Decade and the Next Century, April, New Zealand Law Society, Wellington.

Berke, P., Ericksen, N., Crawford, J. and Dixon, J. (2002), 'Planning and Indigenous People: Human Rights and Environmental Protection in New Zealand', *Journal of Planning Education and Research*, 22: 115-134.

Boast, R.P. (1989), *The Treaty of Waitangi: a Framework for Resource Management Law*. New Zealand Planning Council and Victoria University of Wellington Law Review, Wellington.

Boast, R. and Edmunds, D. (1994), 'The Treaty of Waitangi and Maori Resource Management Issues', *Resource Management*, Volume 1A, Brooker's Ltd, Wellington.

Boston, J., Martin, J., Pallot, J. and Walsh, P. (eds) (1996), *Reshaping the State: New Zealand's Bureaucratic Revolution*, Oxford University Press, Auckland.

Chen, M. and Palmer, G. (1999), *He Waka — Local Government and the Treaty of Waitangi*, Research Monograph Series, Paper no. 8, Local Government New Zealand, Wellington.

Clark, Rt. Hon. H. (2000), *Prime Minister's Speech*, Presented to Local Government New Zealand Conference, 10-12 July, Christchurch.

Durie, M., (1998), *Te Mana, te Kawanatanga: The Politics of Maori Self-determination*, Oxford University Press, Auckland.

Hayward, J. (1999), 'Local Government and Maori: Talking Treaty?', *Political Science*, Vol. 50 (2), pp. 182-94.

Hewison, G. (1997), *Agreements Between Maori and Local Authorities*, Manukau City Council, Manukau City.

James, C. (2000), *A New Paradigm?*, Presentation to Local Government New Zealand Conference, 10-12 July, Christchurch.

Kelly, J. and Marshall, B. (1996), *Atlas of New Zealand Boundaries*, Auckland University Press, Auckland.

Kelsey, J. (1990), *A Question of Honour? Labour and the Treaty, 1984-1989*, Allen and Unwin, Wellington.

King, M., (1997), *Nga Iwi O Te Motu. One Thousand Years of Maori History*, Reed Books, Wellington.

Lee, Hon. S. (2000), *Keynote Address*, Presented to Local Government New Zealand Conference, 10-12 July, Christchurch.

Local Government New Zealand (1997), *Liaison and Consultation with Tangata Whenua: A Survey of Local Government Practice*, Local Government New Zealand, Wellington.

Manukau City Council (2000), *Te Tiriti o Waitangi: Treaty of Waitangi Toolbox 2000*, Manukau City Council, Manukau City.

Matunga, H. (1989), *Local Government: a Maori Perspective*, Unpublished report for the Maori Consultative Group on Local Government Reform, January 1989.

Matunga, H. (2000), 'Decolonising Planning: The Treaty of Waitangi, the Environment and a Dual Planning Tradition', in P.A. Memon and H. Perkins (eds), *Environmental Planning and Management in New Zealand*, Dunmore Press, Palmerston North, pp. 36-47.

McDowell, M. and Webb, D. (1998), *The New Zealand Legal System: Structures, Processes and Legal Theory*, Second Edition, Butterworths, Wellingon, New Zealand.

Ministry for the Environment (1988), *People, Environment and Decision Making: The Government's Proposals for Resource Management Law Reform*, Ministry for the Environment, Wellington.

Ministry for the Environment (1991), *Resource Management – Consultation with Tangata Whenua*, A guide to assist local authorities in meeting the consultation requirements of the RMA 1991, Ministry for the Environment, Wellington.

Ministry for the Environment (1992a), *The Resource Management Act – Kia Matiratira – A Guide for Maori*, Ministry for the Environment, Wellington.

Ministry for the Environment (1992b), *Tangata Whenua and Local Government Planning: A Guide to Effective Participation*, Information pamphlet, Ministry for the Environment, Wellington.

Ministry for the Environment (1993a), *Taking into Account the Principles of the Treaty of Waitangi: Ideas for the Implementation of Section 8 Resource Management Act 1991*, Ministry for the Environment, Wellington.

Ministry for the Environment (1993b), *He Whenua, he Marae, he Tangata – Planning by Maori for Maori. Te Whakatoitunga – Long-term Protection of Land and Water and Maori Land*, Information Pamphlet, Ministry for the Environment, Wellington.

Ministry for the Environment (1993c), *Te Whakatoitunga – Long-term Protection of Land and Water and Maori Land*, Pamphlet, Ministry for the Environment, Wellington.

Ministry for the Environment (1998), *He Tohu Whakamarama: A Report on the Interaction Between Local Government and Maori Organisations in Resource Management Act Processes*, Ministry for the Environment, Wellington.

Ministry for the Environment (1999), *Case Law on Tangata Whenua Consultation*, RMA Working Paper, Ministry for the Environment, Wellington.

Ministry for the Environment (2000a), *Talking Constructively: A Practical Guide for Iwi, Hapu and Whanau on Building Agreements with Local Authorities*, Ministry for the Environment, Wellington.

Ministry for the Environment, (2000b), *Iwi and Local Government Interaction under the Resource Management Act 1991: Examples of Good Practice*, Ministry for the Environment, Wellington.

Ministry for the Environment and Te Puni Kokiri (1996), *The Duty to Consult – A Survey of Iwi Participation in the Resource Management Process*, Ministry for the Environment and Te Puni Kokiri, Wellington.

Ministry of Justice (1999), *Post Election Briefing for Incoming Ministers*, Ministry of Justice, Wellington.

Minister of Maori Affairs (1988a), *He Tirohanga Rangapu: Partnership Perspectives*, Government Printer (Government discussion paper), Wellington.

Minister of Maori Affairs (1988b), *Te Urupare Rangapu: Partnership Response*, Government Printer (Government discussion paper), Wellington.

Moller, H., Horsely, P., Lyver, P., Taiepa, T. and Bragg, M. (2000), 'Co-management by Maori and Pakeha for Improved Conservation in the Twenty-first Century', in A. Memon and H. Perkins (eds), *Environmental Planning and Management in New Zealand*, Dunmore Press, Palmerston North, pp. 156-167.

Nuttall, P. and Ritchie, J. (1995), *Māori Participation in the Resource Management Act: An Analysis of Provision made for Māori Participation in Regional Policy Statements and District Plans produced under the Resource Management Act 1991*, Centre for Māori Studies and Research, University of Waikato, Hamilton.

Officials Co-ordinating Committee on Local Government (1988), *Discussion Document on Reform of Local and Regional Government*, February, Parliament Buildings, Wellington.

Palmer, K. (1993), *Local Government Law in New Zealand*, The Law Book Company, Sydney.

Parliamentary Commissioner for the Environment (1992), *Proposed Guidelines for Local Authority Consultation with Tangata Whenua*, Parliamentary Commissioner for the Environment, Wellington.

Parliamentary Commissioner for the Environment (1998), *Kaitiakitanga and Local Government: Tangata Whenua Participation in Environmental Management*, Parliamentary Commissioner for the Environment, Wellington.

Rennie, H., Thomson, J. and Tutua-Nathan, T. (2000), *Factors Facilitating and Inhibiting Section 33 Transfers to Iwi, Research Report*, University of Waikato, Hamilton.

Rikys, P. (1999), 'A Commentary', in Local Government New Zealand, *He Whaka Taurua – Local Government and the Treaty of Waitangi*, Local Government New Zealand, Wellington, pp. 87-90.

Rosson, L. (2000), *Fit for the Future: Local Government at the Turn of a New Century*, Presidential address to Local Government New Zealand Conference, 10-12 July, Christchurch.

Sunde, C., Taiepa, T., and Horsley, P. (1999), *Exploring Collaborative Management Initiatives Between Whanganui Iwi and The Department of Conservation*, Occasional Paper no. 3, Massey University, School of Resource and Environmental Planning, Palmerston North.

Statistics New Zealand (1998), *Māori*, New Zealand Now Series, Statistics New Zealand, Wellington.

Taiepa, T. (1999), 'Māori Participation in Environmental Planning: Institutional Reform and Collaborative Management', *He Pukenga Kōrero Koanga*, Spring 1999, Vol. 5(1), pp. 34-9.

Te Puni Kokiri (1993), *Mauriora Ki Te Ao: An Introduction to Environmental and Resource Management Planning*, Te Puni Kokiri (Ministry for Maori Development), Wellington.

Te Puni Kokiri (1996), *Sites of Significance Process: a Step by Step Guide to Protecting Sites of Cultural, Spiritual and Historical Significance to Māori*, Te Puni Kokiri, Wellington.

Te Puni Kokiri (1998), *Progress Towards Closing Social and Economic Gaps Between Māori and Non-Māori: A Report to the Minister of Māori Affairs*, Te Puni Kokiri, Wellington.

Te Puni Kokiri (2000), *Progress Towards Closing Social and Economic Gaps Between Māori and Non-Māori: A Report to the Minister of Māori Affairs*, May 2000, Te Puni Kokiri, Wellington.

Wickliffe, C. (1997), 'The Role of the PCE: Maori and the Treaty of Waitangi Issues', in G. Hawke (ed), *Guardians for the Environment*, Proceedings of a Symposium to mark the first decade and to provide directions for the New Zealand Parliamentary Commissioner for the Environment, Institute of Policy Studies, Wellington, pp.17-24.

Williams, J. (1997), *Legal, Technical and Mechanical Issues*, Paper presented at New Zealand Law Society seminar Treaty of Waitangi Issues: the Last Decade and the Next Century, April, New Zealand Law Society, Wellington.

Chapter 6: Regional Councils: Lightweight Policy Statements and Limited Capability

Berke, P., Dixon, J. and Ericksen, N. (1997), 'Coercive and Co-operative Intergovernmental Mandates: a Comparative Analysis of Florida and New Zealand Environmental Plans', *Environment and Planning B: Planning and Design*, Vol. 24, pp. 451-68.

Berke, P., Crawford, J., Dixon, J. and Ericksen, N. (1999), 'Do Co-operative Environmental Planning Mandates Produce Good Plans? Empirical Results from the New Zealand Experience', *Environment and Planning B: Planning and Design*, Vol. 26, pp. 643-64. Pion Limited, London.

Nuttall, P. and Ritchie, J. (1995), *Māori Participation in the Resource Management Act*, Tainui Maaori Trust Board and Centre for Māori Studies and Research, University of Waikato, Hamilton.

Statistics New Zealand (1997), *Table 2: 1996 Census of Population and Dwellings. Usually Resident Population – Changes in Usually Resident Population for Regional Councils, 1986-1996* [Table]. Retrieved May 15, 2000 from the World Wide Web: http://www.stats.govt.nz/domino/external/pas/pascs96.nsf/695873eb162291ec4c256804 00049867/4fed38d308a4c6924c256673000e15ae?OpenDocument.

Chapter 7: District Councils: Mixed Results in Planning and Capability

Becker, J. and D. Johnston (2000), *Planning and Policy for Earthquake Hazard in New Zealand*, Institute of Geological and Nuclear Sciences Limited, Science Report 2000/28.

Berke, P. (1995), 'Evaluating Environmental Quality: the Case of Planning for Sustainable Development in New Zealand', *Journal of Environmental Planning and Management*, Vol. 63(3), pp. 329-44.

Berke, P., Ericksen, N., Crawford, J. and Dixon, J. (2002), 'Planning and Indigenous People: Human Rights and Environmental Protection in New Zealand', *Journal of Planning Education and Research*, 22: 115-134.

Nuttall, P. and J. Ritchie (1995), *Maaori Participation in the Resource Management Act: A Think Piece*, Ministry for the Environment, Wellington, New Zealand.

Parliamentary Commissioner for the Environment (1992), *Proposed Guidelines for Local Authority Consultation with Tangata Whenua*, Parliamentary Commissioner for the Environment, Wellington.

Parliamentary Commissioner for the Environment (1998), *Kaitiakitanga and Local Government: Tangata Whenua Participation in Environmental Management*, Parliamentary Commissioner for the Environment, Wellington.

Ritchie, J. (1992), *Becoming Bicultural*, Huia Publishers, Wellington.

Chapter 8: Influencing Factors — Linking Mandates, Councils, Capability and Quality

Berke, P., Crawford, J., Dixon, J. and Ericksen, N. (1999), 'Do Co-operative Environmental Planning Mandates Produce Good Plans? Empirical Results from the New Zealand Experience', *Environment and Planning B: Planning and Design*, Vol. 26, pp. 643-64. Pion Limited, London.

Burby, R. and May P., with Berke, P., Dalton, L., French, S. and Kaiser, E. (1997), *Making Governments Plan: State Experiments in Managing Land Use,* Johns Hopkins University Press, Baltimore.

Cullingworth J.B. (1994), 'Alternative Planning Systems: is there Anything to Learn from Abroad?', *Journal of the American Planning Association*, Vol. 60(2), pp. 162-72.

Day, M., Backhurst, M. and Ericksen, N., with Crawford, J., Chapman, S., Berke, P., Laurian, L., Dixon, J., Jefferies, R., Warren, T., Barfoot, C., Mason, G., Bennett, M. and Gibson, C. (2003), *District Plan Implementation Under the RMA: Confessions of a Resource Consent*, Second PUCM Report to Government, The International Global Change Institute (IGCI), University of Waikato, Hamilton.

DeGrove, J. (1992), *The New Frontier for Land Use Policy: Planning and Growth Management in the United States*, Lincoln Institute of Land Use Policy, Cambridge, Massachusetts.

Logan, J. and Motoloch, H. (1987), *Urban Fortunes: The Political Economy of Place*, University of California Press, Berkeley, CA.

Rohe, W. and Bates, L. (1984), *Planning the Neighbourhoods*, University of North Carolina Press, Chapel Hill, North Carolina.

Chapter 9: Far North District: Resisting Innovation

Bay of Islands District of Federated Farmers (1992), *Submission on Far North District Discussion Document "C" for Preparation of New District Plan*, Submitted 30 September 1992, Federated Farmers of New Zealand Inc., Northland Province, Whangarei.

Bay of Islands District of Federated Farmers (1994), *Comments on Far North District Council Draft Objectives and Policies for the District Plan*, Submitted 8 December 1994, Federated Farmers of New Zealand Inc., Northland Province, Whangarei.

Bay of Islands District of Federated Farmers (1996), *Comments on the Draft Objectives, Policies, Outcomes and Methods for the District Plan*, Submitted 26 March 1996, Federated Farmers of NZ Inc., Northland Province, Whangarei.

Bay of Islands District of Federated Farmers (1997), *Problems Associated with Far North District Plan and Recommended Actions*, Information provided to Ministers of Conservation (Mr Smith), Environment (Mr Upton), and Local Government (Mr Carter), Kaeo.

Bay of Islands Planning Ltd. (1997a), *Table of Activity Status: Far North District Council Proposed District Plan and Variation No. 1*, Bay of Islands Planning Ltd, Kerikeri.

Bay of Islands Planning Limited (1997b), *Summation of Primary Standard: FNDC Proposed District Plan and Variation No. 1*, Bay of Islands Planning Limited, Kerikeri.

Claridge, M. and Kerr, S. (1998), *A Case Study in Devolution: The Problem of Pressuring Kiwi Habitat in the Far North*, Treasury Working Paper 98/7a, Wellington.

Crampton, P., Salmond, C., Kirkpatrick, R. with Scarborough, R. and Skelly, C. (2000), *Degrees of Deprivation in New Zealand: An Atlas of Socio-economic Difference*, David Bateman, Albany.

Davis, P., and Cocklin, C. (1997), *Who Pays? Habitat Protection on Private Land*, Paper presented to Nature Conservation in Production Environments Conference, 30 November to 5 December 1997, Taupo.

Department of Conservation (1994a), *Letter to Landowners*, September 1994, Northland Conservancy, Department of Conservation, Whangarei.

Department of Conservation (1994b), *News Release*, September 1994, Department of Conservation, Whangarei.

Department of Conservation (1996a), *The Significant Natural Areas of the Far North District Provisional Report*, Northland Conservancy, Department of Conservation, Whangarei.

Department of Conservation (1996b), *Submission on Discussion Paper – Objectives and Policies and Outcomes and Methods for the District Plan*, 15 March 1996, Department of Conservation, Northland Conservancy, Whangarei.

Far North District Council (1992), *Maori Issues Discussion Document for Preparation of New District Plan*, 10 August 1992, Far North District Council, Kaikohe.

Far North District Council (1996), *Proposed Far North District Plan*, Vols. I-III, Far North District Council, Kaikohe.

Far North District Council (1997a), *A Demographic Overview of the Far North District*, Far North District Council, Kaikohe.

Far North District Council (1997b), *Proposed Far North District Plan: Variation No. 1*, Vols. I-III, February 1997, Far North District Council, Kaikohe.

Far North District Council (1997c), *Summary of Submissions*, Proposed Far North District Plan: Variation No. 1, February 1997, Vols. I-III, Far North District Council, Kaikohe, November 1997.

Far North District Council (1998), *Draft Proposed Far North District Plan*, Far North District Council, Kaikohe.

Far North District Council (2000), *Proposed Far North District Plan*, Far North District Council, Kaikohe.

Ministry for the Environment (1992a), *Discussion Document A – Submission*, Submission to Far North District Council, 24 January 1992, Ministry for the Environment, Northern Regional Office, Auckland.

Ministry for the Environment (1992b), *Discussion Document B – Submission*, Submission to Far North District Council, 24 April 1992, Ministry for the Environment, Northern Regional Office, Auckland.

Ministry for the Environment (1994), *District Plan: Draft Objectives and Policies*, Letter to Far North District Council, 29 November 1994, Ministry for the Environment, Northern Regional Office, Auckland.

Ministry for the Environment (1996), *District Plan: Draft Objectives and Policies*, Letter of 26 February 1996, Ministry for the Environment, Northern Regional Office, Auckland.

Northland Regional Council (1992), *Submission on Discussion Document A – Proposed Far North District Plan*, Northland Regional Council, Whangarei.

Northland Regional Council (1993), *Proposed Northland Regional Policy Statement*, Northland Regional Council, Whangarei.

Northland Regional Council (1996), *Comments on District Plan Discussion Paper – Objectives and Policies, Outcomes and Methods for the District Plan*, Northland Regional Council, Whangarei.

Planning Consultants Ltd (1997), *The Future of the Proposed Far North District Plan: Briefing Notes for Council's Workshop*, 4 November 1997, Planning Consultants Ltd, Auckland.

Salmon, G. (1998), 'RMA Debacle in Far North: What future for nature conservation on private land?', *Maruia Pacific*, June 1998, 7-14.

Significant Natural Areas Review Group (1997), *Report to Council*, Far North District Council, Kaikohe.

Te Runanga o Te Rarawa (1992), *Submission: Re Maori Issues Discussion Document*, Submission by iwi/runanga to Far North District Council, 16 October 1992.

Chapter 10: Queenstown Lakes District: Development Meets Environment

Bennett, E., and Davie Lovell-Smith and Partners (1993), *Wanaka-Hawea-Makarora: Planning for Landscape Assessment*, Davie Lovell-Smith and Partners, Christchurch.

Boffa Miskell Partners Ltd (1991), *The Wakatipu Landscape*, Boffa Miskell Partners Ltd, Christchurch.

Boffa Miskell Partners Ltd (1992), *The Upper Wakatipu Landscape*, Boffa Miskell Partners Ltd, Christchurch.

Christchurch City Council (1995), *Proposed Christchurch City Plan*, Christchurch City Council, Christchurch.

Constantine Planners Ltd (1993a), *Heritage Values*, Constantine Planners Ltd, Dunedin.

Constantine Planners Ltd (1993b), *Settlements Strategy*, Constantine Planners Ltd, Dunedin.

Controller and Auditor-General (1999), *Contracting Out Local Authority Regulatory Functions*, Office of the Controller and Auditor-General, Wellington.

Davie Lovell-Smith and Partners (1993), *Public Open Space and Recreation Areas*, Davie Lovell-Smith, Christchurch.

Department of Conservation (1995), *Draft Conservation Management Strategy*, Otago Conservancy, Department of Conservation, Dunedin.

Lucas Associates (1995), *Indigenous Ecosystems: An Ecological Plan Structure for the Lakes District*, Lucas Associates, Christchurch.

North and South (1994), 'Blot on the Bay: Queer Times in Queenstown', November, pp. 89-100.

Otago Regional Council (1993a), *Proposed Regional Policy Statement*, Otago Regional Council, Dunedin.

Otago Regional Council (1993b), *Floodplain Management Report*, Otago Regional Council, Dunedin.

Queenstown Lakes District Council (1993), *Queenstown Lakes District Issues and Options*, Queenstown Lakes District Council, Queenstown.

Queenstown Lakes District Council (1995a), *Queenstown Lakes District Proposed District Plan,* Volumes 1-3, Queenstown Lakes District Council, Queenstown.

Queenstown Lakes District Council (1995b), *Queenstown Lakes District Strategic Plan*, Queenstown Lakes District Council, Queenstown.

Queenstown Lakes District Council, (1995c), *Queenstown Lakes District Transitional District Plan*, Queenstown Lakes District Council, Queenstown.

Queenstown Lakes District Council (1998a), *Queenstown Lakes District Proposed District Plan*, Queenstown Lakes District Council, Queenstown.

Queenstown Lakes District Council (1998b), *Queenstown Lakes District Proposed District Plan: District Planning Maps*, Queenstown Lakes District Council, Queenstown.

Queenstown Lakes District Council (1998c), *Queenstown Lakes District Proposed District Plan Planning Report Issue 51 - Landscape and Visual Amenity Planning Report*, Queenstown Lakes District Council, Queenstown.

Queenstown Lakes District Council (1998d), *Queenstown Lakes District Proposed District Plan Planning Report Issue 51 - Landscape and Visual Amenity*, Decision, August, Queenstown Lakes District Council, Queenstown.

Queenstown Lakes District Council (1998e), *Queenstown Lakes District Proposed District Plan. Issue 27: Areas of Significant Indigenous Vegetation and Habitat of Indigenous Fauna,* Decision, August, Queenstown Lakes District Council, Queenstown.

Queenstown Lakes District Council (1999), *Personal Communication from Finance Manager,* Queenstown Lakes District Council, Queenstown.

Queenstown Lakes District Council, Davie Lovell Smith and Partners, and Thomas Consultants (1995), *The Development Strategy — Frankton,* Queenstown Lakes District Council, Queenstown.

Riddolls Consultants Ltd (1993), *Geology for Resource Management Planning,* Riddolls Consultants Ltd, Queenstown.

Chapter 11: Tauranga District: Policy Coherence on the Coast

Beca Carter Hollings and Ferner Ltd. (1991), *Tauranga Urban Growth Strategy 1991,* Beca Carter Hollings and Ferner Ltd. Consulting Engineers, Tauranga.

Bay of Plenty Regional Council (1995), *Proposed Regional Coastal Environment Plan,* Environment Bay of Plenty, Whakatane, January 1995.

Bay of Plenty Regional Council (1996), *Proposed Bay of Plenty Regional Policy Statement Incorporating Decisions on Submissions,* Clear Copy, Environment Bay of Plenty, Whakatane, August 1996.

Department of Conservation (1994), *New Zealand Coastal Policy Statement,* Department of Conservation, Te Papa Atawhai, Wellington.

Gibb, J. (1996), *Coastal Hazard Risk Assessment between Mauao and Papamoa, Tauranga District, Bay of Plenty,* Project Dune Watch, CR96/1, Prepared for Tauranga District Council, Tauranga.

Kay, R.C., Ericksen, N.J., and Warrick, R.A, with Foster, G.A., Gillgren, D.J., Healy, T.R., and Sheffield, A.T. (1994), *Assessment of Coastal Hazards and Their Management for Selected Parts of the Coastal Zone Administered by Tauranga District Council,* Final Report to the Tauranga District Council and Environment Bay of Plenty by CEARS, University of Waikato, Hamilton.

Statistics New Zealand (1999), *New Zealand Now – Incomes (Census 1996),* Reference Reports. Statistics New Zealand. [Online] 1 May 2003 .http://www.stats.govt.nz/domino/external/pasfull/pasfull.nsf/web/Reference+ReportsNew+Zealand+Now++Incmes+(Census+96)+1999?open.

Tauranga District Council (1990), *Mount Maunganui Borough Council District Planning Scheme: Third Review,* Tauranga District Council, Tauranga.

Tauranga District Council (1991a), *Tauranga County District Scheme: Fourth Review,* Tauranga District Council, Tauranga.

Tauranga District Council (1991b), *Tauranga District Council 1991-92 Annual Plan,* Tauranga District Council, Tauranga.

Tauranga District Council (1993), *Tauranga District Council Transitional District Plan Proposed Change No. 1: Tauranga Urban Growth Strategy – Annotated Copy Following Council Decisions,* Department of Planning and Environment, Tauranga District Council, Tauranga.

Tauranga District Council (1994), *Planning and Environment Committee Minutes,* 14 June 1994, Tauranga District Council, Tauranga.

Tauranga District Council (1995), *Tauranga District Council 1995 Annual Report,* Tauranga District Council, Tauranga.

Tauranga District Council (1996), *The Proposed District Plan Record of Section 32 Assessment,* Prepared by the Tauranga District Council Department of Strategic Planning, November 1996.

Tauranga District Council (1997a), *Tauranga District Council 1996/97 Annual Report*, Tauranga District Council, Tauranga.

Tauranga District Council (1997b), *Tauranga District Proposed District Plan: Part I – Policy Statement*, Tauranga District Council, Tauranga.

Tauranga District Council (1997c), *Tauranga District Proposed District Plan: Part II – Management Rules*, Tauranga District Council, Tauranga.

Tauranga District Council (1997d), *Tauranga District Proposed District Plan: Part III – Planning Maps*, Tauranga District Council, Tauranga.

Warrick, R., Kenny, G. and Wakeling, D. (1993), *Section 32 Analysis – Bay of Plenty Draft Coastal Hazard Policy*, Final Report to the Bay of Plenty Regional Council.

Western Bay of Plenty District Council (1996), *Proposed Western Bay of Plenty District Plan*, Western Bay of Plenty District Council, Tauranga.

Chapter 12: Tasman District: Political Populism

Bush-King, D. (1997), *Integrated Resource Management*, Paper Presented to the 1997 New Zealand Planning Institute Conference, Palmerston North.

Frieder, J. (1997), *Approaching Sustainability: Integrated Environmental Management and New Zealand's Resource Management Act*, paper supported with funding from the Ian Axford New Zealand Fellowship in Public Policy.

Jackson, Moira (2003), Email, Iwi consultant, Auckland.

Nahkies, G. (1998), *A Review of the Structure of the Environment and Planning Department, Tasman District Council*, Unpublished Report, Tasman District Council, Richmond.

Ngati Tama Manawhenua Ki Te Tau Ihu Trust (1997), *Submission on the Proposed Resource Management Plan (Re- Removal of Section 18)*. Nelson: Ngati Tama Manawhenua Ki Te Tau Ihu Trust (27 February 1997), 5 pp.

Tasman District Council (1992), *Proposed Moutere Water Management Plan*, Tasman District Council, Richmond.

Tasman District Council (1994), *Proposed Regional Policy Statement*, Tasman District Council, Richmond.

Tasman District Council (1995a), *Archaeological and Maori Sites Investigation and Assessment*, August 1995, Tasman District Council, Richmond.

Tasman District Council (1995b), *Natural and Built Heritage*, June 1995, Tasman District Council, Richmond.

Tasman District Council (1995c), *Synopsis of Policy Paper Issues (25 Papers)*, Tasman District Council, Richmond.

Tasman District Council (1996a), *Advice to Property Developers and Consultants Affected by Cultural Heritage Area rules (Section 18.2)*, Tasman District Council, Richmond.

Tasman District Council (1996b), *Issues Arising from Cultural Heritage Protection Provisions of the proposed Tasman Resource Management Plan: A paper to council and stakeholders to assist resolving issues*, Tasman District Council, Richmond.

Tasman District Council (1996c), *Plan Drafting Protocol: Tasman Resource Management Plan*, 25 May 1996, Tasman District Council, Richmond.

Tasman District Council (1996d), *Proposed Tasman Resource Management Plan: Volume 1 and Volume 2 Planning Maps*, 25 May 1996,Tasman District Council, Richmond.

Tasman District Council (1996e), *Record of Action Taken and Documentation Prepared in Fulfilling Duties to Consider Alternatives and Assess Benefits and Costs Under Section 32, Resource Management Act 1991*, May 1996, Tasman District Council, Richmond.

Tasman District Council (1996f), *Report to the Tasman District Council from the Natural Heritage Working Party*, October 1996, Tasman District Council, Richmond.

Tasman District Council (1997a), *Proposed Tasman Resource Management Plan: Proposed Variation No. 1,* Tasman District Council, Richmond.

Tasman District Council (1997b), *Summary of Decisions Requested by Submissions on the Proposed Tasman Resource Management Plan (by Topic in Sequence of Plan),* Tasman District Council, Richmond

Conclusion: A Decade On: Unfulfilled Expectations

Carroll, L. (1947), *Alice's Adventures in Wonderland,* First Australian Edition 1945, reprinted 1947, Wilke and Co. Pty Ltd, Melbourne; for Oxford University Press.

Chapman, S., Crawford, J. and Ericksen, N. (2003), *A Guide to Plan-Making in New Zealand: the Next Generation,* International Global Change Institute, University of Waikato, Hamilton.

Department of Prime Minister and Cabinet (2003), *Sustainable Development for New Zealand,* Department of Prime Minister and Cabinet, Wellington.

Ericksen, N., Crawford, J., Berke, P. and Dixon, J. (2001), *Resource Management, Plan Quality, and Governance: A Report to Government,* International Global Change Institute, University of Waikato, Hamilton.

Foundation for Research, Science and Technology (2002), Sustainability Review Report, Foundation for Research, Science and Technology, Wellington.

Godschalk, D., Beatley, T., Berke, P., Brower, D. and Kaiser, E. (1999), *Natural Hazard Mitigation: Recasting Disaster and Policy Planning,* Island Press, Washington D.C.

Hobbs, M. (2003), *Government Progress and Intentions on Planning Under RMA* (inferred title), Keynote address of Marian Hobbs, Minister for the Environment to Vision 50/50, New Zealand Planning Institute Conference, 22-24 May, Hamilton.

Local Government New Zealand (2000), *Devolution – Fact or Fiction? A Report on the Extent of New Functions and Costs Acquired by Local Government Since Reform in 1989,* March 2000, Local Government New Zealand, Wellington.

Ministry for the Environment (2003), 'Ready to Take up the Challenge', *Environz,* March 2003, Ministry for the Environment, Wellington.

Parliamentary Commissioner for the Environment (1999), *Local Government Environmental Management: A Study of Models and Outcomes,* Parliamentary Commissioner for the Environment, Wellington.

Parliamentary Commissioner for the Environment (2002), *Creating Our Future: Sustainable Development for New Zealand,* Parliamentary Commissioner for the Environment, Wellington.

Appendices

Dillman, D.A. (1978), *Mail and Telephone Surveys: the Total Design Method,* Wiley, New York.

May, P., Burby, R., Ericksen, N., Handmer, J., Dixon, J., Michaels, S. and Smith, D.I. (1996), *Environmental Management and Governance: Intergovernmental Approaches to Hazards and Sustainability,* Roultledge Press, London.

Miles, M.B. and Huberman, A.M. (1984), *Analyzing Qualitative Data: A Source Book for New Methods,* Sage, Beverly Hills, CA.

Index